Encyclopaedia of Mathematical Sciences

Volume 10

Editor-in-Chief: R.V. Gamkrelidze

S. G. Gindikin G. M. Khenkin (Eds.)

Several Complex Variables IV

Algebraic Aspects of Complex Analysis

Springer-Verlag
Berlin Heidelberg New York
London Paris Tokyo
Hong Kong

Title of the Russian edition:
Itogi nauki i tekhniki, Sovremennye problemy matematiki,
Fundamental'nye napravleniya, Vol. 10, Kompleksnyĭ
analiz—monogie peremennye 4
Publisher VINITI, Moscow 1986

Mathematics Subject Classification (1980):
32-02, 32CXX, 32GXX, 32M10, 32L10

Library of Congress Cataloging-in-Publication Data
Gindikin, S.G. (Semen Grigor 'evich)
[Kompleksnyĭ analiz-mnogie peremennye 4. English]
Several complex variables IV: algebraic aspects
of complex analysis / S.G. Gindikin, G.M. Khenkin.
p. cm.—(Encyclopaedia of mathematical sciences; v. 10)
Translation of: Kompleksnyĭ analiz-mnogie peremennye 4.
Bibliography : p. Includes index.

ISBN-13: 978-3-642-64766-6 e-ISBN-13: 978-3-642-61263-3
DOI: 10.1007/978-3-642-61263-3

1. Functions of several complex variables. I. Khenkin, G.M.
II. Title. III. Title: Several complex variables 4. IV. Series.
QA331.7.G5613 1990 515.9'4—dc19 89-4119 CIP

Softcover reprint of the hardcover 1st edition 1990

Typesetting: Asco Trade Typesetting Ltd., Hong Kong
2141/3140-543210—Printed on acid-free paper

List of Editors, Contributors and Translators

Editor-in-Chief

R.V. Gamkrelidze, Academy of Sciences of the USSR, Steklov Mathematical Institute, ul. Vavilova 42, 117966 Moscow, Institute for Scientific Information (VINITI), Baltiiskaya ul. 14, 125219 Moscow, USSR

Consulting Editors

S.G. Gindikin, A.N. Belozersky Laboratory of Molecular Biology and Bioorganic Chemistry, Moscow State University, 119899 Moscow GSP-234, USSR
G.M. Khenkin, Academy of Sciences of the USSR, Central Economic and Mathematical Institute, ul. Krasikova 32, 117418 Moscow, USSR

Contributors

D.N. Akhiezer, Moscow Construction Engineering Institute, Department of Applied Mathematics, Yaroslavskoe Shosse 26, 129337 Moscow, USSR
J. Leiterer, Akademie der Wissenschaften der DDR, Karl-Weierstraß-Institut für Mathematik, Mohrenstraße 39, DDR-1086 Berlin
A.L. Onishchik, Yaroslavl University, Sovetskaya ul. 14, 150000 Yaroslavl, USSR
V.P. Palamodov, Moscow State University, Department of Mechanics and Mathematics, Leninskie Gory, 119899 Moscow, USSR

Translators

J. Leiterer, Akademie der Wissenschaften der DDR, Karl-Weierstraß-Institut für Mathematik, Mohrenstraße 39, DDR-1086 Berlin
J. Nunemacher, Ohio Wesleyan University, Department of Mathematics, Delaware, OH 43015, USA

Contents

I. Methods in the Theory of Sheaves and Stein Spaces

A.L. Onishchik

Translated from the Russian
by J. Nunemacher

Contents

Introduction

This article is devoted to cohomological methods in complex analysis, which have undergone intensive development in the course of the last 35 years. The basic object of study here is a complex analytic space or, roughly speaking, a complex analytic manifold with singular points. Analytic spaces belong (as do also analytic and differentiable manifolds, supermanifolds, and algebraic varieties) to the class of mathematical structures which are defined by fixing on a given topological space a certain stock of continuous local functions. An adequate means for describing such a structure is the concept of a sheaf, which is discussed in Chapter 1. In Chapter 2 we consider first the technically convenient concept of a ringed space, i.e., a space endowed with a sheaf of rings (or algebras), and then a particular case of it—the concept of a complex analytic space, which is fundamental for what follows. We also define here a coherent analytic sheaf. In Chapter 3 we discuss the theory of cohomology with values in a sheaf of abelian groups and indicate its simplest applications to problems of analysis, for example, to the solution of the Cousin problems for polycylindrical domains. In Chapter 4 we give a survey of results related to the so-called Stein spaces. These remarkable complex spaces, which can be defined, roughly speaking, as spaces with a very large stock of global analytic functions, emerged historically as the first objects in complex analysis on which the methods of cohomology were tried.

Chapter 1. Sheaves

The concept of a sheaf of abelian groups arose in the forties in the work of Leray (see [65]) on the topology of fiber spaces. Independently in the work of Oka [81] subsheaves of ideals of the sheaf of germs of holomorphic functions on a domain in $\mathbb{C}^n$ were in effect introduced. The definition of a sheaf in the form discussed below was formulated in the Cartan seminar [15] and, as was remarked there, is due to Lazard. The exposition in this section is close to that given in Godement [41] to which we refer the reader for more details.

1. Definition of a Sheaf. Let X be a topological space. A *sheaf of sets* on X is a topological space $\mathscr{F}$ together with a surjective mapping $p: \mathscr{F} \to X$ which is a local homeomorphism. (This means that any point $y \in \mathscr{F}$ has a neighborhood V in $\mathscr{F}$ so that $p(V)$ is open in X and so that $p: V \to p(V)$ is a homeomorphism.) The set $\mathscr{F}_x = p^{-1}(x)$ is called the *stalk* (or *fiber*) of the sheaf $\mathscr{F}$ over the point $x \in X$ and the mapping p the *projection* of the sheaf. We note that every stalk $\mathscr{F}_x$ is a set on which $\mathscr{F}$ induces the discrete topology and which thus itself carries no intrinsic structure. A sheaf of sets can be described as a family of sets $(\mathscr{F}_x)_{x \in X}$ whose disjoint union is given the structure of the topological space $\mathscr{F}$.

The simplest example of a sheaf of sets is the *constant sheaf* defined by a set A: this is the space $\mathscr{F}_A = X \times A$ (we take on A the discrete topology) endowed with the projection of a direct product $p: (x, a) \mapsto x$ for $x \in X$ and $a \in A$. In this case each stalk $\mathscr{F}_x$ can be identified with A. Another example of a sheaf of sets is given by any topological covering of the space X. For more interesting examples see Chapter 1, Section 2.

A *section* (or, more precisely, a *continuous section*) of a sheaf of sets $\mathscr{F}$ on X is a continuous mapping $s: X \to \mathscr{F}$ such that $p \circ s = \mathrm{id}$. The value of a section s at a point $x \in X$ will be denoted by s_x. Any section is a homeomorphism from the space X onto the open set $s(X) \subset \mathscr{F}$. The set of all sections of the sheaf $\mathscr{F}$ is denoted by $\Gamma(X, \mathscr{F})$. For the constant sheaf $\mathscr{F}_A$ the set $\Gamma(X, \mathscr{F}_A)$ is identified naturally with the set of all continuous mappings $X \to A$. Since A is discrete, a mapping $X \to A$ is continuous if and only if it is locally constant, i.e., constant in some neighborhood of an arbitrary point in X.

Along with global sections $s: X \to \mathscr{F}$ of a sheaf $\mathscr{F}$ we may also consider sections over an arbitrary set $Y \subset X$, i.e., continuous mappings $s: Y \to \mathscr{F}$ satisfying the condition $p \circ s = \mathrm{id}$. The collection of all of these mappings is denoted by $\Gamma(Y, \mathscr{F})$. On the other hand, $\mathscr{F}$ defines on Y a sheaf $\mathscr{F} \mid Y = p^{-1}(Y)$ with the same projection p; it is clear that $\Gamma(Y, \mathscr{F}) = \Gamma(Y, \mathscr{F} \mid Y)$.

We note that for a wide class of spaces X (for example, for locally compact spaces which are countable at infinity) every section in $\Gamma(Y, \mathscr{F})$ extends to some neighborhood of the set Y. What is more, the set $\Gamma(Y, \mathscr{F})$ can be identified with the *inductive limit* (or *direct limit*) $\varinjlim_{U \supset Y} \Gamma(U, \mathscr{F})$ of the family of sets $\Gamma(U, \mathscr{F})$, where U runs over all open neighborhoods of the set Y.

Since the inductive limit is a concept which we shall meet again, we recall its definition. Let I be a partially ordered set, and let there be given a functor from I to the category of sets, i.e., we have a family of sets $(S_i)_{i \in I}$ and for each pair (i, j) where $i \geqslant j$ there is defined a mapping $\varphi_{ij}: S_i \to S_j$ so that $\varphi_{ii} = \text{id}$ for all i and $\varphi_{ik} = \varphi_{jk} \circ \varphi_{ij}$ for $i \geqslant j \geqslant k$. We assume that I is *directed*, i.e., for any $i, j \in I$ there exists $k \in I$ so that $k \leqslant i$ and $k \leqslant j$. Then the inductive limit $\varinjlim_{i \in I} S_i$ is defined to be the quotient set of $\bigcup_{i \in I} S_i$ relative to the equivalence relation: $x \sim y$ if $x \in S_i$, $y \in S_j$, and if there exists $k \in I$ so that $k \leqslant i$, $k \leqslant j$ and $\varphi_{ik}(x) = \varphi_{jk}(x)$. In the case mentioned above the set I of all neighborhoods $U \supset Y$ is ordered by inclusion, and the mapping $f_{UV}: \Gamma(U, \mathscr{F}) \to \Gamma(V, \mathscr{F})$ for $U \supset V$ is the restriction mapping.

Most important for applications are those sheaves whose stalks carry some algebraic structure, where the corresponding operations are assumed to be continuous in the topology of the space $\mathscr{F}$. For example, $\mathscr{F}$ is called a *sheaf of abelian groups* if there is defined in every stalk $\mathscr{F}_x$ an operation $(y, z) \mapsto y + z$ which converts $\mathscr{F}_x$ into an abelian group with zero element 0_x such that:

a) the mapping $(y, z) \mapsto y + z$ from the space $\{(y, z) \in \mathscr{F} \times \mathscr{F} \mid p(y) = p(z)\}$ into $\mathscr{F}$ is continuous;

b) the mapping $y \mapsto -y$ from the space $\mathscr{F}$ into itself is continuous;

c) the mapping $x \mapsto 0_x$ from the space X into $\mathscr{F}$ is continuous, i.e., is a section of the sheaf; it is called the *zero section* and is denoted by 0.

The *support of a sheaf* of abelian groups $\mathscr{F}$ is the set

$$\operatorname{supp} \mathscr{F} = \{x \in X \mid \mathscr{F}_x \neq 0\}.$$

Similarly, one defines a *sheaf of algebras* with unity elements 1_x for $x \in X$ over some field k (the mapping $x \mapsto 1_x$ must be a section of the sheaf—the so-called *unity section*; it is denoted by the symbol 1).

It is easy to see that any algebraic operation defined on the stalks of a sheaf $\mathscr{F}$ carries over to the set of all sections $\Gamma(X, \mathscr{F})$, so that $\Gamma(X, \mathscr{F})$ becomes, for example, an abelian group or an algebra with unity.

2. Sheaves of Germs of Functions; Presheaves. Interesting examples of sheaves on X can be constructed by means of local functions.

Example 1.1. Let x be a fixed point in the topological space X. We consider the set of all complex-valued functions f, each of which is defined and continuous in some neighborhood U_f of the point x in X. We define in this set the following equivalence relation: $f \sim g$ if the functions f and g agree in some neighborhood $V \subset U_f \cap U_g$ of the point x. The equivalence class f_x of the function f is called the *germ of the continuous function* f at the point x. The set of all germs $\mathscr{C}_x$ is endowed with the natural structure of an algebra over $\mathbb{C}$ with unity. Let $\mathscr{C} = \bigcup_{x \in X} \mathscr{C}_x$. With every pair (f, U), where f is a continuous function on an open set $U \subset X$, we associate the set $W(f, U) = \{f_x \mid x \in U\}$. The sets $W(f, U)$ form a basis of open sets for a topology on $\mathscr{C}$ (which generally is not Hausdorff). If we define a projection $p: \mathscr{C} \to X$ by the formula $p(f_x) = x$, then $\mathscr{C}$ becomes a sheaf of $\mathbb{C}$-algebras with unity on X. This sheaf is called the *sheaf of germs of* (complex-valued) *continuous functions* on X. Similarly, we can define the sheaf of germs of continuous functions with real values or with values in some topological field.

If X is a domain in $\mathbb{R}^n$ (or in a differentiable manifold), then we may consider the *sheaf* $\mathscr{C}^l$ *of germs of differentiable functions* of some fixed class C^l or the *sheaf* $\mathscr{C}^\omega$ *of real-analytic functions* (in the latter case the manifold must be real-analytic).

It is common to consider the sheaf $\mathscr{C}^G$ of germs of continuous functions with values in some topological group G; it is a sheaf of groups which are abelian if the group G is abelian. The sections of the sheaf $\mathscr{C}$ (or $\mathscr{C}^G$) are continuous functions on X.

Example 1.2. If G is a certain group, algebra, etc., then the constant sheaf $\mathscr{F}_G$ is a sheaf of groups, algebras, etc. The sheaf $\mathscr{F}_G$ can be interpreted as the sheaf of germs of locally constant G-valued functions on X.

Example 1.3. Let D be an open set in $\mathbb{C}^n$. We denote by $\mathcal{O}$ the subset of the sheaf $\mathscr{C}$ on D consisting of the germs of holomorphic functions. Then $\mathcal{O}$ is a sheaf on D, which is called the *sheaf of germs of holomorphic functions*. By expanding a function which is holomorphic in a neighborhood of a point $x \in D$ in a power series centered at the point, we obtain an isomorphism from the stalk $\mathcal{O}_x$ of the sheaf $\mathcal{O}$ to the $\mathbb{C}$-algebra $\mathbb{C}\{z_1, \ldots, z_n\}$ of all convergent power series in $z_1, \ldots, z_n$. We note that the topology of the sheaf $\mathcal{O}$, unlike that of the sheaf $\mathscr{C}$, is Hausdorff. Sections of the sheaf $\mathcal{O}$ are holomorphic functions on D.

It is natural to formulate the idea of analytic continuation in the language of germs of holomorphic functions. Thus, if $g_0 \in \mathcal{O}$ is the germ of a holomorphic function at some point $x_0 = p(g_0) \in D$, then the connected component of the element g_0 in $\mathcal{O}$ is the so-called multi-valued analytic function in D obtained from the germ g_0 via analytic continuation; it is at the same time the natural geometric object on which this function becomes single-valued (its *Riemann domain*).

The construction of a sheaf of germs applies not only to functions but also to other objects—vector and tensor fields, functions of several variables, etc. This construction admits the following formalization. We consider the set $\mathscr{K}_X$ of all open subsets of a topological space X partially ordered by inclusion. A *presheaf* of sets, groups, algebras, etc., on X is a functor from $\mathscr{K}_X$ to the category of sets, groups, algebras, etc., i.e., a correspondence which associates to every $U \in \mathscr{K}_X$ a set (group, etc.) F_U and to every pair $U \supset V$ of sets in $\mathscr{K}_X$ a mapping (homomorphism) $r_{UV}: F_U \to F_V$ so that $r_{UU} = \mathrm{id}$ for $U \in \mathscr{K}_X$ and $r_{UW} = r_{VW} \circ r_{UV}$ for $U \supset V \supset W$. To every presheaf there is associated a sheaf in the following manner. If $x \in X$ then the set of all open neighborhoods of a point x is directed, which allows the inductive limit

$$\mathscr{F}_x = \varinjlim_{U \ni x} F_U.$$

to be defined. If $f \in F_U$ and $x \in U$ then the class of the element f in $\mathscr{F}_x$ is denoted by f_x. The set $\mathscr{F} = \bigcup_{x \in X} \mathscr{F}_x$ is endowed with the topology for which sets of the form $\{f_x | x \in U\}$ for all possible $f \in F_U$ with $U \in \mathscr{K}_X$ form a basis with the natural projection onto X. It is easy to verify that $\mathscr{F}$ is a sheaf of sets, groups, algebras, etc., on X. The sheaves of germs of functions constructed above can be obtained using this construction if we take as F_U the set of all functions of the given class on U and as r_{UV} the restriction mapping.

It is also clear that any sheaf $\mathscr{F}$ can be defined from some presheaf, namely, the presheaf which associates to each $U \in \mathscr{K}_X$ the set of sections $\Gamma(U, \mathscr{F})$ and to each pair $U \supset V$ the restriction mapping $\Gamma(U, \mathscr{F}) \to \Gamma(V, \mathscr{F})$.

On the other hand, if there is given on X an arbitrary presheaf $U \to F_U$ then any element $f \in F_U$ defines a section $\tilde{f} \in \Gamma(U, \mathscr{F})$ of the corresponding sheaf $\mathscr{F}$ by the formula

$$\tilde{f}_x = f_x \quad \text{for } x \in U.$$

However, the mapping $f \mapsto \tilde{f}$ from the set F_U into $\Gamma(U, \mathscr{F})$ is not in general either injective or surjective.

In later examples of presheaves we shall explicitly indicate only the objects F_U; it will be implicit that r_{UV} is the natural restriction mapping.

3. The Simplest Concepts Related to Sheaves of Groups, Algebras, and Modules. Let $\mathscr{F}$ be a sheaf of sets (groups, algebras, etc.) on X. A *subsheaf* of the sheaf $\mathscr{F}$ is a open subset $\mathscr{G} \subset \mathscr{F}$ such that for any $x \in X$ the subset (or subgroup, subalgebra, etc.) $\mathscr{G}_x = \mathscr{F}_x \cap \mathscr{G}$ is nonempty in $\mathscr{F}_x$. If $\mathscr{F}'$ is another sheaf on X (with the same algebraic structure on the stalks as $\mathscr{F}$), then a *homomorphism* from the sheaf $\mathscr{F}$ to $\mathscr{F}'$ is a continuous mapping $h: \mathscr{F} \to \mathscr{F}'$ which induces a homomorphism of stalks $h_x: \mathscr{F}_x \to \mathscr{F}'_x$ for every $x \in X$. The image $\operatorname{Im} h$ of a sheaf homomorphism h is a subsheaf of $\mathscr{F}'$. If $\mathscr{F}$ and $\mathscr{F}'$ are sheaves of groups (or rings), then the kernel $\operatorname{Ker} h$ of the homomorphism $h: \mathscr{F} \to \mathscr{F}'$ is well-defined; it is the subsheaf of $\mathscr{F}$ formed by the union of the kernels $\operatorname{Ker} h_x$ for $x \in X$. The kernel $\operatorname{Ker} h$ is a sheaf of normal subgroups (resp., ideals) in $\mathscr{F}$. An *isomorphism* of sheaves is a homomorphism for which there exists an inverse homomorphism. Sheaves of sets (groups, algebras) over a fixed space X form a category whose morphisms are sheaf homomorphisms.

There is also the concept of a *quotient sheaf*. In what is for us the most important case, when there is given a sheaf of abelian groups $\mathscr{F}$ and a subsheaf $\mathscr{G}$ of $\mathscr{F}$, the quotient sheaf $\mathscr{F}/\mathscr{G}$ is defined as the union $\bigcup_{x \in X} \mathscr{F}_x/\mathscr{G}_x$ endowed with the quotient topology with respect to the natural mapping $\pi: \mathscr{F} \to \mathscr{F}/\mathscr{G}$; this mapping is a homomorphism of sheaves with kernel $\mathscr{G}$. The *homomorphism theorem* in this context is valid. For the case of sheaves of abelian groups it can be formulated as follows: for every surjective homomorphism $\varphi: \mathscr{F} \to \mathscr{F}'$ of sheaves of abelian groups there exists a sheaf isomorphism $\bar{\varphi}: \mathscr{F}/\operatorname{Ker} \varphi \to \mathscr{F}'$ such that $\varphi = \bar{\varphi} \circ \pi$, where $\pi: \mathscr{F} \to \mathscr{F}/\operatorname{Ker} \varphi$ is the natural homomorphism. In particular, a sheaf homomorphism which is both injective and surjective is an isomorphism.

A sequence of sheaves of abelian groups and their homomorphisms

$$\mathscr{F}_1 \xrightarrow{\varphi_1} \mathscr{F}_2 \xrightarrow{\varphi_2} \cdots \xrightarrow{\varphi_{k-1}} \mathscr{F}_k$$

is called *exact* if $\operatorname{Im} \varphi_i = \operatorname{Ker} \varphi_{i+1}$ for all $i = 1, \ldots, k-2$. If we denote the sheaf with zero stalks by 0, then the exactness of the sequence

$$0 \to \mathscr{F}' \xrightarrow{\varphi} \mathscr{F} \xrightarrow{\psi} \mathscr{F}'' \to 0$$

means that φ is injective and ψ is surjective and induces an isomorphism $\mathscr{F}'' \cong \mathscr{F}/\varphi(\mathscr{F}')$.

If $\mathscr{F}$ and $\mathscr{G}$ are sheaves of sets on X with projections p and q, then the fibred product $\mathscr{F} \times_X \mathscr{G} = \{(f,g) \in \mathscr{F} \times \mathscr{G} \mid p(f) = q(g)\}$ is also a sheaf on X, called the *direct product* of the sheaves $\mathscr{F}$ and $\mathscr{G}$; its stalk over a point $x \in X$ is $\mathscr{F}_x \times \mathscr{G}_x$. The direct product of two sheaves of groups is in a natural manner also a sheaf of groups. If $\mathscr{F}$ and $\mathscr{G}$ are sheaves of abelian groups, then their direct product is also called a *direct sum* and is denoted by $\mathscr{F} \oplus \mathscr{G}$. This definition generalizes in the usual fashion to the case of any finite (and even infinite) family of sheaves. In particular, we write $\mathscr{F}^m$ for $\mathscr{F} \oplus \cdots \oplus \mathscr{F}$ (m times).

It is easy to see that $\Gamma(X, \mathscr{F} \times_X \mathscr{G}) = \Gamma(X, \mathscr{F}) \times \Gamma(X, \mathscr{G})$. In particular, if $\mathscr{F}$ and $\mathscr{G}$ are sheaves of abelian groups, then the group $\Gamma(X, \mathscr{F} \oplus \mathscr{G})$ is the direct sum of groups $\Gamma(X, \mathscr{F}) \oplus \Gamma(X, \mathscr{G})$.

Let $\mathscr{A}$ be a sheaf of associative rings with unity on a space X. A sheaf of abelian groups $\mathscr{F}$ on X is called a *sheaf of $\mathscr{A}$-modules* (or simply an *$\mathscr{A}$-module*) if there is a homomorphism of sheaves of sets $\mathscr{A} \oplus \mathscr{F} \to \mathscr{F}$ which defines for every $x \in X$ the structure of a (left unitary) $\mathscr{A}_x$-module on $\mathscr{F}_x$. Sheaves of modules over sheaves of algebras are defined similarly. One also considers the concepts of a sheaf of submodules (a subsheaf of a sheaf of $\mathscr{A}$-modules), homomorphism, exact sequence, and direct sum of sheaves of $\mathscr{A}$-modules. A sheaf of rings $\mathscr{A}$ is in the natural way a sheaf of $\mathscr{A}$-modules; its submodules are sheaves of left ideals in $\mathscr{A}$. The group of sections $\Gamma(X, \mathscr{F})$ of a sheaf of $\mathscr{A}$-modules $\mathscr{F}$ is a module over the ring $\Gamma(X, \mathscr{A})$.

If $\mathscr{A}$ is a sheaf of commutative rings, then one can associate to any two sheaves of $\mathscr{A}$-modules $\mathscr{F}$ and $\mathscr{G}$ their *tensor product* $\mathscr{F} \otimes_{\mathscr{A}} \mathscr{G}$ and the *sheaf of germs of homomorphisms* $\mathscr{H}om_{\mathscr{A}}(\mathscr{F}, \mathscr{G})$. These are defined using presheaves given by the correspondences

$$U \mapsto \Gamma(U, \mathscr{F}) \otimes_{\Gamma(U, \mathscr{A})} \Gamma(U, \mathscr{G})$$

and

$$U \mapsto \operatorname{Hom}_{\mathscr{A}|U}(\mathscr{F}|U, \mathscr{G}|U)$$

(U is an open set in X).

We note that the $\Gamma(X, \mathscr{A})$-module $\Gamma(X, \mathscr{H}om_{\mathscr{A}}(\mathscr{F}, \mathscr{G}))$ is identified naturally with the module $\operatorname{Hom}_{\mathscr{A}}(\mathscr{F}, \mathscr{G})$ of all sheaf homomorphisms from $\mathscr{F}$ to $\mathscr{G}$ (as sheaves of $\mathscr{A}$-modules).

4. The Lifting of Sections. Let $h: \mathscr{F} \to \mathscr{G}$ be a homomorphism of sheaves defined on a space X. Then h defines a natural mapping $\Gamma(X, \mathscr{F}) \to \Gamma(X, \mathscr{G})$; it is denoted by $\Gamma(h)$ and acts according to the formula

$$(\Gamma(h)s)_x = h(s_x) \quad \text{for} \quad s \in \Gamma(H, \mathscr{F}) \quad \text{and} \quad x \in X.$$

If h is a homomorphism of sheaves of groups, rings, modules, etc., then $\Gamma(h)$ is also a homomorphism of groups, rings, modules, etc. Injectivity of the homomorphism h easily implies the injectivity of $\Gamma(h)$, but a surjective homomorphism of sheaves h may induce a nonsurjective mapping $\Gamma(h)$.

The question of the surjectivity of the mapping $\Gamma(h)$ for a surjective homomorphism of sheaves h is known as the *problem of lifting of sections.* Its solution (in the case of sheaves of abelian groups) is given by the theory of cohomology with values in a sheaf (see Chapter 3). For several sufficient conditions for the lifting of sections see also Chapter 1, Section 5. Many analytic and geometric problems reduce to the problem of lifting of sections of a sheaf. We mention here only a few examples.

Example 1.4. Let $\mathcal{O}$ be the sheaf of germs of holomorphic functions on an open set $D \subset \mathbb{C}^n$. We denote by $\mathcal{O}^*$ the sheaf of germs of holomorphic functions on D which vanish nowhere. We consider $\mathcal{O}$ and $\mathcal{O}^*$ as sheaves of abelian groups relative to the operations of addition and multiplication respectively. The exponential function $f \mapsto e^f$ defines a sheaf homomorphism $\exp: \mathcal{O} \to \mathcal{O}^*$. This homomorphism is surjective, since for any point $z_0 \in D$ and any function $g \in \Gamma(U, \mathcal{O}^*)$ defined on a spherical neighborhood $U \subset D$ of the point z_0 there exists a function $f \in \Gamma(U, \mathcal{O})$ such that $g = e^f$. But the mapping $\Gamma(\exp): \Gamma(D, \mathcal{O}) \to \Gamma(D, \mathcal{O}^*)$ having the form $\Gamma(\exp f) = e^f$ may be nonsurjective (it suffices to take $n = 1$ and $D = \mathbb{C} \setminus \{0\}$, then the function $g(z) = z$ in $\Gamma(D, \mathcal{O}^*)$ has no logarithm on D).

Example 1.5. Let $\mathscr{C}^\infty$ be the sheaf of germs of infinitely differentiable functions in an open set $D \subset \mathbb{C}^n$. We consider the homomorphism $h: \mathscr{C}^\infty \to (\mathscr{C}^\infty)^n$ defined by the formula

$$h(f) = \left(\frac{\partial f}{\partial x_1}, \ldots, \frac{\partial f}{\partial x_n}\right).$$

This same formula clearly also defines $\Gamma(h)$. The subsheaf $\operatorname{Im} h$ consists of the germs of those tuples of local functions $(g_1, \ldots, g_n)$ satisfying $\dfrac{\partial g_i}{\partial x_j} = \dfrac{\partial g_j}{\partial x_i}$ for $i, j = 1, \ldots, n$. The homomorphism $\Gamma(h): \Gamma(D, \mathscr{C}^\infty) \to \Gamma(D, \operatorname{Im} h)$ is in general nonsurjectve. For example, let $n = 2$ and $D = \mathbb{R}^2 \setminus \{0\}$. Then $\left(-\dfrac{x_2}{x_1^2 + x_2^2}, \dfrac{x_1}{x_1^2 + x_2^2}\right) \in \Gamma(D, \operatorname{Im} h)$. However the system of differential equations

$$\frac{\partial f}{\partial x_1} = -\frac{x_2}{x_1^2 + x_2^2}, \qquad \frac{\partial f}{\partial x_2} = \frac{x_1}{x_1^2 + x_2^2}$$

does not have a solution in the domain D (this follows easily from the fact that one of the "solutions" of the system is the multi-valued function Arctan x_2/x_1).

The problem we have just considered is the problem of finding a function f whose differential is a given differential form of degree one $\omega = \sum_{i=1}^{n} g_i\,dx_i$. We may also consider the analogous question for differential forms of higher degree. We denote by $\Phi^p(D)$ for $p = 0, 1, \ldots$ the space of all (exterior) *differential forms* of class $\mathscr{C}^\infty$ and degree p on a domain D, i.e., all exterior polynomials of degree p in $dx_1, \ldots, dx_n$ over the ring $\Gamma(D, \mathscr{C}^\infty)$; its elements have the form

$$\omega = \sum_{i_1 < \cdots < i_p} a_{i_1 \cdots i_p}\, dx_{i_1} \wedge \cdots \wedge dx_{i_p},$$

where $a_{i_1, \ldots, i_p} \in \Gamma(D, \mathscr{C}^\infty)$. Setting

$$d\omega = \sum_{i_1 < \cdots < i_p} da_{i_1 \cdots i_p} \wedge dx_{i_1} \wedge \cdots \wedge dx_{i_p},$$

we obtain a linear mapping $d: \Phi^p(D) \to \Phi^{p+1}(D)$ (the exterior *differential*) such that $d \circ d = 0$. It is clear that $\operatorname{Im} d$ is contained in the kernel of the mapping $d: \Phi^{p+1}(D) \to \Phi^{p+2}(D)$. The question of whether $\operatorname{Im} d$ and this kernel coincide is a particular case of the problem of lifting of sections. In oder to see this, let us denote by Φ^p the *sheaf of germs of differential p-forms* on D and by Ψ^p the kernel of the induced homomorphism of sheaves $d: \Phi^p \to \Phi^{p+1}$. Then $d\Phi^p = \Psi^{p+1}$. This follows from the following classical result:

Poincaré Lemma. *If D is an open ball in $\mathbb{R}^n$ and $p \geqslant 0$ then for any form $\omega \in \Phi^{p+1}(D)$ satisfying $d\omega = 0$ there is a form $\varphi \in \Phi^p(D)$ so that $d\varphi = \omega$.*

In our problem the issue is whether or not the mapping $d = \Gamma(d): \Phi^p(D) = \Gamma(D, \Phi^p) \to \Gamma(D, \Psi^{p+1})$ is surjective.

As is well known, differential forms and the differential d can be defined on an arbitrary differentiable manifold M. As in the case above, we denote by Φ^p the sheaf of germs of all p-forms of class $\mathscr{C}^\infty$ on M and by Ψ^p the kernel of the homomorphism $d: \Phi^p \to \Phi^{p+1}$. It follows from the Poincaré lemma that the sheaf mapping $d: \Phi^p \to \Psi^{p+1}$ is surjective for all $p \geqslant 0$.

In order to discuss the next example we need to define the sheaf of germs of meromorphic functions of several complex variables. Let U be a domain (i.e., a connected open set) in $\mathbb{C}^n$. As is well known, the ring $\Gamma(U, \mathcal{O})$ of holomorphic functions on U is an integral domain, i.e., it contains no zero divisors. Therefore $\Gamma(U, \mathcal{O})$ can be imbedded in its field of fractions M_U, which by definition consists of all functions of the form f/g, where $f, g \in \Gamma(U, \mathcal{O})$ and $g \neq 0$. The correspondence $U \mapsto M_U$ defines a presheaf of rings on $\mathbb{C}^n$. The sheaf $\mathscr{M}$ defined by it is a sheaf of fields and is called the *sheaf of germs of meromorphic function* on $\mathbb{C}^n$. The sheaf $\mathcal{O}$ is a subsheaf of $\mathscr{M}$. The stalk $\mathscr{M}_x$ of the sheaf $\mathscr{M}$ at any point x is the field of fractions of the integral domain $\mathcal{O}_x$. Discarding from $\mathscr{M}$ the germs of zero functions we obtain a sheaf of abelian groups $\mathscr{M}^*$ which contains $\mathcal{O}^*$ as a subsheaf.

A *meromorphic function* in an open set $D \subset \mathbb{C}^n$ is a section in $\Gamma(D, \mathcal{M})$. The meromorphic functions in a domain D clearly form a field, which is an extension of the field $\mathbb{C}$.

Example 1.6. Let D be a domain in $\mathbb{C}^n$. Let there be given an open covering $(U_i)_{i \in I}$ of the set D and a meromorphic function f_i on each U_i with the property that for each pair (i, j) with $U_i \cap U_j \neq \emptyset$ the function $f_i - f_j$ is holomorphic in $U_i \cap U_j$. We consider the problem of finding a meromorphic function f on D so that $f - f_i$ is holomorphic in U_i for all $i \in I$. This problem is called the *first* or *additive Cousin problem*, and the collection of functions $(f_i)_{i \in I}$ satisfying the above mentioned consistency requirement is called *additive Cousin data*. It is clear that additive Cousin data defines a section of the quotient sheaf $\mathcal{M}/\mathcal{O}$ on D and that any section of this sheaf can be defined in this manner. Thus the Cousin problem is equivalent to the problem of lifting of sections for the natural sheaf homomorphism $\mathcal{M} \to \mathcal{M}/\mathcal{O}$.

For any point $x \in D$ the quotient space $\mathcal{M}_x/\mathcal{O}_x$ is identified naturally with the space of principal parts of Laurent series for meromorphic functions. Thus it is natural to call the sheaf $\mathcal{M}/\mathcal{O}$ the *sheaf of principal parts*.

In the case $n = 1$ the additive Cousin problem is equivalent to the problem of constructing a meromorphic function on D having prescribed principal parts for Laurent series at all points of some countable set having no limit point in D. This is the so-called Mittag-Leffler problem, which was solved by him in 1884 for an arbitrary domain $D \subset \mathbb{C}$ (see [95]). As Cousin [23] showed, the first Cousin problem is solvable in any polycylindrical domain, i.e., in a direct product of one-dimensional domains (see Chapter 3, Section 6, where an example is given of a domain in which the first Cousin problem has no solution).

Example 1.7. The formulation of the *second* or *multiplicative Cousin problem* in a domain $D \subset \mathbb{C}^n$ is completely analogous. *Multiplicative Cousin data* consists of a collection of meromorphic functions $(f_i)_{i \in I}$, where $f_i \in \Gamma(U_i, \mathcal{M}^*)$ and $(U_i)_{i \in I}$ is an open covering of the domain D, such that $f_i/f_j \in \Gamma(U_i \cap U_j, \mathcal{O}^*)$ whenever $U_i \cap U_j \neq \emptyset$. It is required to find a meromorphic function $f \in \Gamma(D, \mathcal{M}^*)$ so that $f/f_i \in \Gamma(U_i, \mathcal{O}^*)$ for all $i \in I$. Such a collection of functions corresponds naturally to a section of the sheaf $\mathcal{M}^*/\mathcal{O}^*$ (with multiplication as the group operation), and the problem is equivalent to the problem of lifting of sections for the natural sheaf homomorphism $\mathcal{M}^* \to \mathcal{M}^*/\mathcal{O}^*$. The sheaf $\mathcal{M}^*/\mathcal{O}^*$ is called the *sheaf of germs of divisors* and its sections *divisors* (for a geometric interpretation of divisors see Chapter 2, Section 9).

In the case $n = 1$ the multiplicative Cousin problem in equivalent to the problem of constructing a meromorphic function with prescribed zeros and poles (of given multiplicities) which is solvable in any domain $D \subset \mathbb{C}$ (the *Weierstrass Theorem*, see [95]). In the article [23] Cousin claimed that the second Cousin problem is solvable in any polycylindrical domain $D = D_1 \times \cdots \times D_n$, where $D_i \subset \mathbb{C}$ for $i = 1, \ldots, n$ but, as was noted in [52], his proof goes through only in the case when all the domains except possibly one are simply connected. For an explanation of this see Chapter 3, Section 6 and Chapter 4, Section 3.

The second Cousin problem is closely connected with another classical problem of complex analysis, the so-called *Poincaré problem*: is every meromorphic function h in a domain D representable in the form $h = f/g$, where f and g are holomorphic functions in D with $g \neq 0$? In other words, does the field $\Gamma(D, \mathscr{M})$ of all meromorphic functions in D coincide with the field of fractions of the ring of holomorphic functions $\Gamma(D, \mathscr{O})$ (which is clearly an integral domain)? The answer is affirmative in the case of $n = 1$. In 1883 Poincaré proved ([86], see also [70]) that this is true for $D = \mathbb{C}^2$. It turns out that in any domain in which the second Cousin problem is solvable the Poincaré problem is solvable also (see Chapter 2, Section 9).

5. The Extension of Sections. Let $\mathscr{F}$ be a sheaf on a topological space X and Y a subspace of X. Then the restriction mapping $r: \Gamma(X, \mathscr{F}) \to \Gamma(Y, \mathscr{F})$ is defined. We say that a section $s \in \Gamma(Y, \mathscr{F})$ extends to all of X if $s \in \operatorname{Im} r$.

The sheaf $\mathscr{F}$ is called *flabby* if each of its sections over an open set $Y \subset X$ extends to all of X.

Example 1.8. For an arbitrary sheaf $\mathscr{F}$ on X the sheaf of germs of not necessarily continuous sections $\tilde{\mathscr{F}}$ is flabby. Clearly $\mathscr{F}$ is a subsheaf of $\tilde{\mathscr{F}}$.

Example 1.9. If the space X is irreducible, i.e., not representable as a union of two closed subsets different from X, then any constant sheaf on X is flabby. This situation occurs often in algebraic geometry, since an irreducible algebraic variety is by definition irreducible in the Zariski topology.

Flabby sheaves are used as a technical device in the problem of lifting of sections. Let $0 \to \mathscr{F}' \to \mathscr{F} \xrightarrow{\psi} \mathscr{F}'' \to 0$ be an exact sequence of sheaves of abelian groups with $\mathscr{F}'$ a flabby sheaf. Then the homomorphism $\Gamma(\psi): \Gamma(X, \mathscr{F}) \to \Gamma(X, \mathscr{F}'')$ is surjective [41].

Another important class consists of *fine sheaves.* A sheaf of abelian groups $\mathscr{F}$ on X is called *fine* if for any locally finite open covering $(U_i)_{i \in I}$ of the space X there exists a family of endomorphisms $\eta_i: \mathscr{F} \to \mathscr{F}$ for $i \in I$ so that $\sum_{i \in I} \eta_i = \mathrm{id}$ and for any $i \in I$ we have $\eta_i(\mathscr{F}_x) = 0$ for all x in some neighborhood of the set $\overline{X \backslash U_i}$.

If $\mathscr{F}$ is a sheaf of rings with unity, then this property is equivalent to the existence of a partition of unity

$$1 = \sum_{i \in I} \varepsilon_i,$$

where $\varepsilon_i \in \Gamma(U_i, \mathscr{F})$ and $\varepsilon_i(\lambda) = 0$ for all x in some neighborhood of the set $\overline{X \backslash U_i}$. If $\mathscr{A}$ is a fine sheaf of rings with unity, then any sheaf of $\mathscr{A}$-modules is also a fine sheaf.

We consider also the class of *soft sheaves*, i.e., sheaves which admit extension of sections from any closed subset to the whole space. If the space X is paracompact and Hausdorff, then any flabby and any fine sheaf on X is soft, and any soft sheaf of rings with unity is a fine sheaf.

Example 1.10. If X is a domain in $\mathbb{R}^n$ (or any paracompact Hausdorff differentiable manifold of class C^∞), then the sheaf $\mathscr{C}^l$ of germs of functions of class C^l on X for $l = 1, 2, \ldots, \infty$ (see Example 1.1) is fine (this follows from the classical lemma on the existence of a partition of unity of class C^∞). Therefore any sheaf of $\mathscr{C}^l$-modules is also fine, for example, the sheaf Φ^p of germs of p-forms of class C^∞. For an arbitrary compact Hausdorff space X, Urysohn's Lemma implies that the sheaf $\mathscr{C}$ (see Example 1.1) is fine. These sheaves can be flabby or soft only in trivial situations. The sheaf $\mathcal{O}$ on a domain $D \subset \mathbb{C}^n$ for $n > 0$ is neither fine, flabby, nor soft.

We note one further connection which exists between the problems of extension and lifting of sections. Let X be a topological space, Y a locally closed subspace of it, and $\mathscr{F}$ a sheaf of abelian groups on Y. It turns out that there exists a unique sheaf of abelian groups $\mathscr{F}^X$ on X such that $\mathscr{F}^X | Y = \mathscr{F}$ and $(\mathscr{F}^X)_x = 0$ for all $x \in X \setminus Y$. The sheaf $\mathscr{F}^X$ is defined by the presheaf $U \mapsto \Gamma(Y \cap U, \mathscr{F})$; it is called the *trivial extension* of the sheaf $\mathscr{F}$ to X.

Let $\mathscr{F}$ be a sheaf of abelian groups on X and let Y be a closed subspace of X. We denote $\mathscr{F}_Y = (\mathscr{F} | Y)^X$ and $\mathscr{F}_{X \setminus Y} = (\mathscr{F} | X \setminus Y)^X$. Then there exists an exact sequence of sheaves

$$0 \to \mathscr{F}_{X \setminus Y} \to \mathscr{F} \xrightarrow{\psi} \mathscr{F}_Y \to 0.$$

Here $\Gamma(X, \mathscr{F}_Y) \cong \Gamma(Y, \mathscr{F})$ and the mapping $\Gamma(\psi)$ coincides under this isomorphism with the restriction $r : \Gamma(X, \mathscr{F}) \to \Gamma(Y, \mathscr{F})$.

6. Direct and Inverse Images. Up to this point we have considered sheaves over a fixed space X. We now define two operations on sheaves related to a continuous mapping $f : X \to Y$.

Let $\mathscr{F}$ be a sheaf on X. The *direct image* of the sheaf $\mathscr{F}$ under the mapping f is the sheaf $f_* \mathscr{F}$ on Y defined by the presheaf $U \mapsto \Gamma(f^{-1}(U), \mathscr{F})$ for U an open set in Y. For any open set $U \subset Y$ we have the natural isomorphism $\Gamma(U, f_* \mathscr{F}) \cong \Gamma(f^{-1}(U), \mathscr{F})$.

Let $\mathscr{G}$ be a sheaf on Y. The *inverse image* of the sheaf $\mathscr{G}$ under the mapping f is the sheaf $f^* \mathscr{G} \subset X \times \mathscr{G}$ consisting of those pairs (x, γ) with $\gamma \in \mathscr{G}_{f(x)}$. The projection $f^* \mathscr{G} \to X$ is defined by the formula $(x, \gamma) \mapsto x$. Thus $(f^* \mathscr{G})_x = \mathscr{G}_{f(x)}$ for $x \in X$. For any open set $V \subset Y$ there exists a natural mapping $\Gamma(V, \mathscr{G}) \to \Gamma(f^{-1}(V), f^* \mathscr{G})$, which is injective but not in general surjective.

Example 1.11. Let X be a subspace of a topological space Y endowed with the induced topology, and let $i : X \to Y$ be the inclusion mapping. If $\mathscr{F}$ is a sheaf of abelian groups on X and X is locally closed, then $i_* \mathscr{F}$ coincides with the trivial extension $\mathscr{F}^Y$ of the sheaf $\mathscr{F}$ to Y. If $\mathscr{F}$ is a sheaf on Y, then $i^* \mathscr{F}$ is the restriction $\mathscr{F} | X$.

Chapter 2. Complex Spaces

Complex spaces appeared as a result of the attempt to carry over to the case of several variables the concept of the Riemann surface of a multi-valued analytic function. The basic idea, which is due to Riemann, consists in viewing a multi-valued analytic function on a domain in $\mathbb{C}^n$ as a single-valued function on some "covering" of this domain. In the case of $n = 1$ this idea led to the concept of an abstract Riemann surface, i.e., a one-dimensional complex manifold (see, [112], [95]). The Riemann surface of an analytic function is a certain one-dimensional manifold, which is constructed as a branched covering of a plane domain. In the multi-dimensional case there is associated to a multi-valued function its Riemann domain, which is an n-dimensional complex manifold with a locally biholomorphic mapping onto a domain in $\mathbb{C}^n$ (see Example 1.3). This construction, however, does not take account of the branch points of the function. The attempt to complete the Riemann domain with points lying over the branch points takes us beyond the bounds of complex manifolds when $n > 1$. A classical example is the function $\sqrt{z_1 z_2}$ in $\mathbb{C}^2$ whose natural "Riemann surface' is the analytic set in $\mathbb{C}^3$ defined by the equation $w^2 = z_1 z_2$. The point 0 of this set is singular.

In 1951 Behnke and Stein [9] introduced the concept of a branched covering over a domain in $\mathbb{C}^n$ and defined a complex space to be a space whose local model is this branched covering. Soon after this Cartan [15] and Serre [94] introduced a different concept of complex space taking as local model an arbitrary analytic set in a domain in $\mathbb{C}^n$. The relation between these two concepts was finally cleared up in 1958 by Grauert and Remmert [47] who proved that a Behnke–Stein space is a normal complex space in the sense of Cartan-Serre. In this section we consider a generalization of complex space in the sense of Serre which is due to Grauert [44]: the structure sheaf of algebras of such a space is allowed to contain nilpotent elements.

In this article we consider only complex spaces, i.e., analytic spaces over the field $\mathbb{C}$. At the same time, the concept of analytic space can be defined over an arbitrary field k which is complete relative to some nontrivial valuation. If k is an algebraically closed field of characteristic 0, then the results of this section in general remain true. Over fields which are not algebraically closed (the most important examples for applications being the real and p-adic numbers) the theory becomes more complicated, for example, an analytic set over such a field is not always coherent (for an explanation see [17]). We also shall not deal with the theory of Banach analytic spaces (see [28], [88]).

1. Analytic Local Algebras. (see [48]). Let A be an associative and commutative algebra with unity 1 over a field k. The algebra A is called *local* if the set $\mathfrak{m}$ of all noninvertible elements is an ideal in A with $A/\mathfrak{m} \cong k$. In this case $A = k \oplus \mathfrak{m}$ and $\mathfrak{m}$ is the largest ideal in A (i.e., it contains all proper ideals in the algebra A), it is called the maximal ideal of A. If I is an ideal in the local algebra A distinct from A, then the algebra A/I is local and its maximal ideal is $\mathfrak{m}/I$. If $\tilde{A}$ is another local k-algebra with maximal ideal $\tilde{\mathfrak{m}}$, then any algebra homomorphism $f: A \to \tilde{A}$ such that $f(1) = 1$ is local, i.e., $f(\mathfrak{m}) \subset \tilde{\mathfrak{m}}$.

The *nilradical* of a nonzero algebra A is the set $\mathfrak{n}$ consisting of all its nilpotent elements. Clearly $\mathfrak{n} \neq A$. The algebra is called *reduced* if its nilradical is trivial. The quotient algebra $\operatorname{red} A = A/\mathfrak{n}$ of an algebra A relative to it nilradical is reduced; it is called the *reduction* of the algebra A. Any homomorphism of k-algebras $f: A \to \tilde{A}$ sends the nilradical into the nilradical and therefore defines a homomorphism of their reductions $\operatorname{red} f: \operatorname{red} A \to \operatorname{red} \tilde{A}$.

The *radical* of an ideal I in an algebra A is the ideal $\operatorname{rad} I = \{a \in A \mid a^n \in I$ for some natural number $n\}$, which contains I. In case $I = 0$ this coincides with nilradical of A. We have $\operatorname{red}(A/I) = A/\operatorname{rad} I$; in particular, A/I is reduced if and only if $\operatorname{rad} I = I$. We have $\operatorname{rad}(\operatorname{rad} I) = \operatorname{rad} I$ for any ideal I.

Example 2.1. Let A be the algebra $\mathscr{C}_x$ of germs of all $\mathbb{C}$-valued continuous functions at a point x of some topological space X. Then A is a reduced local $\mathbb{C}$-algebra; its maximal ideal $\mathfrak{m}$ consists of the germs of all functions which vanish at the point x. Analogously, the algebra of germs of all functions of some fixed class C^l at a point $x \in \mathbb{R}^n$ is a reduced local $\mathbb{C}$-algebra. Finally, the algebra of germs of all holomorphic functions at the point $0 \in \mathbb{C}^n$ is a reduced local $\mathbb{C}$-algebra; it is the same as the algebra $\mathbb{C}\{z_1, \ldots, z_n\}$ of convergent power series in the variables $z_1, \ldots, z_n$. The maximal ideal again consists of the germs of functions which vanish at the point 0, i.e., power series with no constant term.

Any nonzero quotient algebra of the algebra $\mathbb{C}\{z_1, \ldots, z_n\}$ is called an *analytic local algebra.*

We note the following algebraic properties of analytic local algebras similar to those of finitely generated algebras over a field.

Theorem 2.1. *The algebra* $\mathbb{C}\{z_1, \ldots, z_n\}$ *is a unique factorization domain; when* $n = 1$ *it is a principal ideal domain.*

Theorem 2.2. *Any analytic local algebra is Noetherian.*

The proof as in the algebraic case goes by induction on n and makes use of Gauss' Theorem that the ring of polynomials is a unique factorization domain and Hilbert's Basis Theorem. The inductive step is made possible by two lemmas due to Weierstrass. A *Weierstrass polynomial* of degree k in z_n is an element $h \in \mathbb{C}\{z_1, \ldots, z_n\}$ having the form

$$h = z_n^k + a_1 z_n^{k-1} + \cdots + a_k,$$

where $a_j \in \mathbb{C}\{z_1, \ldots, z_{n-1}\}$ and $a_j(0) = 0$ for $j = 1, \ldots, k$.

Preparation Lemma. *If a germ* $f \in \mathbb{C}\{z_1, \ldots, z_n\}$ *is such that* $f(0, \ldots, 0, z_n)$ *has a zero of order* k *at the point* $z_n = 0$, *then* $f = uh$, *where* h *is a Weierstrass polynomial of degree* k *relative to* z_n, $u \in \mathbb{C}\{z_1, \ldots, z_n\}$, *and* $u(0) \neq 0$. *Also this representation is unique.*

Division Lemma. *If h is a Weierstrass polynomial of degree k relative to z_n, then any element $f \in \mathbb{C}\{z_1,\ldots,z_n\}$ can be represented uniquely in the form $f = hq + r$, where $q \in \mathbb{C}\{z_1,\ldots,z_n\}$ and r is a polynomial in z_n over $\mathbb{C}\{z_1,\ldots,z_{n-1}\}$ of degree $<k$ (or $r = 0$).*

2. Analytic Sets (see [54], [21]). Let D be an open set in $\mathbb{C}^n$. A set $X \subset D$ is called *analytic* in D if for every point $z_0 \in D$ there is a neighborhood U of the point z_0 in D and there are holomorphic functions $f_1, \ldots, f_p$ in U so that $X \cap U = \{z \in U \mid f_1(z) = \cdots = f_p(z) = 0\}$. An analytic set X is always closed in D. If X contains no connected component of the set D, then X is nowhere dense in D. If D is connected, then the set $D \backslash X$ is also connected.

We shall consider *germs of analytic sets* at the point $0 \in \mathbb{C}^n$, i.e., analytic sets in various neighborhoods of the point 0 in $\mathbb{C}^n$ which are defined up to local coincidence. These germs are closely related to ideals in the algebra $\mathbb{C}\{z_1,\ldots,z_n\}$, viewed as the algebra of germs of holomorphic functions at the point 0 in $\mathbb{C}^n$. Every element $f \in \mathbb{C}\{z_1,\ldots,z_n\}$ determines the germ of an analytic set $V(f)$, which is defined in a neighborhood of the point 0 by the equation $f(z) = 0$. If I is an ideal in the algebra $\mathbb{C}\{z_1,\ldots,z_n\}$, then by the Noetherian property of this algebra, I corresponds to the germ of an analytic set $V(I) = \bigcap_{i=1}^{s} V(f_i)$, where $f_1, \ldots, f_s$ are any set of generators for the ideal I. Conversely, to each germ of an analytic set X at the point 0 there corresponds an ideal $J(X) = \{f \in \mathbb{C}\{z_1,\ldots,z_n\} \mid (f|X = 0)\}$ in the algebra $\mathbb{C}\{z_1,\ldots,z_n\}$. Clearly $V(J(X)) = X$ for any germ of an analytic set X and $I \subset J(V(I))$ for any ideal $I \subset \mathbb{C}\{z_1,\ldots,z_n\}$. The following result holds:

Theorem 2.3. (Analytic Version of the Hilbert Nullstellensatz). *For any ideal $I \subset \mathbb{C}\{z_1,\ldots,z_n\}$ we have*

$$J(V(I)) = \operatorname{rad} I.$$

Thus there is a bijective correspondence between germs of analytic sets at the point 0 and ideals I in the algebra $\mathbb{C}\{z_1,\ldots,z_n\}$ with the property that $\operatorname{rad} I = I$.

A nonempty germ of an analytic set at the point 0 is called *irreducible* if it is impossible to represent it as a union of two germs of analytic sets distinct from it. An irreducible germ X is characterized by the fact that $J(X)$ is a prime ideal, i.e., that $\mathbb{C}\{z_1,\ldots,z_n\}/J(X)$ is an integral domain. It follows from the Noetherian property of the algebra $\mathbb{C}\{z_1,\ldots,z_n\}$ that any nonempty germ X can be represented as a union $X = X_1 \cup \cdots \cup X_k$, where the X_i are maximal irreducible germs of analytic sets contained in X; they are called the *irreducible components* of the germ X. Thus the local study of an analytic set reduces to the case when its germ is irreducible. In the irreducible case the local algebra $\mathbb{C}\{z_1,\ldots,z_n\}/J(X)$ is described by the following theorem.

Theorem 2.4. *Let I be a prime ideal in the algebra $\mathbb{C}\{z_1,\ldots,z_n\}$. Then coordinates z_i can be chosen so as to satisfy the following requirements:*

1) $I \cap \mathbb{C}\{z_1,\ldots,z_s\} = 0$ *for some* $s \leqslant n$;

2) *the algebra* $\mathbb{C}\{z_1,\ldots,z_n\}/I$ *is integral over the naturally embedded subalgebra* $\mathbb{C}\{z_1,\ldots,z_s\}$ (*i.e., any element of the algebra* $\mathbb{C}\{z_1,\ldots,z_n\}/I$ *is the root of a polynomial over* $\mathbb{C}\{z_1,\ldots,z_s\}$ *with leading coefficient* 1).

3) *The field of fractions of the algebra* $\mathbb{C}\{z_1,\ldots,z_n\}/I$ *is generated over the field of fractions of the algebra* $\mathbb{C}\{z_1,\ldots,z_s\}$ *by the element* $z_{s+1} + I$.

Geometrically this means that the projection $\pi: \mathbb{C}^n \to \mathbb{C}^s$ which sends $(z_1,\ldots,z_n)$ to $(z_1,\ldots,z_s)$ maps $X = V(I)$, an irreducible analytic set at 0, in a sufficiently small neighborhood of zero Δ onto some open polycylinder $\Delta' \subset \mathbb{C}^s$. The mapping $\pi: X \to \Delta'$ is proper and there exists in Δ' an analytic set $Z \neq \Delta'$ so that $X \backslash Z$ is a connected complex submanifold in Δ and $\pi: X\backslash Z \to \Delta' \backslash Z$ is an l-sheeted covering, where l is the degree of the minimal polynomial of the element $z_{s+1} + I$ over $\mathbb{C}\{z_1,\ldots,z_s\}$.

Thus in a neighborhood of 0 an irreducible analytic set X at 0 is in a neighborhood of 0 a branched covering over a polycylinder in $\mathbb{C}^s$. The integer s is called the *dimension* of the analytic set X (or its germ) at the point 0 and is denoted by $\dim_0 X$. It is equal to half of the topological dimension of the set X at the point 0.

If X is an irreducible germ of an analytic set at $0 \in \mathbb{C}^n$, then the dimension of the germ X at the point 0 is defined by the formula

$$\dim_0 X = \max_i \dim_0 X_i,$$

where the X_i are the irreducible components of the germ X.

A germ $X \subset \mathbb{C}^n$ is called a *complete intersection* at the point 0 if $\dim_0 X_i = s$ for all irreducible components X_i and the ideal defining X has $n - s$ generators.

3. Ringed Spaces. A topological space X with a fixed sheaf of rings $\mathscr{A}$ defined on it is called a *ringed space*. The sheaf $\mathscr{A}$ is called the *structure sheaf* of the space $(X, \mathscr{A})$. We shall assume without mentioning it that the structure sheaf is a sheaf of associative and commutative local algebras over some field k (in what follows k will be the field $\mathbb{C}$ of complex numbers).

Let $(X, \mathscr{A})$ and $(Y, \mathscr{B})$ be two ringed spaces over the same field k. A *morphism* from $(X, \mathscr{A})$ to $(Y, \mathscr{B})$ is a pair $F = (f, f^\#)$, where $f: X \to Y$ is a continuous mapping and $f^\#: f^*\mathscr{B} \to \mathscr{A}$ is a homomorphism of sheaves of k-algebras which sends the unity section to the unity section. The mappings f and $f^\#$ we shall call respectively the *geometric* and *algebraic components* of the morphism F. We note that $f^\#$ defines for every $x \in X$ a homomorphism of k-algebras $f_x^\#: \mathscr{B}_{f(x)} \to \mathscr{A}_x$. Further, the homomorphism $\Gamma(f^\#): \Gamma(X, f^*\mathscr{B}) \to \Gamma(X, \mathscr{A})$ when composed with the natural homomorphism $\Gamma(Y, \mathscr{B}) \to \Gamma(X, f^*\mathscr{B})$ from Chapter 1, Section 6 gives an algebra homomorphism $f^*: \Gamma(Y, \mathscr{B}) \to \Gamma(X, \mathscr{A})$.

Let $G = (g, g^\#): (Y, \mathscr{B}) \to (Z, \mathscr{C})$ be a second morphism of ringed spaces and let $h = g \circ f: X \to Z$. Then the homomorphisms $h_x^\# = f_x^\# \circ g_{f(x)}^\#: \mathscr{C}_{h(x)} \to \mathscr{A}_x$ define a sheaf homomorphism $h^\#: h^*\mathscr{C} \to \mathscr{A}$. The morphism $H = (h, h^\#)$ is called the *composition* of the morphisms G and F and is denoted by $G \circ F$. Ringed spaces over a field k and their morphisms form a category with respect to this operation.

Let $(Y, \mathscr{B})$ be a ringed space and X a subset of Y. By setting $\mathscr{A} = \mathscr{B}|X$, we obtain a ringed space $(X, \mathscr{A})$. Recalling that $\mathscr{A} = i^*\mathscr{B}$, where $i: X \to Y$ is the inclusion, we obtain a morphism $(i, \mathrm{id}): (X, \mathscr{A}) \to (Y, \mathscr{B})$. The space $(X, \mathscr{A})$ will be most often considered when X is open in Y (an *open subspace* of a ringed space).

Let $(X, \mathscr{A})$ be a ringed space. The nilradicals $\mathscr{n}_x$ of the stalks $\mathscr{A}_x$ of the sheaf $\mathscr{A}$ form a subsheaf of ideals $\mathscr{n} \subset A$ called the *nilradical* of the sheaf $\mathscr{A}$. The ringed space $(X, \mathrm{red}\,\mathscr{A})$, where $\mathrm{red}\,A = A/\mathscr{n}$, is called the *reduction* of the space $(X, \mathscr{A})$. A space $(X, \mathscr{A})$ is called *reduced* if $\mathscr{n} = 0$.

The most important ringed spaces in the sequel (in particular, all reduced complex analytic spaces) have structure sheaves consisting of germs of functions. However, also in the general case we can associate with sections of a ringed space $(X, \mathscr{A})$ certain functions on X. Let $\mathscr{m}_x$ be the maximal ideal of the algebra $\mathscr{A}_x$ for $x \in X$, and let $v_x: \mathscr{A}_x \to k$ denote the projection orthogonal to $\mathscr{m}_x$. To any section $s \in \Gamma(X, \mathscr{A})$ there corresponds a function $X \to k$ defined by the formula $x \mapsto v_x(s_x)$ for $x \in X$. It is clear that the section $\Gamma(\mathrm{red})s \in \Gamma(X, \mathrm{red}\,\mathscr{A})$ defines the same function on X as does s.

Example 2.2. Any topological space X can be turned into a reduced ringed space over $\mathbb{C}$ by taking as the structure sheaf the sheaf $\mathscr{C}_X$ of germs of continuous functions. Let $f: X \to Y$ be a continuous mapping into some other topological space Y. Then for any open set $U \subset Y$ a homomorphism $f^*: \Gamma(U, \mathscr{C}_Y) \to \Gamma(f^{-1}(U), \mathscr{C}_X)$ is defined by the formula $(f^*\varphi)(x) = \varphi(f(x))$ for $x \in f^{-1}(U)$. These homomorphisms define homomorphisms $f_x^\#: \mathscr{C}_{Y, f(x)} \to \mathscr{C}_{X,x}$ for $x \in X$ which form a sheaf homomorphism $f^\#: f^*\mathscr{C}_Y \to \mathscr{C}_X$. Thus every continuous mapping $f: X \to Y$ canonically defines a morphism $(f, f^\#): (X, \mathscr{C}_X) \to (Y, \mathscr{C}_Y)$. It is not difficult to deduce from the fact that a homomorphism of local algebras which sends 1 to 1 is local that any morphism of ringed spaces has this form, i.e., the algebraic component of the morphism is completely determined by the geometric component.

Example 2.3. If the spaces are not reduced, then the algebraic component of a morphism is not necessarily determined by the geometric one. For example, the structure of a ringed space on the space $X = \{x\}$ consisting of a single point is defined by any local k-algebra A. Morphisms from the space $(\{x\}, A)$ into itself have the form (id, α) where α is an arbitrary endomorphism of the algebra A such that $\alpha(1) = 1$.

Example 2.4. We consider the reduced ringed space $(D, \mathcal{O})$, where D is an open set in $\mathbb{C}^n$ (see Example 1.3). The morphisms of spaces of this type are holomorphic mappings from open sets in $\mathbb{C}^n$ into themselves. A *complex* (*analytic*) *manifold* of dimension n is defined to be a Hausdorff ringed space $(X, \mathcal{O}_X)$ over $\mathbb{C}$ which is locally isomorphic to a space of the form $(D, \mathcal{O})$, where D is an open set in $\mathbb{C}^n$. This means that any point of the space X must possess a neighborhood U so that the ringed space $(U, \mathcal{O}_X(U))$ is isomorphic to a space of the form $(D, \mathcal{O})$, where D

is an open set in $\mathbb{C}^n$. A complex manifold is by definition reduced. Clearly sections of the sheaf $\mathcal{O}_X$ can be identified with certain functions on X which are known as the *holomorphic* functions. If $(Y, \mathcal{O}_Y)$ is another complex manifold, the morphism $(f, f^\#): (X, \mathcal{O}_X) \to (Y, \mathcal{O}_Y)$ is completely determined by the mapping $f: X \to Y$ and is called a *holomorphic mapping*; the homomorphism $f^\#$ is defined by the induced mapping of functions $\varphi \mapsto \varphi \circ f$. Holomorphic functions on X can be viewed also as holomorphic mappings $X \to \mathbb{C}$. One-dimensional complex manifolds are also called *Riemann surfaces*.

The definition of complex manifold given above is completely equivalent to the usual definition involving charts and atlases. Its advantage over the latter definition consists not only in its brevity but also in the ease of generalization and translation to other situations. Thus, considering ringed spaces which are locally isomorphic to spaces of the form $(D, \mathscr{C}^l)$, where D is an open set in $\mathbb{R}^n$ (see Example 1.1), we obtain the concept of an n-dimensional *differentiable manifold* of class C^l. On the same principle is based the definition of a complex (analytic) space (see Chapter 2, Section 4) and also those of an algebraic variety, scheme, and supermanifold.). We note that already in the book of H. Weyl [112] one-dimensional complex manifolds were essentially defined as ringed spaces.

The next example can be viewed as setting up "local models" for the definition of a complex space.

Example 2.5. Let X be an analytic set in an open set $D \subset \mathbb{C}^n$. A function on an open set $U \subset X$ with values in $\mathbb{C}$ will be called *holomorphic* if it extends to a holomorphic function on some open set $V \subset \mathbb{C}^n$ such that $V \cap X = U$. This concept allows us to define on X the *sheaf of germs of holomorphic functions* $\mathcal{O}_X$ which is a subsheaf of $\mathscr{C}_X$. It is easy to see that $\mathcal{O}_X$ can be represented as a certain quotient sheaf of the sheaf $\mathcal{O}_D$ restricted to X. Indeed, $\mathcal{O}_X \cong (\mathcal{O}_D/\mathscr{J}(X))|X$, where $\mathscr{J}(X)$ is the subsheaf of germs of functions which vanish on X. Thus for any point $x \in X$ the algebra $\mathcal{O}_{X,x} \cong \mathcal{O}_{D,x}/\mathscr{J}(X)_x$ is an analytic local algebra. Therefore $(X, \mathcal{O}_X)$ is a reduced ringed space over $\mathbb{C}$. We note also that $\mathcal{O}_D/\mathscr{J}(X) \cong (\mathcal{O}_X)^D$ (the trivial extension of the sheaf $\mathcal{O}_X$).

4. Coherent Sheaves of Modules. In Example 2.5 we associated to each analytic set X in a domain D in $\mathbb{C}^n$ a certain sheaf of ideals $\mathscr{J}(X) \subset \mathcal{O}_D$. A natural question is: can we characterize those sheaves of ideals which can arise in this way? This question is closely related to the technical problem of transporting to sheaves of modules the classical idea of a finitely generated module. If we fix a point $x_0 \in D$, then as we saw in Chapter 2, Section 2, any ideal in the algebra $\mathcal{O}_{D,x_0}$ is finitely generated and therefore defines the germ of an analytic set at the point x_0. It turns out that finite sets of generators for the ideals $\mathscr{J}(X)_x$ can be chosen in some neighborhood U of the point x_0 in D in a compatible (coherent) fashion, so that they will be germs of certain holomorphic functions $f_1, \ldots, f_p$ in U (here $X \cap U$ is clearly defined by the equations $f_1 = \cdots = f_p = 0$).

The following simple example shows that not every subsheaf of ideals in $\mathcal{O}_D$ possesses this property.

Example 2.6. We define a subsheaf of ideals $\mathscr{I} \subset \mathcal{O}_{\mathbb{C}}$ by putting $\mathscr{I}_x = \mathcal{O}_{\mathbb{C},x}$ if $x \neq 0$ and $\mathscr{I}_0 = 0$. Then in any neighborhood U of the point 0 the only section of the sheaf $\mathscr{I}$ is the zero section, hence $\mathscr{I}|U$ is not generated by its sections.

We now introduce, following Serre [93], the general concept of a coherent sheaf of $\mathscr{A}$-modules on an arbitrary ringed space $(X, \mathscr{A})$. We note that the requirement which we have imposed that the algebras $\mathscr{A}_x$ for $x \in X$ be local is not necessary here.

Let $\mathscr{F}$ be a sheaf of $\mathscr{A}$-modules on a ringed space $(X, \mathscr{A})$ and let $s_1, \ldots, s_p \in \Gamma(X, \mathscr{F})$ be sections of the sheaf $\mathscr{F}$. We consider the homomorphism of sheaves of modules $\varphi : \mathscr{A}^p \to \mathscr{F}$ defined by the formula

$$\varphi(u_1, \ldots, u_p) = \sum_{i=1}^{p} u_i s_{ix} \qquad (u_i \in \mathscr{A}_x, x \in X).$$

Its kernel $\operatorname{Ker} \varphi = R(s_1, \ldots, s_p) \subset \mathscr{A}^p$ is called the *sheaf of relations* among the sections $s_1, \ldots, s_p$, and the image $\operatorname{Im} \varphi = [s_1, \ldots, s_p]_{\mathscr{A}} \subset \mathscr{F}$ the *subsheaf generated by the sections* $s_1, \ldots, s_p$.

A sheaf of modules $\mathscr{F}$ is called *free* of rank p if it is isomorphic to the sheaf $\mathscr{A}^p$ for some integer $p \geqslant 0$, i.e., if it is generated by sections $s_1, \ldots, s_p \in \Gamma(X, \mathscr{F})$ so that $R(s_1, \ldots, s_p) = 0$. The collection of sections $s_1, \ldots, s_p$ is called the *basis* for the free sheaf $\mathscr{F}$. A sheaf $\mathscr{F}$ is called *locally free* if it is free over some neighborhood of every point in X.

A sheaf $\mathscr{F}$ is called a *sheaf of finite type* if each point of the space X has a neighborhood U so that the sheaf $\mathscr{F}|U$ is generated by a finite number of sections in $\Gamma(U, \mathscr{F})$, i.e., if there exists an exact sequence of sheaves of modules over $\mathscr{A}|U$ of the form

$$\mathscr{A}^p|U \to \mathscr{F}|U \to 0,$$

where p depends on U.

In the case when the ringed space is $(D, \mathcal{O}_D)$, where D is an open set in $\mathbb{C}^n$, the following theorems hold, whose proofs are based on the study of analytic local algebras and the local structure of analytic sets.

Theorem 2.5 (Cartan, see [54]). *Let X be an analytic set in an open set $D \subset \mathbb{C}^n$. Then the sheaf of ideals $\mathscr{I}(X) \subset \mathcal{O}_D$ is a sheaf of finite type over $\mathcal{O}_D$.*

Theorem 2.6 (Oka, see [54]). *Let D be an open subset of $\mathbb{C}^n$ and let $s_1, \ldots, s_p \in \Gamma(D, \mathcal{O}_D^q)$ for some natural number q. Then the sheaf of relations $R(s_1, \ldots, s_p) \subset \mathcal{O}_D^p$ is a sheaf of $\mathcal{O}_D$-modules of finite type.*

In [15] a subsheaf of $\mathcal{O}_D$-modules $\mathscr{F} \subset \mathcal{O}_D^p$ is called "coherent" if $\mathscr{F}$ is a sheaf of modules of finite type. In this terminology the theorems of Cartan and Oka say that the sheaves $\mathscr{I}(X)$ and $R(s_1, \ldots, s_p)$ are coherent. In the general case it turns out to be convenient to include in the definition of a coherent sheaf of modules the requirement that the sheaf of relations also be of finite type.

A sheaf of $\mathscr{A}$-modules $\mathscr{F}$ is called *coherent* if it is a sheaf of finite type and if for any sections $s_1, \ldots, s_p \in \Gamma(U, \mathscr{F})$ over some open set U the sheaf of relations

$R(s_1, \ldots, s_p)$ is a sheaf of finite type on U. It follows from this that every point in the space X has a neighborhood U over which there exists an exact sequence of sheaves of modules over $\mathscr{A}|U$ of the form

$$\mathscr{A}^q|U \to \mathscr{A}^p|U \to \mathscr{F}|U \to 0, \tag{1}$$

where p and q depend on U.

In particular, the structure sheaf $\mathscr{A}$ viewed as a sheaf of $\mathscr{A}$-modules is coherent if and only if for any $s_i \in \Gamma(U, \mathscr{A})$ with $i = 1, \ldots, p$ and U open in X the sheaf $R(s_1, \ldots, s_p)$ is a sheaf of finite type on U. In this situation the ringed space $(X, \mathscr{A})$ is called *coherent*. It follows from the theorem of Oka that the sheaf $\mathcal{O}_D$ for an open set $D \subset \mathbb{C}^n$ is coherent (as is also any free $\mathcal{O}_D$-module). Since the concept of coherence is local, any complex manifold $(X, \mathcal{O}_X)$ is also coherent.

The following theorem describes the basic properties of a coherent sheaf of modules.

Theorem 2.7 (see [54], [93]).

1) *If in an exact sequence* $0 \to \mathscr{F}' \to \mathscr{F} \to \mathscr{F}'' \to 0$ *of sheaves of* $\mathscr{A}$*-modules two of the three sheaves* $\mathscr{F}'$, $\mathscr{F}$, $\mathscr{F}''$ *are coherent, then the third sheaf is also coherent. In particular, the direct sum of a finite number of coherent sheaves is a coherent sheaf.*

2) *The kernel, cokernel, and image of a homomorphism from one coherent sheaf of* $\mathscr{A}$*-modules into another are coherent sheaves.*

3) *The tensor product* $\mathscr{F} \otimes_{\mathscr{A}} \mathscr{G}$ *of two coherent sheaves of* $\mathscr{A}$*-modules is a coherent sheaf.*

4) *If* $\mathscr{F}$ *and* $\mathscr{G}$ *are coherent sheaves of* $\mathscr{A}$*-modules, then the sheaf* $\mathscr{H}om_{\mathscr{A}}(\mathscr{F}, \mathscr{G})$ *is coherent, where* $\mathscr{H}om_{\mathscr{A}}(\mathscr{F}, \mathscr{G})_x \cong \mathrm{Hom}_{\mathscr{A}_x}(\mathscr{F}_x, \mathscr{G}_x)$ *for any* $x \in X$.

Corollary 1. *If the space* $(X, \mathscr{A})$ *is coherent, then a subsheaf of the free sheaf* $\mathscr{A}^p$ *is coherent if and only if it is a sheaf of finite type.*

Corollary 2. *If the space* $(X, \mathscr{A})$ *is coherent, then the existence for a sheaf of* $\mathscr{A}$*-modules* $\mathscr{F}$ *of an exact sequence of the form* (1) *over a neighborhood of each point in X is not only necessary but also sufficient for the coherence of the sheaf* $\mathscr{F}$.

Corollary 3. *If the space* $(X, \mathscr{A})$ *is coherent and* $\mathscr{I} \subset \mathscr{A}$ *is a coherent sheaf of ideals, then* $\mathscr{A} \backslash \mathscr{I}$ *is a coherent sheaf of algebras.*

Corollary 4. *If the space* $(X, \mathscr{A})$ *is coherent and* $\mathscr{F}$ *is a coherent sheaf of* $\mathscr{A}$*-modules, then the ideals* $\mathscr{I}_x = \mathrm{Ann}\, \mathscr{F}_x \subset \mathscr{A}_x$ *form a coherent sheaf of ideals* $\mathscr{I} = \mathrm{Ann}\, \mathscr{F} \subset \mathscr{A}$.

5. Analytic Spaces (see [49], [54]). In the reduced case a complex analytic space is a ringed space over $\mathbb{C}$ which is locally isomorphic to a space of the form $(X, \mathcal{O}_X)$ from Example 2.5 (see Chapter 2, Section 3). Since we wish to include the unreduced case also, we first define an unreduced structure on an analytic set.

Let $D \subset \mathbb{C}^n$ be an open set and let $\mathscr{I}$ be a sheaf of ideals in $\mathcal{O}_D$. Then we may consider the subset $V(\mathscr{I}) \subset D$—the *set of zeros of the ideal sheaf* $\mathscr{I}$. It can be defined as the set of those $x \in X$ such that $\varphi(x) = 0$ for all functions φ whose germs belong to $\mathscr{I}$, or equivalently by the formula

$$V(\mathscr{I}) = \{x \in D | \mathscr{I}_x \neq \mathcal{O}_{D,x}\} = \operatorname{supp}(\mathcal{O}_D/\mathscr{I}).$$

If the ideal sheaf $\mathscr{I}$ is coherent (or equivalently a sheaf of finite type), then $V(\mathscr{I})$ is clearly an analytic set. Any analytic set X in D can be obtained in this fashion, since X is the zero set of the ideal sheaf $\mathscr{J}(X)$ which is coherent by the Cartan Theorem. The same X can be the zero set for different coherent ideal sheaves, for example, the sheaves $\mathscr{J}(X)^m$ for any integer m.

If $X = V(\mathscr{I})$ is the zero set of a coherent ideal sheaf $\mathscr{I} \subset \mathcal{O}_D$, then we fix on X the sheaf of algebras $\mathscr{A} = (\mathcal{O}_D/\mathscr{I})|X$. Then $(X, \mathscr{A})$ is a ringed space. Such spaces corresponding to various coherent ideal sheaves $\mathscr{I} \subset \mathcal{O}_D$ are called *analytic subspaces* of D. By the analytic version of the Hilbert Nullstellensatz (see Chapter 2, Section 2) the sheaf $\operatorname{rad} \mathscr{I} = \bigcup_{x \in D} \operatorname{rad} \mathscr{I}_x$ coincides with $\mathscr{J}(X)$ so the reduction of the space $(X, \mathscr{A})$ is the space $(X, \mathcal{O}_X)$ from Example 2.5. In particular, the space $(X, \mathscr{A})$ is reduced if and only if $\mathscr{I} = \mathscr{J}(X)$. It follows from Corollary 3 of Theorem 2.7 that the space $(X, \mathscr{A})$ is coherent.

We are now able to give the general definition of a complex space.

A *complex analytic space* (or simply a *complex space*) is a Hausdorff ringed space $(X, \mathcal{O}_X)$ over $\mathbb{C}$ each of whose points has a neighborhood U so that the open subspace $(U, \mathcal{O}_X | U)$ is isomorphic to an analytic subspace of an open set in $\mathbb{C}^n$. Morphisms of analytic spaces are also called *holomorphic mappings*.

It is clear that complex spaces and their holomorphic mappings form a category.

In the notation for an analytic space we shall sometimes for brevity omit the structure sheaf, i.e., we shall write X instead of $(X, \mathcal{O}_X)$. We shall also omit the algebraic component in the notation for a morphism.

Sections of the structure sheaf $\mathcal{O}_X$ of a complex space X are in natural correspondence with holomorphic mappings $X \to \mathbb{C}$; we shall call them *holomorphic functions* on X. As we saw in Chapter 2, Section 3, every section $f \in \Gamma(X, \mathcal{O}_X)$ defines an actual function $X \to \mathbb{C}$, which we shall denote by the same letter f, but which does not completely determine the section f.

It follows from the definition that for the local study of a complex space the space can be replaced by an analytic subspace of an open set in $\mathbb{C}^n$; this is a so-called *local model* for a complex space. A local isomorphism from a complex space X to its local model is called a *local chart*, and a set of local charts covering all of X is called an *analytic atlas*.

A point $x \in X$ is called a *smooth* (or *nonsingular*, or *simple*) point of the complex space if in a neighborhood of the point x the space admits a local model of the form $(D, \mathcal{O}_D)$, where D is open in $\mathbb{C}^n$. In the alternative case the point x is called *singular*. A complex space all of whose points are smooth (with the same n in all local models) is a complex manifold.

A point $x \in X$ is called a *reduced point* of the complex space X if the algebra $\mathcal{O}_{X,x}$ is reduced, or equivalently if in a neighborhood of the point x there exists a reduced local model (for example, a smooth point is always reduced). On a reduced complex space $(X, \mathcal{O}_X)$ sections of the sheaf $\mathcal{O}_X$ over an open set $U \subset X$ are identified with the corresponding functions on U, which in the local models are represented by holomorphic functions on analytic sets (see Example 2.5). In the general case the reduction $(X, \operatorname{red} \mathcal{O}_X)$ of a space $(X, \mathcal{O}_X)$ is a reduced complex space and the nilradical $\mathfrak{n}$ of the structure sheaf consists of the germs of all sections defining zero functions on X.

Any complex space X is coherent. Sheaves of $\mathcal{O}_X$-modules on X are usually called *analytic sheaves.*

Every complex space X is locally compact and locally arcwise connected. We usually assume in addition that X has a countable basis of open sets. A complex space with a countable basis is always metrizable (in particular, it is paracompact) and is the union of a countable family of compact sets.

Let X be a complex space. A set $Y \subset X$ is called *analytic* if in a neighborhood of each point of the space X the set Y is defined by a set of equations of the form $f_1 = \cdots = f_p = 0$, where the f_i are holomorphic functions in this neighborhood. An analytic set is always closed in X. Any coherent ideal sheaf $\mathcal{I} \subset \mathcal{O}_X$ defines the analytic set $Y = V(\mathcal{I}) = \operatorname{supp}(\mathcal{O}_X/\mathcal{I})$ and the ringed space $(Y, \mathcal{O}_Y)$, where $\mathcal{O}_Y = (\mathcal{O}_X/\mathcal{I})|Y$. This space, as is easy to see, is again a complex space; it is called an *analytic subspace* of the space X. The inclusion $i: Y \to X$ and the natural homomorphism $i^*: \mathcal{O}_X|Y \to \mathcal{O}_Y$ define the holomorphic mapping $(i, i^*): (Y, \mathcal{O}_Y) \to (X, \mathcal{O}_X)$.

We give some examples of analytic sets and subspaces. The reduction $(X, \operatorname{red} \mathcal{O}_X)$ is an analytic subspace corresponding to the coherent subsheaf $\mathfrak{n} \subset \mathcal{O}_X$. The support $\operatorname{supp} \mathcal{F}$ of any coherent analytic sheaf of modules $\mathcal{F}$ on X is an analytic set. In particular, the set of unreduced points is an analytic set in X. Also analytic is the set $\operatorname{Sing} X$ of all singular points; if X is reduced then $\operatorname{Sing} X$ is nowhere dense in X. In particular, $X_{\mathrm{reg}} = X \setminus \operatorname{Sing} X$ is open in X. Connected components of the open subspace $(X_{\mathrm{reg}}, \mathcal{O}_X|X_{\mathrm{reg}})$ are complex manifolds.

Example 2.7. Let $F = (f, f^\#): X \to Y$ be a holomorphic mapping of complex spaces. If $y \in Y$ then the set $X_y = f^{-1}(y)$ is called the *fiber of the mapping f* over the point y. We assume that $X_y \neq \varnothing$ and introduce on X_y the natural structure of an analytic subspace of X. We denote by $\mathfrak{m}(y)$ the sheaf of ideals $\mathcal{I}(\{y\}) \subset \mathcal{O}_Y$; we have $\mathfrak{m}(y)_z = \mathcal{O}_{Y,z}$ for $z \neq y$ and $\mathfrak{m}(y)_y = \mathfrak{m}_y$, the maximal ideal in the algebra $\mathcal{O}_{Y,y}$. Then the sheaf of ideals $\hat{\mathfrak{m}} \subset \mathcal{O}_X$ which is generated by the subsheaf $f^\#(f^*\mathfrak{m}(y))$ is coherent and $X_y = \operatorname{supp}(\mathcal{O}_X/\hat{\mathfrak{m}})$, so $\hat{\mathfrak{m}}$ defines on X_y the structure of an analytic subspace $(X_y, (\mathcal{O}_X/\hat{\mathfrak{m}})|X_y)$.

Let Y be an analytic subspace of X and let $\mathcal{F}$ be an analytic sheaf on Y. Then $\mathcal{F}$ has the natural structure of a sheaf of modules over $\mathcal{O}_X|Y$. Therefore $\mathcal{F}^X$ can be viewed as an analytic sheaf on X.

Theorem 2.8. *If $\mathcal{F}$ is a coherent analytic sheaf on an analytic subspace $Y \subset X$, then $\mathcal{F}^X$ is a coherent analytic sheaf on X.*

The *dimension of a complex space X at a point $x \in X$* is the dimension of the reduction of the corresponding local model at the point corresponding to x; this dimension is denoted by $\dim_x X$. The *dimension of the space X* is given by $\dim X = \sup_{x \in X} \dim_x X$ (it is a nonnegative integer or ∞). If Y is an analytic subspace of X then $\dim_y Y \leqslant \dim_y X$ for all $y \in Y$, and strict inequality occurs for all $y \in Y$ if and only if Y is nowhere dense in X. An analytic subspace $Y \subset X$ is called a *hypersurface* if $\dim_y Y = \dim_y X - 1$ for all $y \in Y$.

The *tangent space* at a point $x \in X$ to a complex space X is the space $T(X)_x = \mathrm{Der}(\mathcal{O}_{X,x}, \mathbb{C})$ of all derivations from the algebra $\mathcal{O}_{X,x}$ into $\mathbb{C}$, i.e., all linear mappings $\tau : \mathcal{O}_{X,x} \to \mathbb{C}$ satisfying the condition $\tau(\varphi\psi) = \varphi(x)\tau(\psi) + \psi(x)\tau(\varphi)$. The space $T(X)_x^*$ is called the *cotangent space* at the point x. To each element $\varphi \in \mathcal{O}_{X,x}$ corresponds its *differential* $d\varphi \in T(X)_x^*$ defined by the formula

$$(d\varphi)(\tau) = \tau(\varphi) \quad (\tau \in T(X)_x).$$

The correspondence $\varphi \mapsto d\varphi$ defines an isomorphism

$$\mathfrak{m}_x/\mathfrak{m}_x^2 \cong T(X)_x^*,$$

where $\mathfrak{m}_x$ is the maximal ideal of the local algebra $\mathcal{O}_{X,x}$. The number

$$\operatorname{im}\dim_x X = \dim T(X)_x$$

coincides with the minimal number of generators of the ideal $\mathfrak{m}_x$ and also with the minimal number n so that X in a neighborhood of the point x has a local model which is an analytic subspace of a domain in $\mathbb{C}^n$; it is called by *tangential dimension* or the *imbedding dimension* at the point x. We have

$$\dim_x X \leqslant \operatorname{im}\dim_x X,$$

with equality if and only if x is a smooth point. We define also the number

$$\operatorname{im}\dim X = \sup_{x \in X} \operatorname{im}\dim_x X.$$

A complex space X is called *irreducible at a point $x \in X$* if $\mathcal{O}_{X,x}$ is an integral domain. This means that X in a neighborhood of the point x has a reduced local model which is irreducible at the corresponding point. There is also the corresponding global concept: a complex space X is called *irreducible* if it cannot be decomposed as a union of two analytic sets, neither of which coincides with X. If X is irreducible, then X is connected and is *pure dimensional*, i.e., $\dim_x X$ does not depend on $x \in X$. An arbitrary complex space X with a countable basis can be decomposed into a union of a countable locally finite family of nonempty irreducible analytic sets X_i. The sets X_i are uniquely defined and are called the *irreducible components* (or *irreducible branches*) of the space X.

6. Normal Spaces (see [49], [54], [77]). An integral domain A is called *normal* if it is integrally closed in its field of quotients M, i.e., if each element of the field

M which is a root of a polynomial over A with leading coefficient 1 belongs to A. A point x of a complex space X is called *normal* if the algebra $\mathcal{O}_{X,x}$ is normal. A space X is called *normal* if all its points are normal. At any normal point a complex space is necessarily reduced and irreducible. Any smooth point is normal.

Example 2.8. Let X be the analytic set in $\mathbb{C}^2$ defined by the equation $z_1^3 - z_2^2 = 0$ (the *semicubical parabola*) endowed with the natural structure of a reduced complex space (see Example 2.5). The element $\varphi = \frac{z_2}{z_1}$ of the field of fractions of the ring $\mathcal{O}_{X,0}$ satisfies the equation $\varphi^2 = z_1$ but does not belong to $\mathcal{O}_{X,0}$. Therefore X is not normal at the point 0. At the other points X is normal.

The set $N(X) \subset \operatorname{Sing} X$ of nonnormal points of a complex space X is analytic. At any normal point $x \in X \backslash N(X)$ we have

$$\dim_x \operatorname{Sing} X \leqslant \dim_x X - 2.$$

In particular, if $\dim_x X = 1$ at all points $x \in X$ and X is normal, then X is a one-dimensional complex manifold.

The most important property of normal spaces is given by the following result which is called the *Riemann Removable Singularity Theorem.*

Theorem 2.9. *Let A be an analytic set in a normal complex space X.*

1) *If $\dim_x A < \dim_x X$ at each point $x \in A$, then any holomorphic function on $X \backslash A$ which is bounded in a neighborhood of each point of A extends to a holomorphic function on X.*

2) *If $\dim_x A \leqslant \dim_x X - 2$ at each point $x \in A$, then any holomorphic function on $X \backslash A$ extends to a holomorphic function on X.*

We note also that if X is reduced and if we require that 1) holds for functions in all open sets of the space X, then X will be normal.

A *normalization* of a reduced complex space X is a holomorphic mapping $f: \hat{X} \to X$ having the following properties: a) $\tilde{X}$ is a normal complex space; b) $f: \tilde{X} \backslash f^{-1}(N(X)) \to X \backslash N(X)$ is an isomorphism of open subspaces; c) f is surjective and proper, and $f^{-1}(x)$ is finite for each $x \in X$.

Theorem 2.10 (see [50], [77]). *For any reduced space X there exists a normalization $f: \tilde{X} \to X$. If $f_1: \hat{X}_1 \to X$ is another normalization of the same space, then there exists an isomorphism $g: \tilde{X} \to \hat{X}_1$ so that $f = f_1 \circ g$.*

For example, a normalization of the space X of Example 2.8 is given by the mapping $f: \mathbb{C} \to X$ defined by the formula

$$f(t) = (t^2, t^3).$$

This bijective holomorphic mapping is not, however, an isomorphism of the complex spaces.

7. Construction of Complex Spaces.

a) Gluing. Let there be given a family of ringed spaces $(X_i, \mathscr{A}_i)$ for $i \in I$ over a field k and for each pair $i, j \in I$ with $i \neq j$ distinguished open sets $X_{ij} \subset X_i$ and $X_{ji} \subset X_j$ and isomorphisms $F_{ij} = (f_{ij}, f_{ij}^{\#}): (X_{ji}, \mathscr{A}_j | X_{ji}) \to (X_{ij}, \mathscr{A}_i | X_{ij})$. Let us assume that the following conditions are satisfied:

1) $F_{ji} = F_{ij}^{-1}$ for all $i, j \in I$ with $i \neq j$.

2) For any distinct $i, j, k \in I$ such that $X_{ij} \cap X_{ik} \neq \varnothing$ we have $f_{ij}(X_{ji} \cap X_{jk}) = X_{ij} \cap X_{ik}$ and $F_{ik} = F_{ij} \circ F_{jk}$ on $X_{kj} \cap X_{ki}$. In this situation it is not difficult to show that there exists a ringed space $(X, \mathscr{A})$ over k which is unique up to isomorphism and possesses the following property: there is an open covering $(U_i)_{i \in I}$ of the space X and isomorphisms $G_i = (g_i, g_i^{\#}): (U_i, \mathscr{A} | U_i) \to (X_i, \mathscr{A}_i)$ so that $g_i(U_i \cap U_j) = X_{ij}$ and $G_i = F_{ij} \circ G_j$ on $U_i \cap U_j$ for any $i \neq j$ so that $U_i \cap U_j \neq \varnothing$. We say that the space $(X, \mathscr{A})$ is obtained from the spaces $(X_i, \mathscr{A}_i)$ by *gluing* with the help of the isomorphisms F_{ij}.

It is clear that by gluing complex spaces we again obtain a complex space. From the definition of a complex space it is evident that it can always be obtained by gluing together analytic subspaces of open sets in $\mathbb{C}^n$.

b) Direct Products. Let X and Y be analytic sets in open sets $U \subset \mathbb{C}^n$ and $V \subset \mathbb{C}^m$ respectively and let $\mathscr{I} \subset \mathscr{O}_U$ and $\mathscr{J} \subset \mathscr{O}_V$ be coherent sheaves of ideals whose zero sets are X and Y. We define the *direct product* of the corresponding analytic subspaces $(X, (\mathscr{O}_U/\mathscr{I})|X)$ and $(Y, (\mathscr{O}_V/\mathscr{J})|Y)$ to be the analytic subspace in $U \times V \subset \mathbb{C}^{n+m}$ corresponding to the ideal sheaf $\mathscr{K} \subset \mathscr{O}_{U \times V}$ which is generated by the sheaves $\mathscr{I}$ and $\mathscr{J}$. Clearly $\mathscr{K}$ is coherent and its zero set coincides with $X \times Y$.

Now let X and Y be complex spaces. We select open coverings $X = \bigcup_{\alpha \in A} U_\alpha$ and $Y = \bigcup_{\beta \in B} V_\beta$ so that the open subspaces U_α and V_β are isomorphic to analytic subspaces of open sets in $\mathbb{C}^r$. Using the construction of direct product of such subspaces in the preceding paragraph, we can endow each set $U_\alpha \times V_\beta \subset X \times Y$ with a certain sheaf of algebras. The complex spaces which are obtained can be glued together in the natural way to form a complex space $(X \times Y, \mathscr{O}_{X \times Y})$ which is called the *direct product* of the original complex spaces.

In the natural way projection morphisms $\Pi_1 : X \times Y \to X$ and $\Pi_2 : X \times Y \to Y$ are defined. If Z is a third complex space and $F: Z \to X$ and $G: Z \to Y$ are holomorphic mappings, then there exists a holomorphic mapping $H = F \times G: Z \to X \times Y$ so that $\Pi_1 \circ H = F$ and $\Pi_2 \circ H = G$; any holomorphic mapping $H: Z \to X \times Y$ can be represented in the form $H = (\Pi_1 \circ H) \times (\Pi_2 \circ H)$.

c) Quotients. Let X be a reduced complex space, R an equivalence relation on X, X/R the quotient space relative to the relation R, endowed with quotient topology, and $p: X \to X/R$ the natural mapping. We consider on X/R the presheaf obtained by associating to each open set $U \subset X/R$ the algebra F_U of all (continuous) functions $\varphi: U \to C$ so that $\varphi \circ p$ is holomorphic in $p^{-1}(U) \subset X$. The corresponding sheaf of algebras $\mathscr{O}_X/R$ turns X/R into a reduced ringed space. The next example shows that the space $(X/R, \mathcal{O}_X/R)$ need not be an analytic space.

Example 2.9. Let $X = \mathbb{C}^2$ and let the relation R be defined by the mapping $h: X \to \mathbb{C}^2$ given by the formula

$$h(z_1, z_2) = (z_1, z_1 z_2).$$

It is easy to see that X/R is not locally compact at the point $p(0)$ and thus cannot be given the structure of an analytic space.

There do exist, however, large classes of equivalence relations which do lead to analytic quotient spaces. We formulate some results of this kind which are due to Cartan [19]. An equivalence relation R is called *proper* if the saturation of any compact set in X relative to R is compact or equivalently if X/R is locally compact space and p is a proper mapping.

Theorem 2.11. *Let R be proper. The ringed space $(X/R, \mathcal{O}_X/R)$ is analytic if and only if each point in X/R has a neighborhood U so that the functions in $\Gamma(U, \mathcal{O}_X/R)$ separate points in U, i.e., for any distinct points $y_1, y_2 \in U$ there is $\varphi \in \Gamma(U, \mathcal{O}_X/R)$ so that $\varphi(y_1) \neq \varphi(y_2)$.*

We consider two particular cases. Let f be a proper holomorphic mapping from a space X into a reduced complex space Y and let R be the equivalence relation defined by the mapping f. Then X/R is a complex space. We note that in this situation $f(X)$ is an analytic set in Y [91]. The natural bijective mapping $X/R \to f(X)$ is holomorphic, and if $f(X)$ is normal, then it is an isomorphism.

Another particular case: R is defined by some group G of automorphisms of the space X possessing the property of *proper discontinuity*: for any compact set $K \subset X$ there exist only finitely many elements $g \in G$ so that $g(K) \cap K \neq \emptyset$. In this case X/R is a complex space.

We note also that in [45] those equivalence relations R in a normal complex space X are characterized which make X/R into a complex space under the hypothesis that R is an analytic set in $X \times X$.

8. Holomorphic Fiber Bundles. Let $Q = (q, q^\#): E \to X$ be a surjective holomorphic mapping of complex spaces. We assume that Q is locally trivial in the following sense: there exists a complex space F so that for some neighborhood U of any point of the space X the open subspace $q^{-1}(U)$ of the space E is isomorphic to the direct product $U \times F$; under this isomorphism the mapping Q is transformed into the projection $\Pi_1: U \times F \to U$. In this situation we say that E is a *holomorphic fiber bundle* with base X and typical fiber F. Any fiber $E_x = q^{-1}(x)$ for $x \in X$ of the bundle E (i.e., of the mapping Q, see Example 2.7) is isomorphic to the space F.

A *holomorphic section of the bundle F* is any holomorphic mapping $S: X \to E$ such that $Q \circ S = \mathrm{id}$. Associating to each open set $U \subset X$ the set of all sections of the bundle $Q: q^{-1}(U) \to U$ and using the restriction mapping, we obtain a presheaf of sections of the bundle E which leads to the *sheaf* $\mathscr{E}$ *of germs of holomorphic sections of the bundle* E, which is a sheaf of sets on X. In the case when $E = X \times F$ is a direct product and $Q = \Pi_1$ the sections are identified with holomorphic mappings $X \to F$ so that $\mathscr{E}$ is the sheaf of germs of holomorphic mappings $X \to F$.

We fix an open covering $(U_i)_{i \in I}$ of the space X so that over each U_i the triviality condition is satisfied, and we set $E_i = q^{-1}(U_i)$. Then for each $i \in I$ there exists an isomorphism $\Phi_i: E_i \to U_i \times F$ so that $Q = \Pi_1 \circ \Phi_i$ on E_i. If $i, j \in I$ and $U_i \cap U_j \neq \emptyset$, then the isomorphism $F_{ij} = \Phi_i \circ \Phi_j^{-1}: (U_i \cap U_j) \times F \to (U_i \cap U_j) \times F$ is defined so that

$$\Pi_1 \circ F_{ij} = \Pi_1. \tag{2}$$

It is clear that F_{ij} satisfies conditions 1) and 2) from Chapter 2, Section 7a, where in our case $X_i = U_i \times F$ and $X_{ij} = X_{ji} = (U_i \cap U_j) \times F$. The space E is clearly obtained by gluing the spaces $U_i \times F$ using the isomorphisms F_{ij}. Conversely, if there are given automorphisms F_{ij} of the spaces $(U_i \cap U_j) \times F$ which satisfy conditions 1) and 2) of Chapter 2, Section 7a and condition (2), then by gluing the spaces $U_i \times F$ using the F_{ij}, we obtain a holomorphic fiber bundle with base X and typical fiber F.

We consider now the case when $F = \mathbb{C}^r$ with the standard (reduced) structure. Let $GL_r(\mathbb{C})$ denote the group of all nonsingular complex matrices of order r and let $\mu: GL_r(\mathbb{C}) \times \mathbb{C}^r \to \mathbb{C}^r$ denote matrix multiplication of the matrix with a column vector. We assume that for any i, j with $i \neq j$ so that $U_i \cap U_j \neq \emptyset$ there is defined a holomorphic mapping $g_{ij} = U_i \cap U_j \to GL_r(\mathbb{C})$ satisfying the conditions:

$$\begin{aligned} &1)\ g_{ij}(x) = g_{ji}(x)^{-1} \quad (x \in U_i \cap U_j);\\ &2)\ g_{ij}(x) g_{jk}(x) = g_{ik}(x) \quad (x_i \in U_i \cap U_j \cap U_k). \end{aligned} \tag{3}$$

We define an automorphism F_{ij} of the space $(U_i \cap U_j) \times \mathbb{C}^r$ by composing the holomorphic mappings

$$(U_i \cap U_j) \times \mathbb{C}^r \xrightarrow{(\mathrm{id} \times g_{ij}) \times \mathrm{id}} (U_i \cap U_j) \times GL_r(\mathbb{C}) \times \mathbb{C}^r \xrightarrow{\mathrm{id} \times \mu} (U_i \cap U_j) \times \mathbb{C}^r.$$

Then the F_{ij} satisfy conditions 1) and 2) of Chapter 2, Section 7a and condition (2). Thus we have defined a holomorphic bundle $Q: E \to X$ with typical fiber $\mathbb{C}^r$.

A bundle constructed in this fashion is called a *holomorphic vector bundle* of rank r over X. Since the automorphism F_{ij} is linear on each fiber $\{x\} \times \mathbb{C}^r$, in each fiber E_x of the bundle E there is defined the structure of an r-dimensional vector space over $\mathbb{C}$. If $Q': E' \to X$ is another vector bundle over X, then a *homomorphism* from the bundle E to the bundle E' is a holomorphic mapping $H: E \to E'$ satisfying the condition $Q' \circ H = Q$ and linear on the fibers. A homomorphism H is called an *isomorphism* if there exists an inverse homomorphism. We shall discuss conditions for the existence of an isomorphism between two vector bundles in Chapter 3, Section 7 (see Example 3.4.).

The sheaf $\mathscr{E}$ of holomorphic sections of a vector bundle E is a sheaf of vector spaces. Moreover, $\mathscr{E}$ is an analytic sheaf. In the case when $E = X \times \mathbb{C}^r$ and

$Q = \Pi_1$ the sheaf $\mathscr{E}$ coincides with $\mathcal{O}_X^r$, i.e., it is a free analytic sheaf of rank r. In the general case the sheaf $\mathscr{E}$ as a sheaf of modules over $\mathcal{O}_X$ is locally free, since it is locally isomorphic to the sheaf $\mathcal{O}_X^r$. It is not difficult to prove that any locally free (more precisely, locally isomorphic to $\mathcal{O}_X^r$) analytic sheaf on a complex space X is the sheaf of germs of sections of some vector bundle of rank r over X which is determined uniquely up to isomorphism. We note that locally free analytic sheaves are coherent.

The above construction for vector bundles can be generalized in the following fashion. Instead of the space $\mathbb{C}^r$ we take as typical fiber an arbitrary complex space F; instead of the group $\mathrm{GL}_r(\mathbb{C})$ an arbitary complex Lie group G; and as μ an arbitrary holomorphic action of the group G on F. Any system of holomorphic mappings $g_{ij}: U_i \cap U_j \to G$ satisfying condition (3) defines a holomorphic bundle with base X and typical fiber F. Such bundles are called *holomorphic fiber bundles with structure group G*. We mention the case of a *principal bundle* with structure group G which is a bundle with fiber $F = G$ on which G acts by left translations.

9. Meromorphic Functions and Divisors. We shall define meromorphic functions on an arbitrary complex space (for a domain in $\mathbb{C}^n$ this was done in Chapter 1, Section 4). Since stalks of the structure sheaf and algebras of sections can contain zero divisors, we need an algebraic construction which generalizes the usual field of fractions.

Let A be an associative and commutative ring with unity $1 \neq 0$. We denote by T the set of all non-zerodivisors in A, i.e., all $a \in A$ so that $ax = 0$ for some $x \in A$ implies that $x = 0$. Clearly T is a subsemigroup of the multiplicative semigroup of the ring A with $1 \in T$. Let S be an arbitrary subsemigroup of T which contains 1. Then we may define a ring A_S containing "fractions" of the form a/b, where $a \in A$ and $b \in S$; which are identified among themselves by the rule: $a/b = a'/b'$ if $ab' = ba'$. The ring A_S is called the *ring of fractions* (or *localization*) of the ring A relative to S. If $S = T$ we obtain the *complete ring of fractions* A_T. The mapping $a \mapsto a/1$ defines an imbedding $A \to A_S$.

Let X be a complex space. For $x \in X$ we denote by T_x the set of all nonzerodivisors in $\mathcal{O}_{X,x}$. We consider on X the presheaf which associates to each open set $U \subset X$ the algebra $\mathscr{M}_U = \Gamma(U, \mathcal{O}_X)_{S_U}$, where $S_U = \{s \in \Gamma(U, \mathcal{O}_X | s_x \in T_x$ for all $x \in U\}$. The corresponding sheaf $\mathscr{M}$ on X is called the *sheaf of germs of meromorphic functions* and its sections are the *meromorphic functions*. For any point $x \in X$ we have $\mathscr{M}_x = (\mathcal{O}_{X,x})_{T_x}$. The sheaf $\mathscr{M}$ contains $\mathcal{O}_X$ as a subsheaf of groups relative to addition. Therefore $\mathscr{M}$ is analytic although not an coherent sheaf. Any analytic subsheaf of finite type in $\mathscr{M}$ is coherent. The analytic sheaf $\mathscr{M}/\mathcal{O}_X$ is called the *sheaf of principal parts*. The *first* (or *additive*) *Cousin problem* on the space X is the problem of lifting of sections for the natural homomorphism of sheaves $\mathscr{M} \to \mathscr{M}/\mathcal{O}_X$.

We denote by $\mathcal{O}_X^* \subset \mathscr{M}^*$ the subsheaves of groups relative to multiplication in $\mathscr{M}$ consisting of the invertible elements in the sheaves $\mathcal{O}_X$ and $\mathscr{M}$ respectively.

The *second* (or *multiplicative*) *Cousin problem* on X is the problem of lifting of sections for the natural homomorphism of sheaves of groups $\mathcal{M}^* \to \mathcal{M}^*/\mathcal{O}_X^*$.

The *Poincaré problem* on the space X is the following question: is every meromorphic function $h \in \Gamma(X, \mathcal{M})$ representable in the form $h = f/g$, where $f \in \Gamma(X, \mathcal{O}_X)$ and $g \in S_X$? In other words, is the natural mapping $M_X \to \Gamma(X, \mathcal{M})$ surjective? This problem is closely related to the second Cousin problem. In particular, the following result holds (see [49], Chapter V):

Theorem 2.12. *Let X be a complex manifold on which the second Cousin problem is always solvable. Then the Poincaré problem is also solvable on X in the following strong sense: every meromorphic function h on X can be represented in the form $h = f/g$, where f and g are holomorphic functions, and at every point $x \in X$ the germs f_x and g_x are relatively prime in the unique factorization domain $\mathcal{O}_{X,x}$ with $g_x \neq 0$.*

The sheaf $\mathcal{D} = \mathcal{M}^*/\mathcal{O}_X^*$ is called the *sheaf of germs of divisors* on the complex space X and its sections are *divisors* on X. In the divisor group $\Gamma(X, \mathcal{D})$ it is usual to employ additive notation. Every invertible meromorphic function defines a divisor; such divisors are called *principal.* The second Cousin problem is the question: is each divisor on X principal? The quotient group of $\Gamma(X, \mathcal{D})$ by the subgroup of principal divisors is called the *group of divisor classes* of the space X and is denoted by $CD(X)$.

Each divisor $d \in \Gamma(X, \mathcal{D})$ can be represented on some open covering $(U_i)_{i \in I}$ of the space X by a set of meromorphic functions $h_i \in \Gamma(U_i, \mathcal{M}^*)$ so that $h_i/h_j \in \Gamma(U_i \cap U_j, \mathcal{O}_X^*)$ for any i, j (*multiplicative Cousin data*). A divisor d is called *positive* or *effective* (written $d \geqslant 0$) if this representation can be chosen so that $h_i \in \Gamma(U_i, \mathcal{O}_X)$ for all $i \in I$.

If X is an n-dimensional complex manifold, then to each positive divisor d on X we may associate a formal linear combination $\sum_j n_j H_j$, where $n_j \in \mathbb{Z}$, $n_j \geqslant 0$, and (H_j) is a locally finite family of reduced irreducible analytic hypersurfaces $H_j \subset X$. The sets H_j are the irreducible components of an analytic set in X defined by the equations $h_i(x) = 0$ for $x \in U_i$, and n_j are the multiplicities of the zeros of the functions h_j on H_j. Analogously, to an arbitrary divisor there corresponds a linear combination $\sum_j n_j H_j$ with arbitrary integers n_j. As a result we obtain an isomorphism between the divisor group $\Gamma(X, \mathcal{D})$ and the group of all integral linear combinations $\sum n_j H_j$ of locally finite families (H_j) of reduced irreducible analytic hypersurfaces (see [111]).

Chapter 3. Cohomology with Values in a Sheaf

Cohomology with values in a sheaf was introduced by Leray [65] in connection with the study of the topology of continuous mappings and fiber spaces. This apparatus was first applied to problems in the theory of functions in the Cartan Seminar ([15], see also [16]). In this section we shall discuss the theory of Čech cohomology, which is the most convenient for applications in complex analysis. This theory works satisfactorily only in the case of paracompact spaces, which suffice for our needs. In Chapter 3, Section 4 we shall discuss a more general approach due to Godement [41] and Grothendieck [53].

1. The Obstruction to the Lifting of a Section. Before beginning a formal exposition of cohomology theory, we shall indicate how one-dimensional cohomology classes arise in the problem of lifting of sections (see Chapter 1, Section 4). Let $h: \mathscr{F} \to \mathscr{G}$ be a surjective homomorphism of sheaves of abelian groups on a topological space X, and choose $t \in \Gamma(X, \mathscr{G})$. The problem consists in constructing a section $s \in \Gamma(X, \mathscr{F})$ so that $\Gamma(h)(s) = t$. It follows easily from the surjectivity of the homomorphism h that the problem can be solved locally, i.e., that a section s with the necessary property exists in some neighborhood of any point of the space X. In other words, there exist an open covering $\mathscr{U} = (U_i)_{i \in I}$ of the space X and a collection of sections $s_i \in \Gamma(U_i, \mathscr{F})$ so that $\Gamma(h)(s_i) = t \mid U_i$ for all $i \in I$. The obstruction to gluing these sections together to form a global section of the sheaf $\mathscr{F}$ is given by the sections $z_{ij} = s_j - s_i$ defined over the nonempty intersections $U_i \cap U_j$ for $i, j \in I$. It is clear that $z_{ij} \in \Gamma(U_i \cap U_j, \mathscr{H})$, where $\mathscr{H} = \operatorname{Ker} h$. Moreover, we have

$$z_{ij} + z_{jk} + z_{ki} = 0 \quad \text{in} \quad U_i \cap U_j \cap U_k \neq \varnothing \tag{4}$$

(here it is understood that the sections are restricted to $U_i \cap U_j \cap U_k$)

A collection of sections $z_{ij} \in \Gamma(U_i \cap U_j, \mathscr{H})$ for $i, j \in I$ with $U_i \cap U_j \neq \varnothing$ which satisfies condition (4) is called a *1-cocycle* for the covering $\mathscr{U}$ with values in the sheaf $\mathscr{H}$; all such 1-cocycles form an abelian group $Z^1(\mathscr{U}, \mathscr{H})$. In this group we distinguish the subgroup $B^1(\mathscr{U}, \mathscr{H})$ of *cocycles cohomologous to zero*, i.e., having the form

$$z_{ij} = a_j - a_i,$$

where $a_i \in \Gamma(U_i, \mathscr{H})$. It is easy to see that for a different choice of sections s_i the cocycle (z_{ij}) is replaced by a cohomologous cocycle, i.e., one which differs by a term in $B^1(\mathscr{U}, \mathscr{H})$. Thus the section t defines a unique cohomology class, i.e., an element ζ of the quotient group $H^1(\mathscr{U}, \mathscr{H}) = Z^1(\mathscr{U}, \mathscr{H})/B^1(\mathscr{U}, \mathscr{H})$. From the following almost obvious lemma it is clear that the class ζ is the natural *obstruction to the lifting of the section t*.

Lemma 3.1. *For the existence of a section $s \in \Gamma(X, \mathscr{F})$ such that $\Gamma(h)(s) = t$ it is necessary and sufficient that $\zeta = 0$.*

As our discussion shows, theorems which give sufficient conditions for the triviality of the cohomology groups $H^1(\mathscr{U}, \mathscr{H})$ are extremely useful for solving the problem of lifting of sections. Theorems of this kind are sometimes called *vanishing theorems for cohomology*. Examples are provided by Lemma 3.2 and Theorem B in Chapter 4.

2. Simplicial Structures and Čech Cohomology (see [41]). We begin with a generalization of the well known combinatorial construction which leads to the classical cohomology groups of a simplicial complex. Let I denote an arbitrary nonempty set. We shall say that a *simplicial structure* with vertex set I is defined if a collection of nonempty finite subsets of I called *simplices* is given which satisfies the following properties:

1) $\{i\}$ is a simplex for any $i \in I$;
2) any nonempty subset of a simplex is again a simplex.

The *dimension* of a simplex $\sigma = \{i_0, i_1, \ldots, i_n\}$, where $i_k \in I$, is the integer n. Every n-dimensional simplex $\{i_0, i_1, \ldots, i_n\}$ defines $(n+1)!$ *ordered simplices* $(i_{w(0)}, \ldots, i_{w(n)})$ for $w \in S_{n+1}$. We shall write

$$s_{j_0 j_1 \ldots j_n} = (j_0, j_1, \ldots, j_n),$$
$$\{j_0, j_1, \ldots, j_n\} = |s_{j_0 j_1 \ldots j_n}|.$$

We say that a *system of coefficients* F is defined on a simplicial structure if to each simplex σ of the structure there is associated an abelian group F_σ and to each inclusion of simplices $\sigma \subset \tau$ a group homomorphism $r_{\sigma\tau}: F_\tau \to F_\sigma$ so that for any collection of three simplices $\rho \subset \sigma \subset \tau$ we have $r_{\rho\tau} = r_{\rho\sigma} \circ r_{\sigma\tau}$. A p-dimensional *cochain* for the simplicial structure S with values in the coefficient system F is an arbitrary family $c = (c_{i_0 \ldots i_p})$, where $(i_0, i_1, \ldots, i_p)$ runs over the set of all ordered p-dimensional simplices of the structure S and $C_{i_0 \ldots i_p} \in F_{\{i_0, \ldots, i_p\}}$. The p-dimensional cochains form an abelian group $C^p(S, F)$ which we may write as a direct product $\prod_s F_{|s|}$, where s runs over the set of all ordered p-dimensional simplices. The formula

$$(\delta c)_{i_0 \ldots i_{p+1}} = \sum_{j=0}^{p+1} (-1)^j r_{\{i_0, \ldots, \hat{i}_j, \ldots, i_{p+1}\}\{i_0, \ldots, i_{p+1}\}} c_{i_0 \ldots \hat{i}_j \ldots i_{p+1}}$$

(the symbol ^ means that the corresponding index has been omitted) defines a collection of homomorphisms $\delta: C^p(S, F) \to C^{p+1}(S, F)$. If we introduce the graded group of cochains

$$C^*(S, F) = \bigoplus_{p \geqslant 0} C^p(S, F),$$

then the homomorphisms δ define an endomorphism δ of the graded group $C^*(S, F)$ having degree 1. Here we have $\delta^2 = 0$, so we obtain a cochain complex in the sense of homological algebra. The graded subgroup $Z^*(S, F) = \operatorname{Ker} \delta$ is called the *group of cocycles*, the graded subgroup $B^*(S, F) = \operatorname{Im} \delta$ the *group of coboundaries*, and the graded group

$$H^*(S, F) = Z^*(S, F)/B^*(S, F)$$

the *cohomology group* for the structure S with values in F. We have

$$H^p(S, F) = Z^p(S, F)/B^p(S, F) \quad (p \geqslant 1),$$
$$H^0(S, F) = Z^0(S, F).$$

Example 3.1. Let P be a polyhedron in the affine space $\mathbb{R}$ endowed with a triangulation. Then it is natural to define a simplicial structure $S(P)$ whose vertex set coincides with the vertex set of the triangulation. Any abelian group A defines a *constant system of coefficients* F_A on $S(P)$ given by the formula $(F_A)_\sigma = A$ for any simplex σ. The group $H^*(S(P), F_A)$ coincides with the classical cohomology group $H^*(P, A)$ of the polyhedron P with values in the group A.

Example 3.2. Let X be a topological space and $\mathcal{U} = (U_i)_{i \in I}$ a covering of it by nonempty open sets U_i. We call a subset $\{i_0, i_1, \ldots, i_p\} \subset I$ a simplex if $U_{i_0 \ldots i_p} = U_{i_0} \cap \cdots \cap U_{i_p} \neq \varnothing$. We obtain a simplicial structure $N_{\mathcal{U}}$ with vertex set I called the *nerve* of the covering $\mathcal{U}$. Any sheaf of abelian groups $\mathcal{F}$ defines on $N_{\mathcal{U}}$ a system of coefficients F according to the formula

$$F_{\{i_0, \ldots, i_p\}} = \Gamma(U_{i_0 \ldots i_p}, \mathcal{F}).$$

We denote the complex $C^*(N_{\mathcal{U}}, \mathcal{F})$ by $C^*(\mathcal{U}, \mathcal{F})$. The corresponding cohomology group $H^*(N_{\mathcal{U}}, \mathcal{F})$ is also denoted by $H^*(\mathcal{U}, \mathcal{F})$.

We have $H^0(\mathcal{U}, \mathcal{F}) = Z^0(\mathcal{U}, \mathcal{F}) = \Gamma(X, \mathcal{F})$. It is easy to verify that the groups $H^1(\mathcal{U}, \mathcal{F})$ coincide with the cohomology groups defined in Chapter 3, Section 1.

The cohomology defined in Example 3.2 depends on the choice of the covering $\mathcal{U}$. However, by "passing to the limit" it is possible to turn these groups into groups which depend only on the space X and the sheaf $\mathcal{F}$.

Let $\mathcal{U} = (U_i)_{i \in I}$ and $\mathcal{V} = (V_j)_{j \in J}$ be two coverings of a topological space X. We say that $\mathcal{V}$ refines $\mathcal{U}$ if a mapping $\lambda : J \to I$ is defined so that $V_j \subset U_{\lambda(j)}$ for any $j \in J$. In this case for any sheaf of abelian groups $\mathcal{F}$ on X there is defined a natural homomorphism of graded groups $\lambda_{\mathcal{U},\mathcal{V}} : H^*(\mathcal{U}, \mathcal{F}) \to H^*(\mathcal{V}, \mathcal{F})$ which does not depend on the choice of the mapping λ. We now consider all possible coverings $\mathcal{U}$ of the space X, i.e., coverings for which all the U_i are open sets. The existence of the homomorphisms $\lambda_{\mathcal{U},\mathcal{V}}$ allows us to define the inductive limit $\varinjlim_{\mathcal{U}} H^*(\mathcal{U}, \mathcal{F})$. The resulting graded group depends only on X and $\mathcal{F}$. It is called the *Čech cohomology group* of the space X with values in $\mathcal{F}$ and is denoted by $H^*(X, \mathcal{F})$.

The homomorphisms $\lambda_{\mathcal{U},\mathcal{V}}$ define for each open covering $\mathcal{U}$ a homomorphism $\lambda_{\mathcal{U}} : H^*(\mathcal{U}, \mathcal{F}) \to H^*(X, \mathcal{F})$. It is natural to ask: in which cases is $\lambda_{\mathcal{U}}$ an isomorphism? A sufficient condition for this will be given in Chapter 3, Section 4. Here we indicate only the following two simple facts:

$$\lambda_{\mathcal{U}} : H^0(\mathcal{U}, \mathcal{F}) = \Gamma(X, \mathcal{F}) \to H^0(X, \mathcal{F}) \text{ is an isomorphism;}$$

$$\lambda_{\mathcal{U}} : H^1(\mathcal{U}, \mathcal{F}) \to H^1(X, \mathcal{F}) \text{ is injective.}$$

In what follows we shall identify $H^0(\mathcal{U}, \mathcal{F}) = \Gamma(X, \mathcal{F})$ and $H^0(X, \mathcal{F})$ via $\lambda_{\mathcal{U}}$.

In the case when $X = P$, a polyhedron endowed with a triangulation, and $\mathcal{F} = \mathcal{F}_A$ is a constant sheaf defined by an abelian group A the cohomology

$H^*(X, \mathscr{F}) = H^*(P, \mathscr{F}_A)$ coincides with the cohomology $H^*(S(P), \mathscr{F}_A)$ of Example 3.1. The proof of this fact is based on the following construction: with each simplex σ of our triangulation we associate its *star* U_σ, i.e., the union of all open simplices of the triangulation whose closure contains σ. Then the stars of the 0-dimensional simplices form an open covering $\mathscr{U}$ of the space P whose nerve $N_{\mathscr{U}}$ is naturally isomorphic to $S(P)$. Thus $H^*(\mathscr{U}, \mathscr{F}_A) \cong H^*(S(P), \mathscr{F}_A)$. When we pass to the limit, it suffices to consider a sufficiently fine barycentric subdivision of the given triangulation relative to which the cohomology $H^*(S(P), \mathscr{F}_A)$, as is well known, does not change. Therefore we obtain the isomorphism $H^*(P, \mathscr{F}_A) \cong H^*(SP, \mathscr{F}_A)$. In what follows the constant sheaf $\mathscr{F}_A$ will often be denoted simply by A.

3. Basic Properties of Cohomology (see [41], [54]). Let $h: \mathscr{F} \to \mathscr{G}$ be a homomorphism of sheaves of abelian groups on X. Then h defines for any open covering $\mathscr{U}$ a homomorphism of complexes $C^*(\mathscr{U}, \mathscr{F}) \to C^*(\mathscr{U}, \mathscr{G})$ and thus a homomorphism of cohomology groups $h^*: H^*(\mathscr{U}, \mathscr{F}) \to H^*(\mathscr{U}, \mathscr{G})$. Passing to the limit relative to the coverings, we obtain a homomorphism $h^*: H^*(X, \mathscr{F}) \to H^*(X, \mathscr{G})$. The homomorphism h^* has the following properties:

$$(h_1 + h_2)^* = h_1^* + h_2^*; \quad (h_1 \circ h_2)^* = h_1^* \circ h_2^*; \quad \mathrm{id}^* = \mathrm{id}.$$

Further, the homomorphism $h^*: H^0(X, \mathscr{F}) \to H^0(X, \mathscr{G})$ is identified with $\Gamma(h)$.

Thus the correspondence $\mathscr{F} \mapsto H^*(X, \mathscr{F})$, $h \mapsto h^*$ is a covariant functor from the category of sheaves of abelian groups on X to the category of graded abelian groups, and the correspondence $\mathscr{F} \mapsto H^0(X, \mathscr{F})$ agrees with the section functor Γ.

Assume now that

$$0 \to \mathscr{A} \xrightarrow{\alpha} \mathscr{B} \xrightarrow{\beta} \mathscr{C} \to 0 \tag{5}$$

is an exact sequence of sheaves of abelian groups.

Theorem 3.1. *Let X be paracompact. Then for any $p \geqslant 0$ there exists a homomorphism $\delta^*: H^p(X, \mathscr{C}) \to H^{p+1}(X, \mathscr{A})$* so that

a) the sequence

$$\begin{gathered} 0 \to H^0(X, \mathscr{A}) \xrightarrow{\alpha^*} H^0(X, \mathscr{B}) \xrightarrow{\beta^*} H^0(X, \mathscr{C}) \xrightarrow{\delta^*} H^1(X, \mathscr{A}) \xrightarrow{\alpha^*} \cdots \\ \xrightarrow{\delta^*} H^p(X, \mathscr{A}) \xrightarrow{\alpha^*} H^p(X, \mathscr{B}) \xrightarrow{\beta^*} H^p(X, \mathscr{C}) \xrightarrow{\delta^*} H^{p+1}(X, \mathscr{A}) \xrightarrow{\alpha^*} \cdots \end{gathered} \tag{6}$$

is exact;

b) for any commutative diagram of exact sequences of sheaves

$$\begin{array}{ccccccccc} 0 & \longrightarrow & \mathscr{A} & \longrightarrow & \mathscr{B} & \longrightarrow & \mathscr{C} & \longrightarrow & 0 \\ & & \downarrow{\scriptstyle\varphi} & & \downarrow & & \downarrow{\scriptstyle\psi} & & \\ 0 & \longrightarrow & \mathscr{A}' & \longrightarrow & \mathscr{B}' & \longrightarrow & \mathscr{C}' & \longrightarrow & 0 \end{array}$$

the diagram

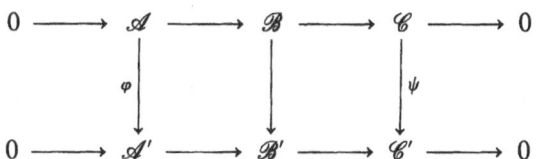

$$\begin{array}{ccc} H^q(X,\mathscr{C}) & \xrightarrow{\delta^*} & H^{q+1}(X,\mathscr{A}) \\ \Big\downarrow{\scriptstyle\psi^*} & & \Big\downarrow{\scriptstyle\varphi^*} \\ H^q(X,\mathscr{C}') & \xrightarrow{\delta^*} & H^{q+1}(X,\mathscr{A}') \end{array}$$

is commutative.

The sequence (6) is called the *long exact cohomology sequence* associated with the sequence (5).

We note that the part of the sequence (6) consisting of the groups H^0 and H^1 exists and is exact for any space X. Further, the exactness of the sequence

$$H^0(X,\mathscr{B}) \xrightarrow{\beta^*} H^0(X,\mathscr{C}) \xrightarrow{\delta^*} H^1(X,\mathscr{A})$$

means that a section $t \in H^0(X,\mathscr{C})$ lifts to a section of the sheaf $\mathscr{B}$ if and only if $\delta^* t = 0$. It is clear that the class $\delta^* t$ is the image of the obstruction defined in Chapter 3, Section 1 under the homomorphism $\lambda_{\mathscr{U}}$, so in fact we have again arrived at Lemma 3.1. In particular, if $H^1(X,\mathscr{A}) = 0$ then the homomorphism $\Gamma(\beta): \Gamma(X,\mathscr{B}) \to \Gamma(X,\mathscr{C})$ is surjective.

Theorem 3.2. *Let Y be a closed subset of a paracompact space X and let $\mathscr{F}$ be a sheaf of abelian groups on Y. Then $H^*(Y,\mathscr{F}) \cong H^*(X,\mathscr{F}^X)$.*

The next result plays an important role in cohomology theory and its applications.

Lemma 3.2. *If $\mathscr{F}$ is a fine sheaf of abelian groups on a paracompact space X and $\mathscr{U}$ is an open covering of X, then $H^p(\mathscr{U},\mathscr{F}) = 0$ for all $p > 0$. In particular $H^p(X,\mathscr{F}) = 0$ for all $p > 0$.*

The definition of cohomology which we have given can be generalized in the following fashion (see [41]). We fix in the space X a certain family of closed sets Φ such that the union of two sets in Φ belongs to Φ and every closed subset of a set in Φ also belongs to Φ. We shall consider only cochains each of which is equal to 0 outside some set in Φ. Then we obtain the *cohomology groups $H^p_\Phi(X,\mathscr{F})$ with supports in the family Φ*. If Φ is the family of all closed subsets of X, then these groups agree with those defined above. Another important case is *cohomology with compact supports* (Φ is the family of all compact subsets of X). In what follows this generalization will not be used.

4. The Calculation of Cohomology Using Resolutions (see [41], [53]). Let $\mathscr{F}$ be a sheaf of abelian groups on X. A *resolution* of the sheaf $\mathscr{F}$ is an exact sequence of sheaves of abelian groups of the form

$$0 \to \mathscr{F} \xrightarrow{j} \mathscr{A}^0 \xrightarrow{d} \mathscr{A}^1 \xrightarrow{d} \cdots \qquad (7)$$

A resolution defines a sheaf of cochain complexes $\mathscr{A}^* = \bigoplus_{p \geqslant 0} \mathscr{A}^p$ with coboundary d, whose sheaf of p-dimensional cohomology is trivial for $p > 0$ and agrees with $\mathscr{F}$ for $p = 0$.

With a resolution (7) of the sheaf $\mathscr{F}$ there is associated the cochain complex

$$\Gamma(X,\mathscr{A}^*) = \bigoplus_{p\geqslant 0} \Gamma(X,\mathscr{A}^p)$$

with coboundary operator $\Gamma(d)$ which for short we shall denote simply by d. We now give a condition which is sufficient for the cohomology groups of this complex $H^p(\Gamma(X,\mathscr{A}^*))$ to be isomorphic to the groups $H^p(X,\mathscr{F})$. A resolution (7) of a sheaf $\mathscr{F}$ is called *acyclic* if $H^p(X,\mathscr{A}^q)=0$ for all $p>0$ and all $q\geqslant 0$. A resolution is called *fine* or *flabby* if all the sheaves $\mathscr{A}^q$ are fine or flabby respectively.

Theorem 3.3. *If X is paracompact and if (7) is a resolution of the sheaf $\mathscr{F}$, then there is a natural homomorphism $H^*(\Gamma(X,\mathscr{A}^*)) \to H^*(X,\mathscr{F})$ which is an isomorphism if the resolution is acyclic.*

The proof of Theorem 3.3 is obtained by applying Theorem 3.1 to the exact sequence of sheaves

$$0 \to \mathscr{Z}^{p-1} \xrightarrow{i} \mathscr{A}^{p-1} \xrightarrow{d} \mathscr{Z}^p \to 0,$$

where $\mathscr{Z}^q \sim (\operatorname{Ker} d) \cap \mathscr{A}^q$ and i is the inclusion for $p \geqslant 1$. The first terms of the long exact cohomology sequence give us an injective homomorphism $H^p(\Gamma(X,\mathscr{A}^*)) = \Gamma(X,\mathscr{Z}^p)/d\Gamma(X,\mathscr{A}^{p-1}) \to H^1(X,\mathscr{Z}^{p-1})$, and further the mappings δ^* define a cochain of homomorphisms

$$H^1(X,\mathscr{Z}^{p-1}) \to H^2(X,\mathscr{Z}^{p-2}) \to \cdots \to H^p(X,\mathscr{Z}^0) = H^p(X,\mathscr{F}).$$

If the resolution is acyclic, then these will all be isomorphisms.

By Lemma 3.2 a fine resolution is acyclic, so we obtain the next result.

Corollary. *If X is paracompact and we are given a fine resolution (7) of the sheaf $\mathscr{F}$, then there is a natural isomorphism*

$$H^*(\Gamma(X,\mathscr{A}^*)) \cong H^*(X,\mathscr{F}).$$

In Chapter 3, Section 5 we shall consider important special cases of this corollary: the theorems of de Rham and Dolbeault.

By using resolutions we may construct a cohomology theory with values in a sheaf which has "good" properties (see Chapter 3, Section 3) for the case of an arbitrary (not necessarily paracompact) space X (see [41]). For this we associate with every sheaf of abelian groups $\mathscr{F}$ on X a *canonical resolution* $0 \to \mathscr{F} \to \mathscr{C}^0 \to \mathscr{C}^1 \to \cdots$ which is simultaneously flabby and fine. The cohomology groups are defined to be the cohomology groups of the complex of sections of this resolution: $H^*(X,\mathscr{F}) = H^*(\Gamma(X,\mathscr{C}^*))$. We shall not discuss here the construction of the canonical resolution: we mention only that it is essential to use the sheaf of germs of arbitrary (not necessarily continuous) sections of the sheaf $\mathscr{F}$ (see Example 1.8).

For a paracompact space X the cohomology groups defined in this way are isomorphic by Lemma 3.2 to the Čech cohomology groups defined in Chapter 3,

Section 2 (for an arbitrary space X also the isomorphisms are valid in dimensions 0 and 1 and the 2-dimensional Čech group is imbedded in $H^2(X, \mathscr{F})$, see [53]). On the other hand, it is not difficult to show that for cohomology groups defined using canonical resolutions the properties listed in Chapter 3, Section 3, are valid for arbitrary spaces. It is necessary to replace Lemma 3.2 with the following result: $H^p(X, \mathscr{F}) = 0$ for all $p > 0$ if $\mathscr{F}$ is a flabby sheaf.

A more algebraic approach is contained in the work of Grothendieck [53] where cohomology groups are defined as derived functors of the functor of sections Γ (this leads essentially to the same canonical resolution as in [41]).

We note also the following *theorem on acyclic coverings*, the initial version being due to Leray.

Theorem 3.4. *Let $\mathscr{U} = (U_i)_{i \in I}$ be an open covering of a topological space X, $\mathscr{F}$ a sheaf of abelian groups on X, and $n \geqslant 0$. If $H^p(U_{i_0 \dots i_q}, \mathscr{F}) = 0$ for all p and q so that $p > 0$ and $p + q \leqslant n$, then $H^n(\mathscr{U}, \mathscr{F}) \cong H^n(X, \mathscr{F})$.*

Corollary. *The homomorphism $\lambda_{\mathscr{U}} : H^n(\mathscr{U}, \mathscr{F}) \to H^n(X, \mathscr{F})$ from Chapter 3, Section 2, is an isomorphism under the hypotheses of Theorem 3.4 if it is assumed in addition that X is a metrizable space or a locally compact space which is countable at infinity.*

5. Complexes of Differential Forms (see [51]). Let X be a paracompact differential manifold and Φ^p the sheaf of germs of p-forms on X of class C^∞. By the Poincaré Lemma (see Chapter 1, Section 4) the sequence

$$0 \to \mathbb{C} \xrightarrow{i} \Phi^0 \xrightarrow{d} \Phi^1 \xrightarrow{d} \cdots, \tag{8}$$

where i is the inclusion, is a resolution of the sheaf $\mathbb{C}$—the so-called *de Rham resolution.* This resolution is fine so that the Corollary to Theorem 3.3 leads to the classical *de Rham Theorem:*

Theorem 3.5. *If X is a paracompact differentiable manifold, then $H^*(\Gamma(X, \Phi^*)) \cong H^*(X, \mathbb{R})$.*

In fact the de Rham Theorem asserts that the isomorphism of Theorem 3.5 can be described in terms of integration of differential forms on the simplices of any smooth triangulation of the manifold X. More precisely, let S denote the simplicial structure corresponding to a triangulation and set $F^p(X) = \Gamma(X, \Phi^p)$ and $F^*(X) = F(X, \Phi^*)$. Every ordered simplex $(i_0, i_1, \dots, i_p)$ from the structure S defines an oriented simplex $\sigma_{i_0 i_1 \dots i_p}$ of the triangulation, which is the image of a standard p-dimensional simplex Δ^p under a smooth mapping $f : \Delta^p \to X$ sending the standardly numbered vertices of the simplex Δ^p to vertices of the triangulation defined by the elements $i_0, i_1, \dots, i_p$ respectively. Then for each form $\omega \in F^p(X)$ the integral

$$\int_{\sigma_{i_0 \dots i_p}} \omega = \int_{\Delta^p} f^* \omega.$$

is defined. The mapping Int: $F^*(X) \to C^*(S, F_{\mathbb{R}})$, which associates to each form $\omega \in F^p(X)$ the cochain

$$(\operatorname{Int} \omega)_{i_0 \ldots i_p} = \int_{\sigma_{i_0 \ldots i_p}} \omega,$$

is a homomorphism of complexes and therefore defines a homomorphism of the graded spaces $\rho: H^*(F^*(X)) \to H^*(S, F_{\mathbb{R}})$. If we take the composition of this homomorphism with the isomorphism $H^*(S, F_{\mathbb{R}}) \to H^*(X, \mathbb{R})$ described in Chapter 3, Section 2, then it turns out that we obtain the isomorphism of Theorem 3.5 (up to sign equal to $(-1)^p$ on elements of degree p). Thus ρ is also an isomorphism.

Let us now assume that X is a complex analytic manifold. Viewing X as a real analytic manifold with twice the dimension, we can construct on X the de Rham resolution (8) of the sheaf $\mathbb{C}$. The complex structure on X allows us to turn the sheaf $\Phi^* = \bigoplus_{p \geqslant 0} \Phi^p$ into a sheaf of bigraded vector spaces. To do this, at each point $x \in X$ we define $\Phi_x^{p,q}$ to be the space of germs at x of forms of type

$$\sum a_{i_1 \ldots i_p j_1 \ldots j_q} dz_{i_1} \wedge \cdots \wedge dz_{i_p} \wedge d\bar{z}_{j_1} \wedge \cdots \wedge d\bar{z}_{j_q},$$

where $z_1, \ldots, z_n$ are local holomorphic coordinates in a neighborhood of the point x on X. The spaces $\Phi_x^{p,q}$ for $x \in X$ do not depend on the choice of local coordinates and define a subsheaf $\Phi^{p,q} \subset \Phi^{p+q}$. The sections of this subsheaf are called *forms of type* (p, q). We have

$$\Phi^* = \bigoplus_{p,q \geqslant 0} \Phi^{p,q}, \quad \Phi^r = \bigoplus_{p+q=r} \Phi^{p,q}, \quad \overline{\Phi^{p,q}} = \Phi^{q,p}.$$

Let $\pi_{p,q}: \Phi^* \to \Phi^{p,q}$ denote the projection. It is easy to see that

$$d\Phi^{p,q} \subset \Phi^{p+1,q} + \Phi^{p,q+1}.$$

Thus there are defined the sheaf homomorphisms

$$\partial = \pi_{p+1,q} \circ d: \Phi^{p,q} \to \Phi^{p+1,q},$$

$$\bar{\partial} = \pi_{p,q+1} \circ d: \Phi^{p,q} \to \Phi^{p,q+1}.$$

They extend by linearity to the whole sheaf Φ^*, where

$$d = \partial + \bar{\partial},$$

which imples that

$$\partial^2 = \bar{\partial}^2 = \partial\bar{\partial} + \bar{\partial}\partial.$$

For example, if $\varphi \in \Phi^{0,0} = \Phi^0$ is the germ of a smooth function, then in local coordinates $z_\alpha = x_\alpha + iy_\alpha$ for $\alpha = 1, \ldots, n$ we have

$$d\varphi = \sum_\alpha \frac{\partial \varphi}{\partial z_\alpha} dz_\alpha + \sum_\alpha \frac{\partial \varphi}{\partial \bar{z}_\alpha} d\bar{z}_\alpha,$$

where by definition

$$\frac{\partial\varphi}{\partial z_\alpha} = \frac{1}{2}\left(\frac{\partial\varphi}{\partial x_\alpha} - i\frac{\partial\varphi}{\partial y_\alpha}\right), \quad \frac{\partial\varphi}{\partial \bar{z}_\alpha} = \frac{1}{2}\left(\frac{\partial\varphi}{\partial x_\alpha} + i\frac{\partial\varphi}{\partial y_\alpha}\right).$$

Thus

$$\partial\varphi = \sum_\alpha \frac{\partial\varphi}{\partial z_\alpha} dz_\alpha, \quad \bar{\partial}\varphi = \sum_\alpha \frac{\partial\varphi}{\partial \bar{z}_\alpha} d\bar{z}_\alpha.$$

Using the Cauchy-Riemann equations, we see that $\operatorname{Ker}\bar{\partial} \cap \Phi^0$ coincides with the sheaf $\mathcal{O}_X$ of germs of holomorphic functions on X and $\operatorname{Ker}\partial \cap \Phi^0$ with the sheaf $\bar{\mathcal{O}}_X$ of antiholomorphic functions. Further, for any p the kernel Ω^p of the homomorphism $\bar{\partial} : \Phi^{p,0} \to \Phi^{p,1}$ consists of germs of forms of type $\sum a_{i_1,\dots,i_p} dz_{i_1} \wedge \cdots \wedge dz_{i_p}$, where $a_{i_1\dots i_p}$ are holomorphic functions. Such forms are called *holomorphic p-forms* and the sheaf $\Omega^p = \operatorname{Ker}\bar{\partial} \cap \Phi^{p,0}$ the *sheaf of germs of holomorphic p-forms*. Analogously, $\operatorname{Ker}\partial \cap \Phi^{0,p}$ coincides with the sheaf $\bar{\Omega}^p$ of germs of antiholomorphic p-forms.

Lemma 3.3. *The sequences*

$$0 \to \Omega^p \xrightarrow{i} \Phi^{p,0} \xrightarrow{\bar{\partial}} \Phi^{p,1} \xrightarrow{\bar{\partial}} \cdots, \tag{9}$$

$$0 \to \bar{\Omega}^p \xrightarrow{i} \Phi^{0,p} \xrightarrow{\partial} \Phi^{1,p} \xrightarrow{\partial} \cdots \tag{10}$$

are fine resolutions of the sheaves Ω^p and $\bar{\Omega}^p$ respectively.

It suffices to do the proof for the sequence (9), since (10) is obtained from (9) via complex conjugation. The only nontrivial point is the existence of a solution of the equation $\bar{\partial}\alpha = \beta$, where $\beta \in \Phi^{p,0}$, $p > 0$ and $\bar{\partial}\beta = 0$. That follows from the following analog of the Poincaré Lemma.

Grothendieck Lemma. *Let D be an open set in $\mathbb{C}^n$ and let U_i be bounded open sets in $\mathbb{C}$ so that $U = U_1 \times \cdots \times U_n$ is a bounded polycylindrical open set in $\mathbb{C}^n$ with $\bar{U} \subset D$. Then for any form $\beta \in \Gamma(D, \Phi^{p,q})$ with $q > 0$ and $\bar{\partial}\beta = 0$ there is a form $\alpha \in \Gamma(U, \Phi^{p,q-1})$ so that $\bar{\partial}\alpha = \beta$.*

The most important step in the proof is the existence in $U = U_1$ of a solution for the equation

$$\frac{\partial\varphi}{\partial\bar{z}} = \psi,$$

where ψ is a given function of class C^∞ in $D \subset \mathbb{C}$. One verifies that one of the solutions is given by the function

$$\varphi(z) = \frac{1}{2\pi i}\iint_U \frac{\psi(\zeta)}{\zeta - z} d\zeta \wedge d\bar{\zeta}.$$

The resolution (9) of the sheaf Ω^p is called the *Dolbeault resolution*. It defines the cochain complex $\Gamma(X, \Phi^{p,*}) = \bigoplus_{q\geqslant 0} \Gamma(X, \Phi^{p,q})$ with coboundary $\bar{\partial}$. The following theorem holds, which is analogous to Theorem 3.3.

Theorem 3.6 (Dolbeault). *If X is a paracompact complex manifold, then* $H^*(\Gamma(X, \Phi^{p,*})) \cong H^*(X, \Omega^p)$.

In the sequel we shall denote by $H^{p,q}(X)$ the group $H^q(\Gamma(X, \Phi^{p,*}))$, which is isomorphic by Theorem 3.6 to the group $H^q(X, \Omega^p)$.

We note that the sheaves Ω^p give a resolution (though not a fine one!) of the sheaf $\mathbb{C}$, in which the coboundary is $d = \partial$:

$$0 \to \mathbb{C} \xrightarrow{i} \Omega^0 \xrightarrow{d} \Omega^1 \xrightarrow{d} \cdots \tag{11}$$

It is called the *holomorphic de Rham resolution.* The sheaves Ω^p are locally free and, in particular, coherent analytic sheaves.

6. The Cousin Theorems. In this section we shall show how the methods of cohomology theory work on the example of the classical Cousin theorems.

Theorem 3.7. *The additive Cousin problem can be solved in any polycylindrical open set in* $\mathbb{C}^n$.

We shall employ Dolbeault cohomology. The next result is basic for the proof.

Lemma 3.4. *If U is a polycylindrical open set in $\mathbb{C}^n$, then $H^{0,1}(U) = 0$.*

For the proof we construct a sequence of bounded open polycylindrical sets $\{U^{(m)}|_{m=1,2,\ldots}\}$ in $\mathbb{C}^n$ such that $\bar{U}^{(m)} \subset U^{(m+1)} \subset U$ for any m and $U = \bigcup_{m=1}^{\infty} U^{(m)}$. Let $\beta \in \Gamma(U, \Phi^{0,1})$ be a form such that $\bar{\partial}\beta = 0$. It follows from the Grothendieck Lemma that in each $U^{(m)}$ there exists a function f_m of class C^∞ so that $\bar{\partial} f_m = \beta$. By using the Runge Approximation Theorem for functions of one complex variable (see [39]), we may assume that the range of the restriction mapping $\Gamma(U^{(m+1)}, \mathcal{O}) \to \Gamma(U^{(m)}, \mathcal{O})$ is dense in $\Gamma(U^{(m)}, \mathcal{O})$ in the topology of compact convergence for any m. Using this fact, we "glue" the functions f_m together to form a global solution of the equation $\bar{\partial} f = \beta$. The function $h_m = f_{m+1} - f_m$ is holomorphic in $U^{(m)}$ for any m. Using approximation on $U^{(m)}$, we can arrange that $\sup_{U^{(m-1)}} |h_m| < 1/2^{m-1}$. Then for any p the function $u_p = \sum_{m=p}^{\infty} h_m$ is holomorphic in $U^{(p-1)}$. It is easy to see that the function f on U which is equal to $f_p + u_p$ in $U^{(p-1)}$ for any p is well defined and satisfies the condition $\bar{\partial} f = \beta$.

It follows from Lemma 3.4 and Theorem 3.6 that $H^1(U, \mathcal{O}) = 0$. If we now apply Theorem 3.1 to the exact sequence of sheaves $0 \to \mathcal{O} \to \mathcal{M} \to \mathcal{M}/\mathcal{O} \to 0$ (see Example 1.6), we obtain Theorem 3.7. We note that the deduction of Theorem 3.7 from Lemma 3.4 is in fact completely elementary and requires no technical results aside from Lemma 3.2, which is based on partitions of unity. Indeed, it suffices to prove that $H^1(\mathcal{U}, \mathcal{O}) = 0$ for any open covering U of the manifold U (see Lemma 3.1). Let $\mathcal{U} = (U_i)_{i \in I}$ and $(z_{ij}) \in Z^1(\mathcal{U}, \mathcal{O})$. We construct using Lemma 3.2 sections $u_i \in \Gamma(U_i, \mathscr{C}^\infty)$ so that $u_j - u_i = z_{ij}$ in $U_i \cap U_j$. Then the form $\beta = \Gamma(U, \Phi^{0,1})$ equal to $\bar{\partial} u_i$ in U_i for $i \in I$ is well defined. By Lemma 3.4

we have $\beta = \bar{\partial} f$, where $f \in \Gamma(U, \mathscr{C}^\infty)$. Then $h_i = u_i - f$ is a holomorphic function in U_i so that $z_{ij} = h_j - h_i$ in $U_i \cap U_j$.

Example 3.3. We consider the domain $X = \mathbb{C}^2 \setminus \{0\} \subset \mathbb{C}^2$. It can be covered by the two bicylindrical domains $U_1 = \mathbb{C}^* \times \mathbb{C} = \{z_1 \neq 0\}$ and $U_2 = \mathbb{C} \times \mathbb{C}^* = \{z_2 \neq 0\}$ with $U_1 \cap U_2 = \mathbb{C}^* \times \mathbb{C}^*$. By what has been proven above $H^1(U_i, \mathcal{O}) = 0$ for $i = 1, 2$ so we have $H^1(X, \mathcal{O}) = H^1(\mathscr{U}, \mathcal{O})$, where $\mathscr{U} = (U_1, U_2)$ (see the Corollary to Theorem 3.4). Using an expansion into Laurent series, it is not difficult to show that the space $H^1(\mathscr{U}, \mathcal{O})$ is infinite dimensional and isomorphic to the space of all holomorphic functions on $\mathbb{C}^* \times \mathbb{C}^*$ of the form $\sum_{k,l<0} C_{kl} z_1^k z_2^l$. If we define on U_1 the meromorphic function $1/(z_1 z_2)$ and on U_2 the function 0, then the corresponding additive Cousin problem has no solution on X. At the same time it can be proven by analogous means that $H^1(\mathbb{C}^n \setminus \{0\}, \mathcal{O}) = 0$ if $n > 2$.

We move now to the multiplicative Cousin problem. As is clear from Example 1.7, the obstruction to its solution on any complex space X lies in the group $H^1(X, \mathcal{O}_X^*)$, which is called the *Picard group* and denoted by $\operatorname{Pic} X$. Indeed, the exact cohomology sequence associated with the exact sequence of sheaves

$$0 \to \mathcal{O}_X^* \to \mathcal{M}^* \to \mathcal{M}^*/\mathcal{O}_X^* \to 0, \tag{12}$$

gives a homomorphism $\Gamma(X, \mathscr{D}) \to \operatorname{Pic} X = H^1(X, \mathcal{O}_X^*)$, whose kernel consists of the divisors of meromorphic functions. So the group of divisor classes $\mathrm{CD}(X)$ is imbedded in $\operatorname{Pic} X$ as a subgroup.

The following exact sequence of sheaves is useful for the calculation of the Picard group:

$$0 \to \mathbb{Z} \xrightarrow{i} \mathcal{O}_X \xrightarrow{e} \mathcal{O}_X^* \to 0,$$

where i in the inclusion mapping and $e(\varphi) = e^{2\pi i \varphi}$ (in the unreduced case the exponential function can be defined, for example, using the exponential series). We consider the first part of the corresponding long exact cohomology sequence:

$$\begin{aligned} 0 \to \Gamma(X, \mathbb{Z}) \xrightarrow{i} \Gamma(X, \mathcal{O}_X) \xrightarrow{e} \Gamma(X, \mathcal{O}_X^*) \xrightarrow{\delta^*} H^1(X, \mathbb{Z}) \xrightarrow{i^*} H^1(X, \mathcal{O}_X) \\ \xrightarrow{e^*} H^1(X, \mathcal{O}_X^*) \xrightarrow{\delta^*} H^2(X, \mathbb{Z}) \xrightarrow{i^*} H^2(X, \mathcal{O}_X). \end{aligned} \tag{13}$$

It is clear from the exactness at the term $\Gamma(X, \mathcal{O}_X^*)$ that if $H^1(X, \mathbb{Z}) = 0$ then $e : \Gamma(X, \mathcal{O}_X) \to \Gamma(X, \mathcal{O}_X)$ is surjective, i.e., every invertible holomorphic function on X possesses a holomorphic logarithm on X. Further, if the groups $H^1(X, \mathcal{O}_X)$ and $H^2(X, \mathbb{Z})$ are trivial then the group $\operatorname{Pic} X$ is trivial also. Using Lemma 3.4, we obtain the following Cousin theorem.

Theorem 3.8. *The multiplicative Cousin problem can be solved on any polycylindrical open set $U = U_1 \times \cdots \times U_n \subset \mathbb{C}^n$, where $U_i \subset \mathbb{C}$ with at most one exception are simply connected open sets.*

The homomorphism $\delta^* : \operatorname{Pic} X \to H^2(X, \mathbb{Z})$ from the sequence (13) is extremely important for the applications of cohomology theory, and it is therefore useful to indicate its explicit form in terms of coverings. For this it is necessary to choose a covering $\mathscr{U} = (U_i)_{i\in I}$ of the manifold X so that $H^1(U_i \cap U_j, \mathbb{Z}) = 0$ for all $i, j \in I$. Let $(h_{ij}) \in Z^1(\mathscr{U}, \mathcal{O}_X^*)$ be a cocycle representing the class $h \in H^1(X, \mathcal{O}_X^*)$. Then $h_{ij} = e(u_{ij})$, where $u_{ij} \in \Gamma(U_i \cap U_j, \mathcal{O}_X)$, and $\delta^* h$ is represented by the cocycle $c = (c_{jkl}) \in Z^2(\mathscr{U}, \mathbb{Z})$, where

$$c_{jkl} = u_{kl} - u_{jl} + u_{jk} = \frac{1}{2\pi i}(\log h_{kl} - \log h_{jl} + \log h_{jk}).$$

We now find a closed 2-form ω which represents (by Theorem 3.5) the image of the class $\delta^* h$ under the natural homomorphism $H^2(X, \mathbb{Z}) \to H^2(X, \mathbb{C})$. Using Lemma 3.2, we can construct a family of functions $a_j : u_j \to \mathbb{R}$ of class C^∞ so that

$$|h_{ij}|^2 = \frac{a_j}{a_i} \quad \text{in} \quad U_i \cap U_j \neq \varnothing.$$

A simple calculation shows that we may take as ω the form which in each neighborhood U_j is given by

$$\omega = \frac{1}{2\pi i} \bar{\partial} \partial a_j.$$

Thus ω is a form of type $(1, 1)$ and $\bar{\omega} = \omega$. It turns out that the converse result also holds: any closed form $\omega = \bar{\omega}$ of type $(1, 1)$ represents the image of some class $h \in \operatorname{Pic} X$ (see [20]).

7. Cohomology with Values in a Sheaf of Nonabelian Groups. Let $\mathscr{F}$ be a sheaf of (not necessarily abelian) groups on a topological space X relative to an operation which we shall write multiplicatively. In analogy with the abelian case we define the cohomology of the space X with values in $\mathscr{F}$ in diminsions 0 and 1 (it is more complicated to define 2-dimensional cohomology, and we shall not consider this).

The group $H^0(X, \mathscr{F})$ is defined by the equation $H^0(X, \mathscr{F}) = \Gamma(X, \mathscr{F})$. Further, let $\mathscr{U} = (U_i)_{i\in I}$ be an open covering of the space X. As in Chapter 3, Section 2 we define the groups of cochains $C^p(\mathscr{U}, \mathscr{F})$ for $p = 0, 1$. The subsets of cocycles $Z^p(\mathscr{U}, \mathscr{F}) \subset C^p(\mathscr{U}, \mathscr{F})$ for $p = 0, 1$, are defined by the formulas

$$Z^0(\mathscr{U}, \mathscr{F}) = \{c \in C^0(\mathscr{U}, \mathscr{F}) | c_i^{-1} c_j = e \quad \text{in} \quad U_i \cap U_j \neq \varnothing\},$$

$$Z^1(\mathscr{U}, \mathscr{F}) = \{c \in C^1(\mathscr{U}, \mathscr{F}) | c_{ij} c_{jk} = c_{ik} \quad \text{in} \quad U_i \cap U_j \cap U_k \neq \varnothing\}.$$

It is clear that $Z^0(\mathscr{U}, \mathscr{F})$ can be identified with the group$\Gamma(X, \mathscr{F}) = H^0(X, \mathscr{F})$. The set $Z^1(\mathscr{U}, \mathscr{F})$ is not in general a subgroup of $C^1(\mathscr{U}, \mathscr{F})$. It contains a cocycle e defined by the formula $e_{ij} = e$ for any i, j with $U_i \cap U_j \neq \varnothing$. This cocycle is called *trivial*. Further, the group $C^0(\mathscr{U}, \mathscr{F})$ acts on $C^1(\mathscr{U}, \mathscr{F})$ by the mapping $(g, c) \mapsto g \,\square\, c$ $(g \in C^0(\mathscr{U}, \mathscr{F}), c \in C^1(\mathscr{U}, \mathscr{F}))$ defined by the formula

$$(g \,\square\, c)_{ij} = g_i^{-1} c_{ij} g_j \quad \text{in} \quad U_i \cap U_j \neq \varnothing.$$

Under this action $Z^1(\mathscr{U}, \mathscr{F})$ maps to itself. The corresponding quotient set, i.e., the set of orbits of the group $C^0(\mathscr{U}, \mathscr{F})$ in $Z^1(\mathscr{U}, \mathscr{F})$ is called the set of 1-dimensional cohomology classes $H^1(\mathscr{U}, \mathscr{F})$. Passing to the inductive limit relative to coverings, we obtain the cohomology set $H^1(X, \mathscr{F})$. This set in the general case carries no natural group structure. However it possesses a distinguished point ε resulting from the trivial cocycles of the coverings. For any open covering $\mathscr{U}$ we have the natural injective mapping $\lambda_{\mathscr{U}} : H^1(U, \mathscr{F}) \to H^1(X, \mathscr{F})$.

Example 3.4. Let X be a complex space and let $\mathscr{F} = \mathscr{O}^{\mathrm{GL}_r(\mathbb{C})}$ be the sheaf of germs of holomorphic mappings from the space X into the matrix group $\mathrm{GL}_r(\mathbb{C})$. As we saw in Chapter 2, Section 8, each cocycle in $Z^1(\mathscr{U}, \mathscr{F})$ defines a holomorphic vector bundle of rank r with base X. It is not hard to prove (see [103]) that two cocycles (associated to different coverings) define isomorphic vector bundles over X if and only if they correspond to the same element of the set $H^1(X, \mathscr{F})$. Thus the set $H^1(X, \mathscr{O}^{\mathrm{GL}_r(\mathbb{C})})$ can be interpreted as the set of isomorphism classes of holomorphic vector bundles of rank r over X; the element ε corresponds to the trivial bundle $X \times \mathbb{C}^r$. This discussion can be generalized, replacing $\mathrm{GL}_r(\mathbb{C})$ with an arbitrary complex Lie group. The topological and differentiable cases can be considered analogously.

Example 3.5. We consider the sheaf from Example 3.4 in the simplest case of $r = 1$. Since $\mathrm{GL}_1(\mathbb{C}) = \mathbb{C}^*$, the sheaf $\mathscr{F}$ in this situation coincides with $\mathscr{O}_X^*$, and we have again arrived at the Picard group $\operatorname{Pic} X = H^1(X, \mathscr{O}_X^*)$. It is clear from Example 3.4 that $\operatorname{Pic} X$ can be interpreted as the group of isomorphism classes of holomorphic vector bundles of rank 1—the so-called *complex line bundles*. The homomorphism δ^* from (13) associates to each element $h \in \operatorname{Pic} X$ the class $\delta^* h \in H^2(X, \mathbb{Z})$ which is called the *Chern class* (or *characteristic class*) of the corresponding bundle. This is the class which we calculated in Chapter 3, Section 6.

The homomorphism $\Gamma(X, \mathscr{D}) \to \operatorname{Pic} X$ resulting from the exact sequence (12) shows that to each divisor there corresponds a complex line bundle. Let X be a complex manifold of dimension n. As we saw in Chapter 2, Section 9, each divisor on X determines as integral linear combination of irreducible $(n-1)$-dimensional analytic sets in X. The corresponding homology class of dimension $2n - 2$ turns out to be dual in the sense of Poincaré to the Chern class of the bundle defining the divisor (see [51], Chapter 1).

We shall not discuss the analogs of exact sequences for sheaves of nonabelian groups and their cohomology or the applications of this machinery (see [53], [83]). We note only an analog of the Dolbeault Theorem for the sheaf $\mathscr{O}^{\mathrm{GL}_r(\mathbb{C})}$ which enables us to describe part of the set $H^1(X, \mathscr{O}^{\mathrm{GL}_r(\mathbb{C})})$ for a complex manifold X in terms of matrix-valued differential forms on X.

We consider on X the space $\Gamma(X, \Phi^{0,1}) \otimes M_r(\mathbb{C})$ of $(0, 1)$-forms with values in the space $M_r(\mathbb{C})$ of all square matrices of order r. We distinguish in this space the set Ψ of forms ω satisfying the equation $\bar{\partial}\omega = \omega \wedge \omega$. The group S of all mappings $X \to \mathrm{GL}_r(\mathbb{C})$ of class C^∞ acts on $\Gamma(X, \Phi^{0,1}) \otimes M_r(\mathbb{C})$ via the *gauge transformations*:

$$s \cdot \omega = (\bar{\partial} s)s^{-1} + s\omega s^{-1}.$$

Here Ψ maps into itself.

Theorem 3.9 (see [66], [83]). *Let X be a complex manifold. The elements of the set $H^1(X, \mathcal{O}^{GL_r(\mathbb{C})})$ for which the corresponding vector bundles are topologically trivial are in one-to-one correspondence with the orbits of the group S on the set of forms Ψ.*

The correspondence in Theorem 3.9 is established in the following manner. Each form $\omega \in \Psi$ can be represented locally as $\omega = f^{-1}\bar{\partial} f$, where f is a function of class C^∞ with values in $GL_r(\mathbb{C})$. Let $\mathcal{U} = (U_i)_{i \in I}$ be an open covering of the manifold X so that $\omega = f_i^{-1}\bar{\partial} f_i$ in each U_i, where $f_i : U_i \to GL_r(\mathbb{C})$ is a mapping of class C^∞. Then the mappings $g_{ij} = f_i f_j^{-1} : U_i \cap U_j \to GL_r(\mathbb{C})$ are holomorphic and define a cocycle in $Z^1(\mathcal{U}, \mathcal{O}^{GL_r(\mathbb{C})})$.

In a somewhat more complicated fashion the holomorphic vector bundles which are topologically isomorphic to a given bundle can also be classified. For this use is made of the concept of a connection on the bundle. There is also a generalization in which $GL_r(\mathbb{C})$ is replaced by an arbitrary complex Lie group [83].

Chapter 4. Stein Spaces

In 1951 Stein [106] introduced a remarkable class of complex manifolds, which subsequently became known as *Stein manifolds*. This class arises from the following question: which complex manifolds possess a sufficiently large collection of analytic functions, or more precisely, as large a collection of analytic functions as domains possess which lie in the one-dimensional complex plane? Stein manifolds and their analogs in complex spaces (*Stein spaces* or *holomorphically complete spaces*) then became the object of intensive study. In this section we give a survey of some of the results which have been obtained concerning them. A considerable portion of these results is discussed in the classical monograph [49], which we have used in writing this article. A relatively complete survey of results on Stein spaces up to 1977 can be found in [84] and [85].

1. Definition and Examples of Stein Spaces. We begin with three classical existence theorems from function theory of one complex variable. The first of these asserts that for any domain $D \subset \mathbb{C}$ there is a holomorphic function $f \in \Gamma(D, \mathcal{O})$ for which all boundary points of the domain D are singular, i.e., so that f cannot be holomorphically coutinued to any larger domain in the plane $\mathbb{C}$ than D (see [110]). As Hartogs discovered in 1906, the analogous theorem for a domain $D \subset \mathbb{C}^n$ when $n > 1$ is false; for example, every analytic function in the domain $D = \mathbb{C}^n \backslash \{0\}$ when $n > 1$ extends analytically to the entire space $\mathbb{C}^n$. (This is evident from the Laurent expansions of these functions at the point 0.) In general, if D is a domain in $\mathbb{C}^n$ with $n \geqslant 2$ and K is a compact set lying in D such that $D \backslash K$ is connected, then every holomorphic function in $D \backslash K$ extends to a holomorphic function in D (see [49]).

The following concept arises in connection with this phenomenon: a domain $D \subset \mathbb{C}^n$ is called a *domain of holomorphy* if there exists in D a holomorphic function which cannot be continued to any larger domain in the space $\mathbb{C}^n$ than D. For this to be the case, it is in fact necessary and sufficient that for any boundary point $Z_0 \in \partial D$ there exists a holomorphic function in D which is unbounded in a neighborhood of the point Z_0 (see [110]). The simplest classes of domains of holomorphy in $\mathbb{C}^n$ are the classes of polycylindrical and convex domains.

Two other existence theorems—the Mittag-Leffler Theorem, which asserts the existence in any domain $D \subset \mathbb{C}$ of a meromorphic function with prescribed principal parts, and the Weierstrass Theorem, which asserts the existence in D of a meromorphic function with prescribed zeros and poles—also do not generalize to arbitrary domain in $\mathbb{C}^n$ when $n > 1$. They lead to the Cousin problems. The results of Cousin on polycylindrical domains, which were discussed in Chapter 3, Section 6, were generalized by Oka [79], [80]. He proved that the Cousin problems are solvable in any domain of holomorphy D (in the multiplicative case one needs the additional hypothesis that $H^2(D, \mathbb{Z}) = 0$).

From these remarks it is evident that domains of holomorphy are the natural generalization (from the point of view of analytic function theory) of domains in $\mathbb{C}$ to the case of several complex variables. The original definition of *Stein manifold* in [106] generalized the concept of domain of holomorphy and imposed on a complex manifold X the following requirements:

1) *Holomorphic separability*: for any $x, y \in X$ with $x \neq y$ there exists a holomorphic function f on X so that $f(x) \neq f(y)$.

2) *Holomorphic convexity*: for any discrete sequence $(x_n) \subset X$ there exists a holomorphic function f on X so that $\sup_n |f(x_n)| = \infty$.

3) *Uniformizability*: for any point $x_0 \in X$ there exist in X holomorphic functions $f_1, \ldots, f_n$ satisfying the condition $\det\left(\frac{\partial f_\alpha}{\partial z_\beta}(z_0)\right) \neq 0$, where $z_1, \ldots, z_n$ are local holomorphic coordinates in a neighborhood of the point z_0.

4) The existence of a countable basis of open sets.

Condition 2) can be replaced by the following conditions:

2′) For any compact set $K \subset X$ its *holomorphic hull* $K = \{x \in X \mid |f(x)| \leqslant \max_{y \in K} |f(y)|$ for all $f \in \Gamma(X, \mathcal{O}_X)\}$ is also compact.

In the definition of *Stein space* conditions 1), 2), 4), and 2′) (equivalent to 2) have the same form, and the uniformizability condition 3) is formulated as follows:

3′) For any point $x_0 \in X$ there exist functions $f_j \in \Gamma(X, \mathcal{O}_X)$ for $j = 1, \ldots, n = \operatorname{im}\dim_{x_0} X$, so that the germs $(f_j)_{x_0}$ for $j = 1, \ldots, n$ at the point x_0 generate the maximal ideal $m_{x_0} \subset \mathcal{O}_{X,x_0}$.

In fact it turns out (see [42]) that conditions 3) and 3′) and also 4) are superfluous, i.e., for the definition of Stein space we may take only conditions 1) and 2) (or 2′). It is proven in [49] that when condition 4) holds, conditions 1) and 2) can be weakened as follows:

1′) Any compact analytic set in X is finite.

2″) *Weak holomorphic convexity*: any compact set $K \subset X$ possesses an open neighborhood W so that $\hat{K} \cap W$ is compact, where $\hat{K}$ is the holomorphic hull of the set K.

We note that any holomorphic function on a connected compact complex space is constant (this follows easily from the Maximum Principle). Therefore $1) \Rightarrow 1')$.

Example 4.1. A domain $D \subset \mathbb{C}^n$ is a Stein manifold if and only if D is a domain of holomorphy. In general, an open subset U of any Stein space is a Stein space if and only if U is holomorphically convex.

We note that for a domain $D \subset \mathbb{C}^n$ the following simple theorem holds, which sometimes allows us to prove by induction that D is a domain of holomorphy (see [49]). It suffices for a domain D to be a domain of holomorphy that it satisfy at least one of the following conditions: for any complex hyperplane $H \subset \mathbb{C}^n$ the open subset $D \cap H$ of H is Stein and the restriction mapping $\Gamma(D, \mathcal{O}_D) \to \Gamma(D \cap H, \mathcal{O}_{D \cap H})$ is surjective; for any complex line $E \subset \mathbb{C}^n$ the restriction mapping $\Gamma(D, \mathcal{O}_D) \to \Gamma(D \cap E, \mathcal{O}_{D \wedge E})$ is surjective. As is evident from Example 4.2 and Theorem 4.1 from Chapter 4, Section 2, below these conditions are also necessary for the domain D to be Stein.

For conditions on a domain in a Stein space to be Stein see also Chapter 4, Section 2, and Chapter 4, Section 3.

A *Riemann domain* over a complex manifold Y is a reduced complex space X together with an open holomorphic mapping $\pi: X \to Y$ whose fibers are not more than countable. A Riemann domain is called *unramified* with π is a local homeomorphism (in this case X is a complex manifold).

Example 4.2. Let X be a Riemann domain over $\mathbb{C}^n$. If X is a Stein space, then X is (in the natural way) a domain of holomorphy, but the converse assertion is false [46]. At the same time, if a Riemann domain of holomorphy X is unramified, then X is a Stein manifold [82].

Example 4.3. Any analytic subspace of a Stein space is a Stein space. In particular, any analytic subspace of $\mathbb{C}^n$ is Stein and any affine algebraic variety over $\mathbb{C}$ is a Stein space.

Example 4.4. Any compact irreducible component of a Stein space consists of a single point (see condition 1′), so a connected Stein manifold of dimension $n > 0$ cannot be compact. In the case $n = 1$ the converse is also true: any connected noncompact Riemann surface is a Stein manifold [39]. The analogous assertion when $n > 1$ is false. For example, if a point is removed from any connected compact complex manifold of dimension $n \geqslant 2$, the result by Theorem 2.9 is a connected noncompact manifold on which all holomorphic functions are constant.

Example 4.5. The direct product of two Stein spaces is a Stein space.

2. The Theorems of Cartan. By definition a Stein space possesses an ample store of holomorphic functions. It turns out that on any Stein space there is also an abundance of other analytic objects—holomorphic differential forms, holomorphic vector and tensor fields, etc. This fact is closely connected with the fact that on a Stein space the cohomological obstructions to the construction of analytic objects with prescribed properties are absent (for example, obstructions to the extension of holomorphic functions from a closed analytic subspace to the whole space). In the most general form these properties of a Stein space are reflected in the so-called Theorems A and B which are formulated as follows.

Theorem A. *Let X be a Stein space and $\mathscr{F}$ a coherent analytic sheaf on X. For any point $x \in X$ the stalk $\mathscr{F}_x$ is generated by the values s_x of the global sections $s \in \Gamma(X, \mathscr{F})$.*

Theorem B. *Under the same hypotheses we have $H^p(X, \mathscr{F}) = 0$ for all $p > 0$.*

Theorems A and B for Stein manifolds were first published in the works of the Cartan Seminar [15]. As was noted in [16], all the main ideas for the proof of Theorem A and also of Theorem B for $p = 1$ are contained already in [14] but without using the language of cohomology. The idea to study the case of any $p > 0$ and the formulation of Theorem B are due to Serre.

We consider one application of Theorem B.

Theorem 4.1. *Let X be a Stein space, $\mathscr{I}$ a coherent sheaf of ideas, and $Y \subset X$ an analytic subspace defined by the sheaf $\mathscr{I}$. Then the restriction mapping $\Gamma(X, \mathcal{O}_X) \to \Gamma(Y, \mathcal{O}_Y)$ is surjective, i.e., any holomorphic function on Y extends to a holomorphic function on X.*

The theorem follows easily from the equation $H^1(X, \mathscr{I}) = 0$ and the exact sequence of sheaves $0 \to \mathscr{I} \to \mathcal{O}_X \to \mathcal{O}_Y \to 0$.

The proof of Theorem 4.1 shows that any complex space for which Theorem B holds is a Stein space. More precisely, the following result is true.

Theorem 4.2. *A complex space X with a coutable basis is a Stein space if and only if $H^1(X, \mathscr{I}) = 0$ for any coherent sheaf of ideals $\mathscr{I} \subset \mathcal{O}_X$ with a discrete zero set.*

The necessity of the condition $H^1(X, \mathscr{I}) = 0$ follows from Theorem B. We prove its sufficiency. As we have just seen, it follows from our hypothesis that a holomorphic function defined on a discrete set in X can be extended to a holomorphic function on X. By applying this consideration, it is easy to prove that X is holomorphically convex and holomorphically separable. We verify condition 3′). Let $\mathscr{I} \subset \mathcal{O}_X$ be a coherent sheaf of ideals defined by the equations: $\mathscr{I}_x = \mathcal{O}_{X,x}$ for $x \neq x_0$ and $\mathscr{I}_{x_0} = m_{x_0}^2$. Using the exact sequence $0 \to \mathscr{I} \to \mathcal{O}_X \to \mathcal{O}_X/\mathscr{I} \to 0$, it is easy to prove that for any $\varphi \in \mathcal{O}_{X,x_0}$ there exists a function $f \in \Gamma(X, \mathcal{O}_X)$ so that $f_{x_0} - \varphi \in m_{x_0}^2$. Using Nakayama's Lemma (see [48]), it follows that there exist functions $f_1, \ldots, f_n \in \Gamma(X, \mathcal{O}_X)$ so that $(f_1)_{x_0}, \ldots, (f_n)_{x_0}$ is a set of generators for the ideal m_{x_0}.

In view of the importance of the Cartan Theorems we pause a bit to consider their proofs. First of all, it is useful to note that Theorem A follows from Theorem B. For this we first prove that for any $\varphi \in \mathscr{F}_{x_0}$ there is a section $s \in \Gamma(X, \mathscr{F})$ so that $\varphi \in \mathscr{F}_{x_0}$ (cf. the verification of condition 3′) in the proof of Theorem 4.2). If this procedure is applied to the set of generators $\varphi_1, \ldots, \varphi_p$ of the $\mathcal{O}_{X,x_0}$-module $\mathscr{F}_{x_0}$, we obtain sections $s_1, \ldots, s_p \in \Gamma(X, \mathscr{F})$ whose values $(s_i)_{x_0}$ at the point x_0 generate the stalk $\mathscr{F}_{x_0}$.

The proof of Theorem B breaks down into several steps.

It is first proven that $H^p(\bar{Z}, \mathscr{F}) = 0$ for $p > 0$, where Z is a bounded polycylinder in $\mathbb{C}^n$ and $\mathscr{F}$ is a coherent analytic sheaf in a neighborhood of the compact set $\bar{Z}$. In the case of $\mathscr{F} = \mathcal{O}$ this can be done using the Dolbeault Theorem (Theorem 3.6) or by the more traditional method originating with Cousin using the Cauchy integral. The case of an aribitrary coherent sheaf $\mathscr{F}$ reduces to the case of a free sheaf by means of a lemma of Cartan[1] and induction on $\dim Z$. Here we discover that the sheaf $\mathscr{F}$ is generated over $\bar{Z}$ by a finite number of global sections, i.e., there exists a surjective homomorphism $h: \mathcal{O}^p \to \mathscr{F}$ over $\bar{Z}$. Using the previously proven triviality of the cohomology group $H^1(\bar{Z}, \operatorname{Ker} h)$, we deduce the surjectivity of the homomorphism $\Gamma(h)$: $\Gamma(\bar{Z}, \mathcal{O}^p) \to \Gamma(\bar{Z}, \mathscr{F})$. This allows the introduction of a seminorm in the space $\Gamma(\bar{Z}, \mathscr{F})$.

We now consider an arbitrary Stein space X. An open set $W \subset X$ is called an *Oka-Weil domain* if $\overline{W}$ is compact and if there exists a holomorphic mapping $X \to \mathbb{C}^n$ which maps some neighborhood of the set $\overline{W}$ isomorphically onto an analytic subspace of an open set in $\mathbb{C}^n$ with W mapped to an analytic subspace of a bounded polycyliner $Z \subset \mathbb{C}^n$.

Let $\mathscr{F}$ be a coherent analytic sheaf on X. For any Oka-Weil domain W we have $H^p(\overline{W}, \mathscr{F}) = 0$ for $p > 0$. Indeed, by Theorem 3.2 we have $H^p(\overline{W}, \mathscr{F}) \cong H^p(\bar{Z}, \hat{\mathscr{F}})$, where $\hat{\mathscr{F}}$ is the coherent (by Theorem 2.8) analytic sheaf in a neighborhood of the set $\bar{Z}$ obtained from $\mathscr{F}$ by trivial extension under the inclusion mapping of the set $\overline{W}$ into $\mathbb{C}^n$. There is a seminorm defined on the space $\Gamma(\overline{W}, \mathscr{F}) \cong \Gamma(\bar{Z}, \hat{\mathscr{F}})$. It follows from the classical Runge Theorem that any section on $\Gamma(\overline{W}, \mathscr{F})$ can be approximated by sections of $\Gamma(X, \mathscr{F})$ relative to this seminorm.

[1] See Theorem 2.2 in Leiterer's article in this volume.

Using the definition of a Stein space, we can prove that for any compact set $K \subset X$ which coincides with its holomorphic hull $\hat{K}$ and any neighborhood U of the compact set K there exists an Oka-Weil domain W in X so that $K \subset W \subset \overline{W} \subset U$. It follows that X can be represented as the union of an infinite sequence of Oka-Weil domains W_n for $n = 1, 2, \ldots$ so that $\overline{W}_n \subset W_{n+1}$ for every n. Here $H^p(\overline{W}_n, \mathscr{F}) = 0$ for all n and all $p > 0$. The concluding step of the proof consists in deducing from this the equation $H^p(X, \mathscr{F}) = 0$ for any $p > 0$. For $p > 1$ this turns out to be completely simple, and for the case of $p = 1$ the argument is similar to the classical Mittag-Leffler procedure which was used in the proof of Lemma 3.4. It is essential to use the seminorms mentioned above in $\Gamma(\overline{W}_n, \mathscr{F})$ and the fact that sections over $\overline{W}_n$ can be approximated relative to them by global sections.

Various versions of the proofs of Theorems A and B can be found in [15], [40], [5], [49], [54], [57], and [98].

It follows from Theorem B, in particular, that for any Stein space X we have

$$H^p(X, \mathcal{O}_X) = 0 \quad \text{if} \quad p > 0, \tag{14}$$

and for any Stein manifold X

$$H^{p,q}(X) = 0 \quad \text{if} \quad p \geqslant 0, \quad q > 0. \tag{15}$$

Property (14) is also possessed by some compact complex manifolds, for example, by complex projective spaces $\mathbb{CP}^n$, so it is not sufficient for the Steinness of a space X. It is not known whether there are any non-Stein complex manifolds which possess property (15). The following result does hold.

Theorem 4.3. *If D is an unramified Riemann domain over an n-dimensional Stein manifold X which satisfies the condition $H^p(D, \mathcal{O}_D) = 0$ for $1 \leqslant p \leqslant n - 1$, then D is a Stein manifold.*

(For a domain $D \subset X$ see [62] and for the general case [77], where it is also proven that for the Steinness of a Riemann domain D it suffices that the dimension of the spaces $H^p(D, \mathcal{O}_D)$ for $1 \leqslant p \leqslant n - 1$ be not more than countable.)

Numerous applications of Theorems A and B to the theory of Stein spaces will be given below. At this point we want to mention two general applications of Theorem B to arbitrary complex spaces. Let X be a complex space with a countable basis and $\mathscr{F}$ a coherent analytic sheaf on X. Then in the stalks $\mathscr{F}_x$ for $x \in X$ there is a topology arising from the topology of compact convergence in the ring $\mathbb{C}\{z_1, \ldots, z_n\}$.

Corollary 1 (see [49]). *In the space $\Gamma(X, \mathscr{F})$ there exists a unique Frechet topology such that for any point $x \in X$ the mapping $s \mapsto s_x : \Gamma(X, \mathscr{F}) \to \mathscr{F}_x$ is continuous. In particular, if X is reduced and $\mathscr{F} = \mathcal{O}_X$, then this topology coincides with the topology of compact convergence.*

We note that the existence of a topology in the space of sections allows the introduction of natural topology also in the cohomology spaces $H^p(X, \mathscr{F})$, but if $p > 0$ then this topology may be not Hausdorff.

An open covering $\mathscr{U} = (U_i)_{i \in I}$ of a space X is called *Stein* if it is locally finite and if all U_i are Stein spaces.

Corollary 2. *If $\mathscr{U}$ is a Stein covering, then the homomorphism $\lambda_{\mathscr{U}}: H^p(\mathscr{U}, F) \to H^p(X, \mathscr{F})$ is an isomorphism for all $p \geqslant 0$.*

Corollary 2 follows easily form Theorem 3.4, Theorem B, and the following simple result: the intersection of two Stein open sets of X is again a Stein space.

3. Further Examples and Constructions of Stein Spaces. We note that two standard constructions—reduction and normalization—preserve the Steinness or non-Steinness of a complex space.

Theorem 4.4. *A complex space X is Stein if and only if its reduction $(X, \operatorname{red} \mathcal{O}_X)$ is Stein* [44], [49]. *A reduced complex space is Stein if and only if its normalization is Stein* [76].

The transition from X to $(X, \operatorname{red} \mathcal{O}_X)$ here is clear, and for the inverse transition use is made of the surjectivity of the mapping $\operatorname{red}^*: \Gamma(X, \mathcal{O}_X) \to \Gamma(X, \operatorname{red} \mathcal{O}_X)$ which follows from Theorem B.

We next consider the following question: let X be a Stein space and Y an analytic set in X; under what circumstance is the open subspace $X \setminus Y$ a Stein space? By using Theorems 4.2, 4.3, and 2.9 it can be proven that if $X \setminus Y$ is a Stein space, then Y is an analytic hypersurface [49]. The converse assertion is true if X is normal and $\dim X = 2$ [96] but not true for normal spaces of larger dimension [46]. The following result is however true.

Theorem 4.5 (see [49]). *If Y is an analytic hypersurface in a Stein space X which is the zero set of a sheaf of principal ideals, then $X \setminus Y$ is a Stein space.*

For further results in this direction see [12].

It is known that a covering of a Stein space is a Stein space [107] and that a complex space which admits a finite mapping (i.e., closed and with finite fibers) into a Stein space is a Stein space [49]. The next result is a generalization of these facts.

Theorem 4.6 [64]. *Let $f: Y \to X$ be a holomorphic mapping with X a Stein space. Assume that every point in X has a neighborhood U such that for any connected component W of the set $f^{-1}(U)$ the mapping $f: W \to U$ is finite. Then Y is a Stein space.*

In the article [101] it is proven that any Stein analytic subspace of a complex space X has a Stein neighborhood in X. From this the following result can be deduced, which shows that a reduced Stein space can have arbitrary singularities:

Theorem 4.7 [11]. *Let (Y_i, y_i) for $i = 1, 2, \ldots$ be a sequence of germs of analytic sets in various $\mathbb{C}^{n_i}$, all irreducible components of which have the same dimension $k \geqslant 1$. Then there exists a reduced irreducible Stein space X with a sequence of points (x_i) for $i = 1, 2, \ldots$ so that in a neighborhood of each point x_i there is a local isomorphism from the space X onto Y_i sending x_i to y_i. Here* $\operatorname{im}\dim X = \sup_i \operatorname{im}\dim Y_i$, *and if y_i are isolated points in* $\operatorname{Sing} Y_i$, *then* $\operatorname{Sing} X \subseteq \{x_i | i = 1, 2, \ldots\}$.

Let X be a complex space which can be represented as a union $X = \bigcup_{i=1}^{\infty} X_i$ of a sequence $X_1 \subset X_2 \subset \cdots$ of open sets, where each X_i is relatively compact in X_{i+1}. We say that X *is exhausted* by the sequence (X_i). If X is reduced, all the open subspaces X_i are Stein, and each pair (X_i, X_{i+1}) is a *Runge pair* (i.e., the image of the restriction mapping $\Gamma(X_{i+1}, \mathcal{O}_X) \to \Gamma(X_i, \mathcal{O}_X)$ is dense in $\Gamma(X_i, \mathcal{O}_X)$), then X is a Stein space [107]. In the case when X is a domain in $\mathbb{C}^n$ (or even an unramified Riemann domain over $\mathbb{C}^n$), the approximation condition is superfluous [8]. At the same time, there exist examples of non-Stein complex manifolds of any dimension $\geqslant 2$ which can be exhausted by a sequence of Stein open domains [34]. In the general case the following result holds.

Theorem 4.8 [67]. *A reduced complex space X which is exhausted by a sequence of Stein open subspaces is a Stein space if and only if $H^1(X, \mathcal{O}_X) = 0$ (or when the space $H^1(X, \mathcal{O}_X)$ is Hausdorff).*

Let X be a Stein space and D an open set in X. We return again to the question of characterizing those sets D which are open Stein subspaces (see Chapter 4, Section 1 and Theorem 4.3). If D is Stein, then $U \cap D$ is a Stein space for any Stein open set $U \subset X$. Cartan [16] posed the following problem. Assume that D is *locally Stein*, i.e., that every boundary point in ∂D possesses a neighborhood U in X such that $U \cap D$ is a Stein space. Will the open subspace D be Stein? Thus the problem is to characterize open Stein domains in terms of a local condition on their boundaries. In the case of $X = \mathbb{C}^n$ this problem is equivalent to the so-called *Levi problem* (which we shall not touch on in this survey) and a positive solution has been known for a long time (see [54]). It also has a positive answer for any Stein manifold X [27] and for a Stein space X with isolated singularities [4], but in the general case the question remains open. We note the following cohomological result:

Theorem 4.9 [60]. *Let D be a locally Stein open set in a Stein space. Then for any coherent analytic sheaf $\mathcal{F}$ on D we have $H^p(D, \mathcal{F}) = 0$ for all $p \geqslant 2$. The space D is Stein if and only if $H^1(D, \mathcal{O}_D) = 0$ (or when $H^1(D, \mathcal{O}_D)$ is finite-dimensional or Hausdorff).*

It is also known that if all irreducible components of a space X are Stein, then X itself is also Stein [4].

4. The Problems of Cousin and Poincaré. The next result follows from Theorem B applied to the structure sheaf of a Stein space together with the discussion contained in Chapter 3, Section 6.

Theorem 4.10. *The first Cousin problem can always be solved on a Stein space.*

For a domain in $\mathbb{C}^2$ (and in general in a two-dimensional reduced Stein space) the converse is also true [13], [10].

Theorem 4.11 [49]. *A domain $D \subset \mathbb{C}^n$ is a domain of holomorphy if and only if any first Cousin problem in D can be solved and for any complex hyperplane H the intersection $H \cap D$ is a Stein manifold.*

Theorem B and the exact sequence (13) imply the following:

Theorem 4.12. *If X is a Stein space, then* $\operatorname{Pic} X \cong H^2(X, \mathbb{Z})$. *In particular, under the hypothesis $H^2(X, \mathbb{Z}) = 0$ the second Cousin problem on a Stein space X can always be solved.*

Using Theorem 2.12, we obtain from this

Corollary. *If X is a Stein manifold and $H^2(X, \mathbb{Z}) = 0$, then the Poincaré problem can be solved on X in the strong sense.*

We have also

Theorem 4.13. *On any Stein space X the Poincaré problem can be solved.*

In the case when X is reduced it suffices to apply Theorem A to the sheaf of ideals $\mathscr{I} \subset \mathcal{O}_X$ consisting of the stalks $\mathscr{I}_x = \{f \in \mathcal{O}_{X,x} | f_x h_x \in \mathcal{O}_{X,x}\}$, where $h \in \Gamma(X, \mathscr{M}^*)$ is the given meromorphic function on X. For the unreduced case see [1].

We recall that the group of divisor classes $CD(X)$ of a complex space X can be viewed as a subgroup of $\operatorname{Pic} X$.

Theorem 4.14. *For any reduced Stein space X we have* $CD(X) = \operatorname{Pic} X \cong H^2(X, \mathbb{Z})$.

For the proof we consider an element $\zeta \in \operatorname{Pic} X$ as a bundle E of complex lines over X (see Example 3.5). Applying Theorem A to the sheaf of germs of sections $\mathscr{E}$ of the bundle E, we can construct a holomorphic section s of the bundle E over all of X such that $s(x) \neq 0$ on a dense set of points x in X. If z is a cocycle defining ζ on some covering $(U_i)_{i \in I}$, then s is given by a collection of holomorphic functions $s_i \in \Gamma(U_i, \mathcal{O}_X)$, whose germs are not zero divisors at any point. These functions satisfy the conditions $s_i = z_{ij} s_j$ in $U_i \cap U_j$, i.e., they define a positive divisor whose class is identical to ζ.

Corollary. *If on a reduced Stein space X every second Cousin problem is solvable, then $H^2(X, \mathbb{Z}) = 0$.*

Theorem 4.15 (see [22], [29]) *If on a Stein manifold X the Poincaré problem is solvable in the strong sense, then any second Cousin problem is also solvable on X, so $H^2(X, \mathbb{Z}) = 0$.*

Example 4.5 (see [92]). Let X be the domain in $\mathbb{C}^3$ defined by the inequality $|z_1^2 + z_2^2 + z_3^2 - 1| < 1$. Then as a complex manifold X is isomorphic to $D \times Q$, where $D = \{u \in \mathbb{C} \mid |u| < 1\}$ is the unit disc and $Q \subset \mathbb{C}^3$ is the affine quadric defined by the equation $z_1^2 + z_2^2 + z_3^2 = 1$. The desired isomorphism is given by the formula

$$z = (z_1, z_2, z_3) \mapsto \left(z_1^2 + z_2^2 + z_3^2 - 1, \frac{z}{\sqrt{z_1^2 + z_2^2 + z_3^2}} \right),$$

where $\sqrt{w}$ is a single-valued branch of the square root in the disc $|w - 1| < 1$. Therefore X is a Stein manifold with $H^2(X, \mathbb{Z}) \cong \mathbb{Z}$. Thus X gives an example of a (simply connected) domain of holomorphy in $\mathbb{C}^3$ on which not every second Cousin problem can be solved and on which the Poincaré problem is not solvable in the strong sense.

5. Topological Properties. Let X be a Stein manifold. It follows from Theorem B that the holomorphic de Rham resolution (1) of the sheaf $\mathbb{C}$ on X is acyclic. Therefore from Theorem 3.3 follows

Theorem 4.16. *If X is a Stein manifold, then $H^p(X, \mathbb{C}) \cong H^p(\Gamma(X, \Omega^*))$ for $p \geqslant 0$.*

Corollary. *If X is an n-dimensional Stein manifold, then $H^p(X, \mathbb{C}) = 0$ for $p > n$ and the homology groups $H_p(X, \mathbb{Z})$ are periodic for $p > n$.*

This last result can be generalized for any Stein space and can be significantly strengthened.

Theorem 4.17 [78]. *If X is an n-dimensional Stein space, then $H_p(X, \mathbb{Z}) = 0$ for $p > n$, and the group $H_n(X, \mathbb{Z})$ has no torsion.*

For an n-dimensional Stein space X it has been proven also [72], [55] that X has the homotopy tupe of an n-dimensional cell complex. Any n-dimensional paracompact analytic manifold over $\mathbb{R}$ can arise as such a complex [43]. We note also that for any countable abelian group G and any integer $q \geqslant 1$ there exists a domain of holomorphy $D \subset \mathbb{C}^{2q+3}$ for which $H_q(X, \mathbb{Z}) \cong G$ [78].

Analogs of Theorems 4.14 and 4.15 are also known for relative homology (see [61], [108]).

6. Imbeddings into Affine Space. As was remarked in Example 4.2, any analytic subspace of the space $\mathbb{C}^n$ is Stein. It turns out that under natural restrictions the converse is also valid.

Theorem 4.18 (see [90], [54]). *A complex manifold X is a Stein manifold if and only if X is isomorphic to an analytic subspace without singular points of some space $\mathbb{C}^N$.*

Theorem 4.19 (see [113], [2]). *A complex space X is a Stein space and satisfies the condition* im dim $X < \infty$ *if and only if X is isomorphic to an analytic subspace of some space $\mathbb{C}^N$.*

In fact much more precise results have been proven. If X is an n-dimensional Stein manifold, then in Theorem 4.18 we may take $N = 2n + 1$; moreover, the imbeddings $X \to \mathbb{C}^{2n+1}$ form a dense set in the space $\Gamma(X, \mathcal{O}_X)^{2n+1}$ [74]. If $n > 1$ then X admits an imbedding (i.e., an isomorphism onto an analytic subspace) into $\mathbb{C}^{2n}$ [37], [38]. The unit disc and the annulus (except, perhaps, the punctured disc) admit imbeddings into $\mathbb{C}^2$ [63], [104]. There is a conjecture (due to Forster) that X can always be imbedded in $\mathbb{C}^{n+[n/2]+1}$ and an interesting collection of results to support this conjecture. It has also been proven that for any $n \geqslant 1$ there exists an n-dimensional Stein manifold which does not admit an imbedding into $\mathbb{C}^{n+[n/2]}$ [37], [38].

Now let X be a reduced Stein space with finite dimension $n = \dim X$. Then the set of all proper holomorphic injective mappings $X \to \mathbb{C}^{2n+1}$ which are regular at each smooth point is dense in $\Gamma(X, \mathcal{O}_X)^{2n+1}$ [74]. If X is a (not necessarily reduced) Stein space, $n = \dim X$, and $m = \operatorname{im}\dim X < \infty$, then the set of all imbeddings $X \to \mathbb{C}_N$, where $N = \max(2n + 1, n + m)$ is dense in the space of all holomorphic mappings $X \to \mathbb{C}^N$ [113], [2].

7. Holomorphic Fiber Bundles with Stein Base and Fiber. In 1953 Serre posed the following problem [92]: will the space E of a holomorphic fiber bundle whose base X and fiber F are Stein manifolds always be a Stein manifold? He noted that the answer is affirmative, for example, in the case when E is a holomorphic vector bundle. This *Serre problem* (which is often without real basis called the "Serre hypothesis") has generated much research and was solved only in 1977, when Skoda [102] constructed the first example of a non-Stein bundle with fiber $\mathbb{C}^2$, whose base is a p-connected ($p \geqslant 2$) planar domain, with gluing automorphisms locally constant on X and defined by automorphisms of the space $\mathbb{C}^2$ having exponential growth. The following example is simpler.

Example 4.6 [24]. Let E be the three-dimensional complex manifold which is obtained from $\mathbb{C} \times \mathbb{C}^2$ by taking the quotient relative to the cyclic group of automorphisms generated by the automorphism

$$\alpha(x, z) + (z + 2\pi i, g(z)) \quad (x \in \mathbb{C}, z \in \mathbb{C}^2),$$

where $g(z_1, z_2) = (z_1^k - z_2, z_1)$ for $k \geqslant 2$. Then E is clearly the total space of a holomorphic fiber bundle with base $X = \mathbb{C}^* = \mathbb{C}\backslash\{0\}$ and fiber $F = \mathbb{C}^2$. It is not difficult to define this bundle using gluing automorphisms (see Chapter 2, Section 8). To do this we set

$$U_0 = \{x \in \mathbb{C}^* | \arg x \neq \pi\}, \quad U_1 = \{x \in \mathbb{C}^* | \arg x \neq 0\}$$

and define $F_{10} : (U_0 \cap U_1) \times \mathbb{C}_2 \to (U_0 \cap U_1) \times \mathbb{C}^2$ by the formula

$$F_{10}(x, z) = \begin{cases} (x, g(z)) & (\operatorname{Im} x < 0), \\ (x, z) & (\operatorname{Im} x > 0). \end{cases}$$

It turns out that all holomorphic functions on E are constant on the fibers, so E is not holomorphically separable.

In another example of a holomorphic fiber bundle all of whose holomorphic functions are constant on the fibers (see [24], [25]), the base is $\mathbb{C}$ or the disc and the fiber $\mathbb{C}^2$ with the gluing automorphisms having exponential growth on the fibers. In these examples the results of Lelong on the growth of holomorphic and plurisubharmonic functions in $\mathbb{C}^n$ with $n \geqslant 2$ are used in an essential way.

At the same time, in many special cases the Serre problem has a positive solution. We indicate the main results which have been obtained in this direction. All complex spaces are assumed to be reduced with a countable basis.

Theorem 4.20. [71], [31]. *Let E be a holomorphic fiber bundle with base X and fiber F, whose structure group is a complex Lie group G, such that the subgroup of elements of G which send a connected component of the fiber into itself has a finite number of connected components. If X and F are Stein spaces, then E is also a Stein space.*

Let Y be a complex space and G a group of automorphisms of Y. The space Y is called a *Banach-Stein space* (relative to G) if there exists a subspace H in $\Gamma(Y, \mathcal{O}_Y)$ having the structure of a Banach space and possessing the following properties: H separates points of the space Y; for any discrete sequence $M \subset Y$ there is a function in H which is not bounded on M; the inclusion $H \to \Gamma(Y, \mathcal{O}_Y)$ is continuous; $g^*(H) = H$ and $g^*|H$ is continuous for all $g \in G$; and if a holomorphic family of automorphisms $\Phi : S \to \operatorname{Aut} Y$ is given, where S is a complex space and $\Phi(S) \subset G$, then the mapping $s \mapsto \Phi(s)^*|H$ is holomorphic.

Theorem 4.21 ([32], [3]). *If the base of a holomorphic fiber bundle E is a Stein space and the fiber F is a Banach-Stein space relative to the structure group*[2], *then E is a Stein space.*

Examples of Banach-Stein spaces (relative to the full automorphism group) are: any domain in $\mathbb{C}$ [99], any relatively compact bounded domain in a Riemann surface [73], any convex or strictly pseudoconvex bounded domain in $\mathbb{C}^n$ [56], any homogeneous bounded domain in $\mathbb{C}^n$ [32], and any bounded domain of holomorphy in $\mathbb{C}^n$ with compact automorphism group [32].

The answer to the Serre problem is also positive in cases when the fiber is a space of one of the following types: a one-dimensional complex space without compact irreducible components [73], an analytic subspace of a convex bounded domain in $\mathbb{C}^n$ [105], a bounded domain of holomorphy in $\mathbb{C}^n$ with smooth boundary [26], or a bounded domain of holomorphy in $\mathbb{C}^n$ with trivial first Betti number [100]. For arbitrary bounded domains of holomorphy the question remains open.

---

[2] In Theorem 4.21 the structure group is to be understood in a wider sense than in Chapter 2, Section 8—it is a subgroup $G \subset \operatorname{Aut} F$ containing all automorphisms of the fiber which arise in gluing together the bundle from the direct products. In particular, G need not be a complex Lie group.

We note that the answer to the Serre problem can be given a cohomological formulation. We consider a more general situation. A holomorphic mapping $p: E \to X$ of (not necessarily reduced) complex spaces is called *Stein* if p is surjective and if each point of the space X has a neighborhood U so that $p^{-1}(U)$ is a Stein space.

Theorem 4.22 ([58], [59]). *Let $p: E \to X$ be a Stein holomorphic mapping with X a finite-dimensional Stein space. Then for any coherent analytic sheaf $\mathscr{F}$ on E we have $H^p(E, \mathscr{F}) = 0$ for all $p \geqslant 2$, and the space $H^1(E, \mathscr{F})$ is either trivial or infinite-dimensional and non-Hausdorff. The space E is a Stein space if and only if $H^1(E, \mathcal{O}_E) = 0$.*

8. Stein Algebras. Let X be a Stein space. In this section we shall give a survey of certain properties of the $\mathbb{C}$-algebras $\Gamma(X, \mathcal{O}_X)$ of holomorphic functions on X, which are reminiscent of properties of the algebra of polynomial functions on an affine algebraic variety. The main result consists in the fact that a finite dimensional Stein space is completely determined by its algebra of holomorphic functions.

Theorem 4.23 ([35], [6], [49]). *Let X and Y be Stein spaces with* $\dim X < \infty$. *Then any algebra homomorphism $h: \Gamma(X, \mathcal{O}_X) \to \Gamma(Y, \mathcal{O}_Y)$ such that $h(1) = 1$ has the form $h = f^*$, where $f: Y \to X$ is some (uniquely determined) holomorphic mapping. In particular, finite-dimensional Stein spaces X and Y are isomorphic if and only if $\Gamma(X, \mathcal{O}_X) \cong \Gamma(Y, \mathcal{O}_Y)$ as abstract $\mathbb{C}$-algebras. The category of finite-dimensional Stein spaces is dual (anti-equivalent) to a full subcategory of the category of $\mathbb{C}$-algebras.*

The proof of Theorem 4.23 is based on the study of characters of the algebra of holomorphic functions. For an arbitrary $\mathbb{C}$-algebra H we denote by $X(H)$ the set of all *characters* of H, i.e., algebra homomorphisms $H \to \mathbb{C}$ which send 1 to 1. We introduce a topology on $X(H)$: a basis of neighborhoods for a point $\chi \in X(H)$ consists of sets of the form $\{\gamma \in X(H) | \gamma(f_i) - \chi(f_i)| < \varepsilon, i = 1, \ldots, s\}$, where $\varepsilon > 0$ and $\{f_1, \ldots, f_s\}$ runs over all finite subsets of H. We assume that $H = \Gamma(X, \mathcal{O}_X)$, where X is some complex space. With every point $p \in X$ there is associated the character $\chi_p \in X(H)$ defined by the formula $\chi_p(f) = f(p)$. The mapping $\xi: X \to X(H)$ which sends p to χ_p is continuous. Clearly ξ is injective if and only if the space X is holomorphically separable. If ξ is a homeomorphism, then X is a Stein space. Conversely, if X is a finite-dimensional Stein space, then ξ is a homeomorphism. It is then possible to prove that the sheaf $\mathcal{O}_X$ on a finite-dimensional Stein space X can be completely described in terms of the algebra H.

The main point of the proof is the surjectivity of the mapping ξ, i.e., the fact that any character $\chi \in X(H)$ has the form $\chi(f) = f(p)$ for some point $p \in X$. This fact can be proven under weaker hypotheses that finite-dimensionality. For example, it is proven in [30] that if a Stein space X admits a filtration

$X = Y_0 \supset Y_1 \supset \cdots \supset Y_n = \varnothing$ of analytic sets Y_i such that $Y_i \setminus Y_{i+1}$ decomposes for any i into finite-dimensional connected components, then each character χ of the algebra $\Gamma(X, \mathcal{O}_X)$ is continuous in the canonical topology. The equality $\chi = \chi_p$ follows from the following general theorem ([36], [49]).

Theorem 4.24. *Let X be a Stein space and M a maximal ideal in the algebra $H = \Gamma(X, \mathcal{O}_X)$. The following conditions are equivalent: M is closed in H; $M = \operatorname{Ker} \chi_p$ for some $p \in X$; M is finitely generated.*

In the study of the characters of the algebra $H = \Gamma(X, \mathcal{O}_X)$ use is made of the following property of Stein spaces, the so-called *weak Nullstellensatz*: if $f_1, \ldots, f_p \in H$ have no common zeros on X, then there exist $g_1, \ldots, g_p \in H$ so that $\sum_{i=1}^{p} g_i f_i = 1$. It turns out that this property can be used to characterize Stein open subspaces of a Stein space.

Theorem 4.25 ([7], [68]). *Let X be an open set in a Stein space Y. Then the following conditions are equivalent: X is Stein; the weak Nullstellensatz holds in X; any closed maximal ideal in $\Gamma(X, \mathcal{O}_X)$ has the form $\operatorname{Ker} \chi_p$ for some $p \in X$; any finitely generated maximal ideal in $\Gamma(X, \mathcal{O}_X)$ has the form $\operatorname{Ker} \chi_p$ for $p \in X$.*

Again let X be Stein space and $H = \Gamma(X, \mathcal{O}_X)$. To each coherent analytic sheaf $\mathcal{F}$ on X there corresponds the H-module $\Gamma(X, \mathcal{F})$. Thus we obtain an equivalence between the category of coherent analytic sheaves on X and a full subcategory of the category of H-modules whose objects admit an explicit characterization [36], [109]. If X is connected and $\dim X < \infty$, then under this equivalence locally free sheaves correspond to projective H-modules.

9. Holomorphically Convex Spaces. Related to the theory of Stein spaces in a natural fashion is the theory of *holomorphically convex* complex spaces with a countable basis, i.e., spaces which satisfy conditions 2) (or 2′)) and 4) from Chapter 4, Section 1. It is proven in [49] that 2′) can be replaced by the condition of weak holomorphic convexity 2″). The connection between holomorphically convex spaces and Stein spaces is evident from the following result.

Theorem 4.26 ([90], [18]). *A complex space X is holomorphically convex if and only if it admits a proper holomorphic mapping π onto some Stein space $\tilde{X}$ which induces an isomorphism $\pi^* : \Gamma(\tilde{X}, \mathcal{O}_{\tilde{X}}) \to \Gamma(X, \mathcal{O}_X)$. The mapping π (the holomorphic reduction of the space X) is uniquely defined and has connected compact fibers.*

The space $\tilde{X}$ is constructed as a quotient space X/R, where $R = \{(x, y) \in X \times Y \mid f(x) = f(y)$ for all $f \in \Gamma(X, \mathcal{O}_X)\}$ (see Theorem 2.11). The fibers of the mapping π are the level sets of the holomorphic functions on X.

Holomorphically convex spaces admit the following cohomological characterization.

Theorem 4.27 ([87], [89], [69]). *If $\mathcal{F}$ is a coherent analytic sheaf on a holomorphically convex space X, then all the spaces $H^p(X, \mathcal{F})$ for $p \geqslant 0$ are*

Hausdorff. Conversely, if for each coherent sheaf of ideals $\mathscr{I} \subset \mathcal{O}_X$ the space $H^1(X, \mathscr{I})$ is Hausdorff and if all level sets of holomorphic functions on X are compact, then X is holomorphically convex.

If X is a holomorphically convex space, then the union Z of all its compact analytic subspaces of positive dimension is an analytic set. If Z is compact, then the space X is called *strongly pseudoconvex*. Any such space can be obtained from a Stein space by "blowing up" a finite number of points. We have the following result.

Theorem 4.28 ([75]). *A complex space X is strongly pseudoconvex if and only if for any coherent analytic sheaf $\mathscr{F}$ on X we have* $\dim H^p(X, \mathscr{F}) < \infty$ *for all $p > 0$. Here $H^p(X, \mathscr{F}) \cong H^p(Z, \mathscr{F}|Z)$ for $p > 0$.*

Bibliography*

1. Acquistapace, F., Broglia, F.: Problemi di Cousin e di Poincaré per spazi di Stein non ridotti. Ann. Sc. Norm. Super. Pisa, Cl. Sci. Fis. Mat., III. Ser. 27 (1973), 889–904 (1974). Zbl. 303.32014
2. Acquistapace, F., Broglia, F., Tognoli, A.: A relative embedding theorem for Stein spaces. Ann. Sc. Norm. Super. Pisa, Cl. Sci., IV. Ser. 2, 507–522 (1975). Zbl. 313.32020
3. Ancona, V., Speder, J.-P.: Espaces de Banach-Stein. Ann. Sc. Norm. Super. Pisa, Cl. Sci. Fis. Mat., III. Ser. 25, 683–690 (1972). Zbl. 233.32012
4. Andreotti, A., Narasimhan, R.: Oka's Heftungslemma and the Levi problem for complex spaces. Trans. Am. Math. Soc. 111, 345–366 (1964). Zbl. 134.60
5. Andreotti, A., Vesentini, E.: Les thèorèmes fondamentaux de la théorie des espaces holomorphiquement complets. Cah. Topologie Géom. Différ. 4 (1962–1963), Inst. H. Poincaré, Paris (1963)
6. Bănică, C., Stănăşilă, O.: A result on section algebras over complex spaces. Atti Accad. Naz. Lincei, Rend. Cl. Sci. Fis. Mat. Nat. 47, 233–235 (1970). Zbl. 193.39
7. Bănică, C., Stănăşilă, O.: Quelques conditions pour q'un espace complexe soit un espace de Stein. Semin. Inst. Math. Acad. R.S.R. 159–164 (1971). Zbl. 222.32006
8. Behnke, H., Stein, K.: Konvergente Folgen von Regularitätsbereichen und die Meromorphiekonvexität. Math. Ann. 116, 204–216 (1938). Zbl. 20,378
9. Behnke, H., Stein, K.: Modifikation komplexer Mannigfaltigkeiten und Riemannscher Gebiete. Math. Ann. 124, 1–16 (1951). Zbl. 43.303
10. Berg, G.: On 2-dimensional Cousin I-spaces. Math. Ann. 248, 247–248 (1980). Zbl. 411.32012
11. Bingener, J., Flenner, H.: Steinsche Räume zu vorgegebenen Singularitäten, Arch. Math. 32, 34–37 (1979). Zbl. 384.32004
12. Bingener, J., Storch, U.: Resträume zu analytischen Mengen in Steinschen Räumen. Math. Ann. 210, 33–53 (1974). Zbl. 275.32006
13. Cartan, H., Les problèmes de Poincaré et de Cousin pour les fonctions de plusieurs variables complexes. C.R. Acad. Sci. Paris 199, 1284–1287 (1934). Zbl. 10.309

*For the convenience of the reader, references to reviews in Zentralblatt für Mathematik (Zbl.), compiled using the MATH database, and Jahrbuch über die Fortschritte der Mathematik (Jrb.) have, as far as possible, been included in this bibliography.

14. Cartan H., Idéaux et modules de fonctions analytiques de variables complexes. Bull. Soc. Math. Fr. 78, 29–64 (1950). Zbl. 38.237
15. Cartan, H.: Sémin. Ec. Norm. Supér., 4. Fonctions analytiques de plusieurs variables complexes. Paris (1951–1952)
16. Cartan, H.: Variétès analytiques complexes et cohomologie, Colloque sur les fonctions de plusieurs variables complexes tenu à Bruxelles, G. Thone, Masson et C-ie, Paris, 41–55 (1953).
17. Cartan, H.: Variétès analytiques réeles et variétès analytiques complexes. Bull. Soc. Math. Fr. 85, 77–99 (1957). Zbl. 83.305
18. Cartan, H., Sur les fonctions de plusiers variables complexes. Les espace analytiques. Proc. Intern. Congress Mathematicians, Edinburgh, 1958, Camb. Univ. Press 33–52 (1960). Zbl. 117.46
19. Cartan, H., Quotients of complex analytic spaces. Contrib. Function Theory. Internat. Colloquium Bombay, Jan. 1960. Tata Inst. Fundament. Res., Bombay. 1–15 (1960). Zbl. 122.87
20. Chern, S.S.: Complex manifolds. Univ. of Chicago Press (1955–1956). (see Zbl. 74,302 and Zbl. 74,303)
21. Chirka, E.M.: Complex analytic sets. Itogi Nauki Tekh., Ser. Sovrem. Probl. Mat., Fundam. Napravleniya 7, 125–166 (1985)
22. Coen, S.: Una nota sul problema di Poincaré. Atti Accad. Naz. Lincei, VIII. Ser., Rend., Cl. Sci. Fis. Mat. Nat. 57, 342–345 (1974). Zbl. 338.32004
23. Cousin, P.: Sur les fonctions de *n* variables complexes. Acta Math. 19, 1–62 (1895). Jrb. 26,456
24. Demailly, J.-P.: Différents exemples de fibrés holomorphes non de Stein. Séminaire P. Lelong, H. Skoda (Analyse), Année, 1976/77, (Springer) Lecture Notes Math. 694, 15–41 (1978). Zbl. 418.32011
25. Demailly, J.-P.: Un exemple de fibré holomorphe non de Stein à fibre $\mathbb{C}^2$ ayant pour base le disque ou le plan. Invent. Math. 48, 293–302 (1978). Zbl. 372.32012
26. Diederich, K., Fornaess, J.E.: Exhaustion functions and Stein neighborhoods for smooth pseudoconvex domains. Proc. Natl. Acad. Sci. USA 72, 3279–3280 (1975). Zbl. 309.32012
27. Docquier, F., Grauert, H.: Levisches Problem und Rungescher Satz fur Teilgebiete Steinscher Mannigfaltigkeiten. Math. Ann. 140, 94–123 (1960). Zbl. 95,280
28. Douady, A., Le problème des modules pour les sous-espaces analytiques compacts d'un espace analytique donné., Ann. Inst. Fourier 16, No. 1–95 (1966). Zbl. 146.311
29. Ephraim, R.: Stein manifolds on which the Strong Poincaré Problem can be solved, Proc. Am. Math. Soc. 70, 136–138 (1978). Zbl. 395.32003
30. Ephraim, R.: Multiplicative linear functionals on Stein algebras. Pac. J. Math. 78, 89–93 (1978). Zbl. 412.32018
31. Fischer, G.: Holomorph-vollständige Faserbündel. Math. Ann. 180, 341–348 (1969). Zbl. 167.368
32. Fischer, G.: Fibrés holomorphes au-dessus d'un espace de Stein. Espaces analytiques., Sémin. Inst. Math. Acad. Bucarest, Romania 1969 57–69 (1971). Zbl. 219,32008
33. Fischer, G.: Complex analytic geometry. Lect. Notes in Math. 538 Berlin–Heidelberg–New York: Springer–Verlag (1976). 201p. Zbl. 343.32002
34. Fornaess, J.E.: 2-dimensional counterexamples to generalizations of the Levi problem. Math. Ann. 230, 169–173 (1977). Zbl. 379.32016
35. Forster, O.: Uniqueness of topology in Stein algebras. Function Algebras Proc. internat. Sympos. Tulane Univ. 1965, 157–163 (1966). Scott-Foresman, Chicago (1966). Zbl. 151.97
36. Forster, O.: Zur Theorie der Steinschen Algebren und Moduln, Math. Z. 97, 376–405 (1967). Zbl. 148.322
37. Forster, O.: Plongements des variétés de Stein. Comment. Math. Helv. 45, 170–184 (1970). Zbl. 191.94
38. Forster, O.: Topologische Methoden in der Theorie Steinscher Räume. Actes Congr. Int. Mathématiciens 1970 Nice/France, Vol. 2, 613–618 Gauthier-Villars, Paris (1971). Zbl. 245.32006

39. Forster, O.: Riemannsche Flächen, Springer, Berlin (1977). Heidelberger Taschenbücher, Band 184. Zbl. 381.30021
40. Fuks, B.A.: Introduction to the theory of analytic functions of several complex variables. Fizmatgiz, Moscow (1962); English transl.: Amer. Math. Soc., Providence, Rhode Island (1963); Special Chapters in the theory of several complex variables. Fizmatgiz, Moscow (1963); English transl.: Amer. Math. Soc., Providence, Rhode Island (1965). Zbl. 146.308
41. Godement, R.: Topologie algébrique et théorie des faisceaux, Hermann, Paris (1958). Zbl. 80.162
42. Grauert, H.: Charakterisierung der holomorph-vollständigen komplexen Räume, Math. Ann. 129, 233–259 (1955). Zbl. 64.326
43. Grauert, H.: On Levi's problem and the imbedding of real-analytic manifolds, Ann. Math. 68, 460–472 (1958). Zbl. 108.78
44. Grauert, H.: Ein Theorem der analytischen Garbentheorie und Modulräume komplexer Strukturen, Inst. Haut. Etud. Sci., Publ. Math., No. 5, 5–64 (1960). Zbl. 100.80
45. Grauert, H.: Set theoretic complex equivalence relations, Math. Ann. 265, 137–148 (1983). Zbl. 504.32007
46. Grauert, H., Remmert, R.: Singularitäten komplexer Mannigfaltigkeiten und Riemannsche Gebiete, Math. Z. 67, 103–128 (1957). Zbl. 77.289
47. Grauert, H., Remmert, R.: Komplexe Räume. Math. Ann. 136, 245–318 (1958). Zbl. 87.290
48. Grauert, H., Remmert, R.: Analytische Stellenalgebren, Springer, Berlin (1971). Die Grundlehren der Mathematischen Wissenschaften Band 176. Zbl. 231.32001
49. Grauert, H., Remmert, R.: Theorie der Steinschen Räume, Springer, Berlin (1977). Die Grundlehren der Mathematischen Wissenschaften Band 227. Zbl. 379.32001
50. Grauert, H., Remmert, R.: Coherent analytic sheaves, Springer, Berlin (1984). Die Grundlehren der Mathematischen Wissenschaften Band 265. Zbl. 537.32001
51. Griffiths, Ph., Harris, J.: Principles of algebraic geometry. John Wiley & Sons, New York (1978) Zbl. 408.14001
52. Gronwall, T.H.: On the expressibility of a uniform function of several complex variables as a quotient of two functions of entire character, Trans. Am. Math. Soc. 18, 50–64 (1917). Jrb. 46,529
53. Grothendieck, A.: Sur quelques points d'algèbre homologique. Tôhoku Math. J. 9, 119–221 (1957). Zbl. 118.261
54. Gunning, R.C., Rossi, H.: Analytic functions of several complex variables, Prentice-Hall, Englewood Cliffs (1965). Zbl. 141.86
55. Hamm, H.A.: Zum Homotopietyp Steinscher Räume, J. Reine Angew. Math. 338, 121–135 (1983). Zbl. 491.32010
56. Hirschowitz, A.: Domaines de Stein et fonctions holomorphes bornées, Math. Ann. 213, 185–193 (1975). Zbl. 284.32011
57. Hörmander, L.: An introduction to complex analysis in several variables, Van Nostrand, Princeton (1966). Zbl. 138.62
58. Jennane, B.: Groupes de cohomologie d'un fibré holomorphe à base et à fibre de Stein, Invent. Math. 54, 75–79 (1979). Zbl. 427.32016
59. Jennane, B.: Problème de Levi et morphisme localement de Stein, Math. Ann. 256, 37–42 (1981). Zbl. 491.32012
60. Jennane, B.: Remarques sur les ouverts localements de Stein. Math. Ann. 263, 371–375 (1983). Zbl. 522.32010
61. Kaup, L.: Eine topologische Eigenschaft Steinscher Räume, Nachr. Akad. Wiss. Gött., Math.-Phys. Kl. 213–224 (1967). Zbl. 146.313
62. Laufer, H.B.: On sheaf cohomology and envelopes of holomorphy, Ann. Math. 84, 102–118 (1966). Zbl. 143.302
63. Laufer, H.B.: Imbedding annuli in $\mathbb{C}^2$. J. Anal. Math. 26, 187–215 (1973). Zbl. 286.32017
64. Le Barz, P.: A propos des revêtements ramifiés d'espace de Stein, Math. Ann. 222, 63–69 (1976). Zbl. 328.32012

65. Leray, J.: L'anneau spectral et l'anneau filtré d'homologie d'un espace localement compact et d'une application continue, J. Math. Pures Appl. 29, 1–139. (1950). Zbl. 38.363
66. Malgrange, B.: Lectures on the theory of functions of several complex variables. Tata Inst. Fundament. Res., Bombay (1958).
67. Markoe, A.: Runge families and inductive limits of Stein spaces. Ann. Inst. Fourier 27, No. 3 117–127 (1978). Zbl. 323.32014
68. Markoe, A.: Holomorphic convexity and the Corona property. Rev. Roum. Math. Pures Appl. 23, 67–70 (1978). Zbl. 418.32010
69. Markoe, A.: A holomorphically convex analogue of Cartan's Theorem B, Ann. Math. Stud. 100, 291–298 (1981). Zbl. 491.32009
70. Markushevich, A.I.: Introduction to the classical theory of Abelian functions (Russian) Nauka, Moscow (1979). Zbl. 493.14023
71. Matsushima, Y., Morimoto, A.: Sur certains espaces fibrés holomorphes sur une variété de Stein. Bull. Soc. Math. Fr. 88, 137–155 (1960). Zbl. 94.281
72. Milnor, J.: Morse Theory. Ann. Math. Stud., No. 51, Princeton Univ. Press, Princeton (1963) Zbl. 108.104
73. Mok, N.: The Serre problem on Riemann surfaces. Math. Ann. 258, 145–168 (1981). Zbl. 497.32013
74. Narasimhan, R.: Imbedding of holomorphically complete complex spaces, Am. J. Math. 82, 917–934 (1960). Zbl. 104.54
75. Narasimhan, R.: The Levi problem for complex spaces. II. Math. Ann. 146, 195–216 (1962). Zbl. 131.308
76. Narasimhan, R.: A note on Stein spaces and their normalizations. Ann. Sc. Norm. Super. Pisa, Sci. Fis. Mat., III. Ser. 16, 327–333 (1962). Zbl. 112,311
77. Narasimhan, R.: Introduction to the theory of analytic spaces, Springer Lect. Notes Math. 25 (1966). Zbl. 168.60
78. Narasimhan, R.: On the homology groups of Stein spaces. Invent. Math. 2, 377–385 (1967). Zbl. 148.322
79. Oka, K.: Sur les fonctions analytiques de plurieurs variables. II: Domaines d'holomorphie. J. Sci. Hiroshima Univ., Ser. A 7, 115–130 (1937). Zbl. 17,122
80. Oka, K.: Sur les fonctions analytiques de plurieurs variables. III: Deuxième problème de Cousin, J. Sci. Hiroshima Univ., Ser. A 9, 7–19 (1939).
81. Oka, K.: Sur les fonctions analytiques de plurieurs variables. VII: Sur quelques notions arithmétiques. Bull. Soc. Math. Fr. 78, 1–27 (1950). Zbl. 36.52
82. Oka, K.: Sur les fonctions analytiques de plusieurs variables. IX: Domaines finis sans point critique intérieur. Jap. J. Math. 23, 97–155 (1953). Zbl. 53.243
83. Onishchik, A.L.: Some concepts and applications of nonabelian cohomology theory. Tr. Mosk. Mat. O.-va 17, 45–88 (1967); English transl.: Trans. Mosc. Math. Soc. 17, 49–98 (1967). Zbl. 191.539
84. Onishchik, A.L.: Stein spaces. Itogi Nauki Tekh., Ser. Algebra Topologiya Germ. 11, 125–151 (1974). Zbl. 304.32010; English transl.: J. Sov. Math. 4, 540–554 (1976). Zbl. 339.32007
85. Onishchik, A.L.: Pseudoconvexity in the theory of complex spaces. Itogi Nauki Tekh., Ser. Algebra Topologiya. Germ. 15, 93–171 (1977); English transl.: Sov. Math. 14, 1363–1407 (1980). Zbl. 449.32020
86. Poincaré, H.: Sur les fonctions de deus variables. Acta Math. 2, 97–103 (1883). Jrb. 15,358
87. Prill, D.: The divisor class groups of some rings of holomorphic functions. Math. Z. 121, 58–80 (1971). Zbl. 208.352
88. Ramis, J.-P.: Sous-ensembles analytiques d'une variétè banachique complexe. Springer, Berlin (1970). Zbl. 212.428
89. Ramis, J.-P.: Théorèmes de séparation et de finitude pour l'homologie et la cohomologie des espaces (p,q)-convex-concaves. Ann Sc. Norm. Super. Pisa, Sci. Fis. Mat., III. Ser. 27, 933–997 (1974). Zbl. 327.32001

90. Remmert, R.: Sur les espaces analytiques holomorphiquement séparables et holomorphiquement convexes. C. R. Acad. Sci. Paris 243, 118–121 (1956). Zbl. 70.304
91. Remmert, R.: Holomorphe and meromorphe Abbildungen komplexer Räume. Math. Ann. 133, 328–370 (1957).
92. Serre, J-P.: Quelques problèmes globaux relatifs aux variétiès de Stein. Colloque sur les fonctions analytiques de plusieurs variables complexes tenu à Bruxelles, Liège, G. Thone, Masson et C-ie, 57–68 Paris (1953). Zbl. 53.53
93. Serre, J-P.: Faisceaux algébriques cohérents, Ann. Math. 61, 197–278 (1955). Zbl. 67.162
94. Serre. J-P.: Géométrie algébrique et géométrie analytique. Ann. Inst. Fourier 6, 1–42 (1956). Zbl. 75.304
95. Shabat, B.V.: Introduction to complex analysis. Part I: Functions of one variable (Russian) Nauka, Moscow (1985) Zbl. 574.30001, Part II: Functions of several variables (Russian) Nauka, Moscow (1985), Zbl. 578.32001
96. Simha, R.R.: On the complement of a curve on a Stein space of dimension two. Math. Z. 82, 63–66 (1963). Zbl. 112.311
97. Siu, Yum-Tong: Non-countable dimensions of cohomology groups of analytic sheaves and domains of holomorphy. Math. Z. 102, 17–29 (1967). Zbl. 167.68
98. Siu, Yum-Tong: A proof of Cartan's theorems A and B. Tôhoku Math. J. 20, 207–213 (1968). Zbl. 176.38
99. Siu, Yum-Tong: All plane domains are Banach-Stein. Manuscr. Math. 14, 101–105 (1974).
100. Siu, Yum-Tong: Holomorphic fiber bundles whose fibers are bounded Stein domains with zero first Betti number. Math. Ann. 219, 171–192 (1976). Zbl. 318.32010
101. Siu, Yum-Tong: Every Stein subvariety admits a Stein neighborhood. Invent. Math. 38, 89–100 (1976).
102. Skoda, H.: Fibrés holomorphes à base et à fibre de Stein. Invent. Math. 43, 97–107 (1977). Zbl. 365.32018
103. Steenrod, N.E.: The topology of fibre bundles. Princeton Univ. Press, Princeton (1951). Zbl. 54.71
104. Stehlé, J.-L., Plongements du disque dans $\mathbb{C}^2$. Seminaire P. Lelong (analyse), année 1970/1971. Springer Lect. Notes Math. 275, 119–130 (1972). Zbl. 245.32007
105. Stehlé, J.-L.: Fonctions plurisousharmoniques et convexité holomorphe de certains fibrés analytique. Séminaire P. Lelong (Analyse), année 1973/1974, Springer Lect. Notes Math. 474, 155–179 (1975). Zbl. 309.32011
106. Stein, K.: Analytische Funktionen mehrerer komplexer Veränderlichen zu vorgegebenen Periodizitätsmoduln und das zweite Couinsche Problem. Math. Ann. 123, 201–222 (1951).
107. Stein, K.: Uberlagerungen holomorph-vollständiger komplexer Räume. Arch. Math. 7, 354–361 (1956). Zbl. 72.80
108. Succi, F. Il teorema di de Rham olomorpho nel caso relativo. Atti Accad. Naz. Lincei Rend. Cl. Sci. Fis. Mat., Nat. 43, 784–791 (1968).
109. Tognoli, A.: Una caratterizzazione dei moduli della sezioni globali di un fascio coerente su uno spazio di Stein. Boll. Unione Mat. Ital. 8, 181–197 (1973). Zbl. 291.32021
110. Vladimirov, V.S.: Methods of the theory of functions of several complex variables. Nauka, Moscow (1964); English transl.: MIT Press, Cambridge (1966). Zbl. 125.319
111. Weil, A.: Introduction à l'étude des variétés Kählériennes. Hermann, Paris (1958). Zbl. 137.411
112. Weyl, H.: Die Idee der Riemannschen Fläche, 3rd ed., Teubner, Stuttgart (1955). Zbl. 68.60
113. Wiegmann, K.W.: Einbettungen komplexer Räume in Zahlenräume. Invent. Math. 1, 229–242 (1966). Zbl. 148.320

II. Holomorphic Vector Bundles and the Oka-Grauert Principle

J. Leiterer

Contents

In 1939 K. Oka [49] proved the following theorem. *Let $D \subseteq \mathbb{C}^n$ be a domain of holomorphy, let $\{U_i\}_{i \in I}$ be an open covering of D, and let $c_i : U_i \to \mathbb{C}^1 \backslash 0$, $i \in I$, be a family of continuous functions such that the functions c_j/c_i are holomorphic on $U_i \cap U_j$. Then there exists a family of holomorphic function $h_i : U_i \to \mathbb{C}^1 \backslash 0$ such that $h_j/h_i = c_j/c_i$ on $U_i \cap U_j$.*

Proof (using solvability of additive Cousin problems on domains of holomorphy): After passing to a refinement we can assume (cf. 1.9 below) that all the

U_i are balls. Then we can find continuous functions $d_i: U_i \to \mathbb{C}^1$ with $\exp(d_i) = c_i$. Since all the c_j/c_i are holomorphic, all the $d_i - d_j$ are holomorphic. Since D is a domain of holomorphy, there are holomorphic functions $f_i: U_i \to \mathbb{C}^1$ such that $d_j - d_i = f_j - f_i$. Taking $h_i := \exp(f_i)$ solves the problem.

In 1957 H. Grauert [21, 22, 23] proved that this theorem of Oka is valid also for functions with values in an arbitrary complex Lie group G. Moreover, H. Grauert proved that the topological and the holomorphic classifications of G-principal bundles (and thus, in particular, of complex vector bundles) over Stein analytic spaces coincide. In §§ 3–6 of the present paper we give the precise formulation and the proof of Grauert's theorem. § 2 can be considered as a preparatory exercise for this. There we prove some elementary facts on holomorphic vector bundles which can be obtained by means of the Implicit Function Theorem in Banach spaces. In particular, special cases of Grauert's theorem are obtained.

In § 7 we present the generalization of Grauert's theorem to so-called Oka pairs given by O. Forster and K.J. Ramspott [15] (1966). In § 8 some applications of Grauert's theorem as well as its generalization to Oka pairs are given. § 9 deals with infinite-dimensional generalizations, and § 8 contains some remarks concerning holomorphic bundles which admit a continuous extension to the boundary. In § 11 we collect some results which show that domains of holomorphy in $\mathbb{C}^n$ can be characterized by Oka's principle. §§ 7–11 contain no proofs.

§ 12 is devoted to holomorphic vector bundles over the Riemann sphere $\mathbb{P}^1$. A proof of Grothendieck's splitting theorem is given.

In § 13 we consider holomorphic vector bundles over $\mathbb{P}^n$, $n \geqslant 2$. We do not give a survey of this theory, which can be found in the book [50], but restrict ourselves to a special point of view: the study of holomorphic vector bundles on $\mathbb{P}^n$ (or certain domains in $\mathbb{P}^n$) by means of the Radon–Penrose transform.

§ 1. Preliminaries

1.1. If Y is a subset of a topological space X, then we denote by $\bar{Y}$ the closure of Y in X. $Y \subset\subset X$ means that Y is a relatively compact subset of X. The notion of a neighborhood will be used only for *open* neighborhoods.

1.2. $\mathbb{C}^n$ $(n = 1, 2, \ldots)$ is the space of n-tuples of complex numbers.

By an *analytic space* we always mean a *reduced* complex space with countable topology. For this notion and the notion of a *Stein analytic space* we refer, for instance, to [26, 27, 29] and to the first article in this volume.

1.3. Let $\mathscr{F}$ be a sheaf over the topological space X. If $U \subseteq X$ is an open set, then we denote by $\mathscr{F}(U)$ the set of sections of $\mathscr{F}$ over U. For $x \in X$, $\mathscr{F}_z$ denotes the stalk of $\mathscr{F}$ over z.

If X is an analytic (topological) space, then we denote by $\mathcal{O} = \mathcal{O}_X$ ($\mathcal{C} = \mathcal{C}_X$) the sheaf of germs of local holomorphic (continuous) functions on X.

By an *analytic sheaf* over an analytic space we mean a sheaf of $\mathcal{O}$-modules. For this notion and the notion of a *coherent* analytic sheaf, we refer, for instance, to [26, 27, 29] and the first article in this volume.

1.4. By a *vector bundle* we always means a vector bundle with characteristic fibre $\mathbb{C}^r$. To point out the distinction with *holomorphic* vector bundles, vector bundles will be called also *continuous* vector bundles.

Let E be a vector bundle over a topological space X. Then we denote by $\mathcal{C}^E = \mathcal{C}_X^E$ the sheaf of germs of local continuous sections of E. If X is an analytic space and E is holomorphic, then $\mathcal{O}^E = \mathcal{O}_X^E$ denotes the sheaf of germs of local holomorphic sections of E over X.

1.5. $L(r)$ $(r = 1, 2, \ldots)$ denotes the algebra of complex $r \times r$-matrices, and $\mathrm{GL}(r)$ denotes the group of invertible matrices in $L(r)$. If X is a topological (analytic) space and $M \subseteq L(r)$, then we denote by $\mathcal{C}^M = \mathcal{C}_X^M$ ($\mathcal{O}^M = \mathcal{O}_X^M$) the sheaf of germs of local continuous (holomorphic) M-valued maps over X.

1.6. We use some facts from Čech cohomology with coefficients in sheaves or presheaves of groups (see, for instance, [35, 36, 26, 29]. In particular, if $\mathcal{F}$ is a presheaf of abelian groups over a topological space X and $\mathcal{U} = \{U_i\}_{i \in I}$ is an open covering of X, then:

$C^q(\mathcal{U}, \mathcal{F})$ is the group of families $\{f_{i_0 \ldots i_q}\}_{i_0, \ldots, i_q \in I}$ with $f_{i_0 \ldots i_q} \in \mathcal{F}(U_{i_0} \cap \cdots \cap U_{i_q})$;

$\delta^q : C^q(\mathcal{U}, \mathcal{F}) \to C^{q+1}(\mathcal{U}, \mathcal{F})$ is the coboundary operator, i.e. $(\delta^q f)_{i_0 \ldots i_{q+1}} = \sum_{k=0}^{q+1} (-1)^k f_{i_0 \ldots \hat{i}_k \ldots i_{q+1}}$;

$H^q(\mathcal{U}, \mathcal{F}) = \mathrm{kernel}(\delta^q)/\mathrm{image}(\delta^{q-1})$.

1.7. If $\mathcal{F}$ is a presheaf of (in general non-abelian) groups over a topological space X and $\mathcal{U} = \{U_i\}_{i \in I}$ is an open covering of X, then:

$C^0(\mathcal{U}, \mathcal{F})$ is the set of all families $\{f_i\}_{i \in I}$ with $f_i \in \mathcal{F}(U_i)$; $f \in C^1(\mathcal{U}, \mathcal{F})$ is called a *cocycle* iff $f_{ij} f_{jk} = f_{ik}$, the set of all cocycles in $C^1(\mathcal{U}, \mathcal{F})$ will be denoted by $Z^1(\mathcal{U}, \mathcal{F})$;

for $f \in H^1(\mathcal{U}, \mathcal{F})$ and $g \in C^0(\mathcal{U}, \mathcal{F})$, $g \,\square\, f$ denotes the cocycle defined by $(g \,\square\, f)_{ij} = g_i^{-1} f_{ij} g_j$;

two cocycles f, $g \in Z^1(\mathcal{U}, \mathcal{F})$ will be called $\mathcal{F}$*-equivalent* (or *equivalent*) iff $h \,\square\, f = g$ for some $h \in C^0(\mathcal{U}, \mathcal{F})$;

$f \in Z^1(\mathcal{U}, \mathcal{F})$ is called $\mathcal{F}$*-trivial* (or *trivial*) iff f is $\mathcal{F}$-equivalent to the cocycle $\{e_{ij}\}$, where e_{ij} is the neutral element in $\mathcal{F}(U_i \cap U_j)$, the cocycle $\{e_{ij}\}$ will be denoted also by 1;

$H^1(\mathcal{U}, \mathcal{F})$ denotes the set of equivalence classes into which $Z^1(\mathcal{U}, \mathcal{F})$ is divided by $\mathcal{F}$-equivalence.

1.8. Let E be a vector bundle over a topological space X, let $\mathcal{U} = \{U_i\}_{i \in I}$ be an open covering of X, and let $f \in Z^1(\mathcal{U}, \mathcal{C}^{\mathrm{GL}(r)})$. We say E is $\mathcal{C}$*-associated* with f

iff there is a family of continuous vector bundle isomorphisms $h_i : E|U_i \to U_i \times \mathbb{C}^r$ such that f is the corresponding cocycle of transition functions. If X is an analytic space, E is holomorphic and $f \in Z^1(\mathcal{U}, \mathcal{O}^{\mathrm{GL}(r)})$, then we say E is *$\mathcal{O}$-associated* with f iff this family h_i can be chosen to be holomorphic.

1.9. Let $\mathcal{F}$ be a presheaf of (abelian or non-abelian) groups over the topological space X, let $\mathcal{U} = \{U_i\}_{i \in I}$, $\mathcal{V} = \{V_s\}_{s \in S}$ be open coverings of X such that $\mathcal{V}$ is a refinement of $\mathcal{U}$, and let $a : S \to I$ be a map with $V_s \subseteq U_{a(s)}$, $s \in S$. For $f \in Z^1(\mathcal{U}, \mathcal{F})$, $a(f) \in Z^1(\mathcal{V}, \mathcal{F})$ is defined by $(a(f))_{sr} = f_{a(s)a(r)}| V_s \cap V_r$, $r, s \in S$. Recall the following simple

Lemma. *Let $b : S \to I$ be a second map with $V_s \subseteq U_{b(s)}$. Then f and g in $Z^1(\mathcal{U}, \mathcal{F})$ are $\mathcal{F}$-equivalent precisely when $a(f)$ and $b(g)$ are $\mathcal{F}$-equivalent.*

If we say "after passing to a refinement we can assume ...", then we mean "in view of this lemma, after passing to a refinement we can assume ...".

§2. The Implicit Function Theorem and Stability of the Holomorphic Structure of Vector Bundles

To give the general idea we begin with A. Douady's proof of H. Cartan's lemma.

2.1. Notation. Denote by $L(r)$, $r = 1, 2, \ldots$, the algebra of complex $r \times r$-matrices endowed with a norm $\|\cdot\|$ such that $L(r)$ becomes a normed algebra. Let $\mathrm{GL}(r)$ be the group of invertible matrices in $L(r)$. The unit matrix in $L(r)$ will be denoted by 1.

2.2. Theorem [9]. *Let $R_1, R_2 \subset\subset \mathbb{C}^1$ be two open rectangles such that $R_1 \cup R_2$ is also a rectangle. Then every continuous map $A : \overline{R_1 \cap R_2} \to \mathrm{GL}(r)$ which is holomorphic in $R_1 \cap R_2$ can be written $A = A_1 A_2$, where $A_j : \bar{R}_j \to \mathrm{GL}(r)$ are continuous on $\bar{R}_j$ and holomorphic in R_j.*

Note. If A is holomorphic in a neighborhood of $\overline{R_1 \cap R_2}$, this theorem is due to H. Cartan [5] and is known as *Cartan's lemma.*

Proof. It suffices to prove the following two statements:
(i) A can be approximated uniformly on $\overline{R_1 \cap R_2}$ by holomorphic maps on $\mathbb{C}^1$ with values in $\mathrm{GL}(r)$.
(ii) There exists $\varepsilon > 0$ such that the conclusion of the theorem is valid for all A with $\|1 - A(z)\| < \varepsilon$, $z \in \overline{R_1 \cap R_2}$.

Proof of (i). Let $0 \in R_1 \cap R_2$ and choose numbers $0 = t_0 < t_1 < \cdots < t_m = 1$ such that $\|A(t_j z) A(t_{j-1} z)^{-1} - 1\| < 1$, $z \in \overline{R_1 \cap R_2}$. Then $X_j(z) := \ln[A(t_j z) A(t_{j-1} z)^{-1}]$ is defined on $\overline{R_1 \cap R_2}$ and

$$A(z) = [\exp(X_m(z))] \dots [\exp(X_1(z))][A(0)]^{-1} \qquad \text{for } z \in \overline{R_1 \cap R_2}.$$

It remains to approximate the maps X_j uniformly on $\overline{R_1 \cap R_2}$ by holomorphic maps on $\mathbb{C}^1$ with values in $L(r)$.

Proof of (ii). Let B_1, B_2, B_{12} be the Banach spaces of $L(r)$-valued maps which are continuous on $\bar{R}_1, \bar{R}_2, \overline{R_1 \cap R_2}$ and holomorphic in $R_1, R_2, R_1 \cap R_2$. Then $\Phi: B_1 \oplus B_2 \to B_{12}$ defined by $\Phi(X_1, X_2) = (1 + X_1)(1 + X_2)$ is holomorphic and has the derivative $\Phi_0'(X_1, X_2) = X_1 + X_2$ at $0 \in B_1 \oplus B_2$. Choose C^∞-functions f_1, f_2 on $\mathbb{C}^1$ with $f_j = 0$ on $R_j \backslash (R_1 \cap R_2)$ and $f_1 + f_2 = 1$ on $\mathbb{C}^1$. Then by

$$T_j(X)(z) = f_j(z)X(z) + \frac{1}{2\pi i} \int_{R_1 \cap R_2} \frac{\partial f_j(y)}{\partial \bar{y}} \frac{X(y)}{y - z} d\bar{y} \wedge dy, \, z \in R_j,$$

bounded linear operators $T_j: B_{12} \to B_j$ are defined with $\Phi_0'(T_1(X), T_2(X)) = X$ for all $X \in B_{12}$. Hence the assertion follows from the Implicit Function Theorem in Banach spaces.

In the remainder of the present section this proof will be generalized, and we obtain some elementary special results on the stability of the holomorphic structure of vector bundles.

2.3. Definition. *A compactified complex manifold* is a compact Hausdorff space X together with a closed subspace ∂X called the boundary of X ($\partial X = \varnothing$ is possible) such that $X \backslash \partial X$ is a complex manifold and X is the closure of $X \backslash \partial X$. Let U be an open subset of a compactified complex manifold X, and let $Y = L(r)$ or $\mathrm{GL}(r)$. Then $\mathscr{A}^Y(U)$ denotes the set of all continuous maps $f: U \to Y$ which are holomorphic in $U \backslash \partial X$, and $\bar{\mathscr{A}}^Y(U)$ denotes the subset of all maps in $\mathscr{A}^Y(U)$ which admit a continuous extension to the closure $\bar{U}$ of U. Observe that $\mathscr{A}^Y$ is a sheaf whereas $\bar{\mathscr{A}}^Y$ is only a presheaf.

2.4. Definition. If $\mathscr{U} = \{U_i\}_{i \in I}$ is a finite open covering of a compactified complex manifold X, then (see 1.6 and 1.7):

(i) $C^q(\mathscr{U}, \bar{\mathscr{A}}^{L(r)})$ will be considered as Banach space with the norm

$$\|f\| := \max_{i_0, \dots, i_q \in I, z \in \overline{U_{i_0} \cap \dots \cap U_{i_q}}} \|f_{i_0 \dots i_q}(z)\|.$$

(ii) Let $f, g \in Z^1(\mathscr{U}, \bar{\mathscr{A}}^{\mathrm{GL}(r)})$. It is easy to see that f and g are $\bar{\mathscr{A}}^{\mathrm{GL}(r)}$-equivalent if and only if f and g are $\mathscr{A}^{\mathrm{GL}(r)}$-equivalent. If this is the case, then we say f and g are *$\mathscr{A}$-equivalent.*

(iii) Let $g \in Z^1(\mathscr{U}, \bar{\mathscr{A}}^{\mathrm{GL}(r)})$. We define bounded linear operators $\delta^q_{\mathrm{Ad}(g)}$ from $C^q(\mathscr{U}, \bar{\mathscr{A}}^{L(r)})$ into $C^{q+1}(\mathscr{U}, \bar{\mathscr{A}}^{L(r)})$ by setting

$$(\delta^q_{\mathrm{Ad}(g)} f)_{i_0 \dots i_{q+1}} = g_{i_0 i_1} f_{i_1 \dots i_{q+1}} + \sum_{s=1}^{q} (-1)^s f_{i_0 \dots \hat{i}_s \dots i_{q+1}} + (-1)^{q+1} f_{i_0 \dots i_q} g_{i_q i_{q+1}}.$$

Then $\delta^q_{\mathrm{Ad}(g)} \circ \delta^{q-1}_{\mathrm{Ad}(g)} = 0$ and we set

$$H^q(\mathscr{U}, \bar{\mathscr{A}}^{\mathrm{Ad}(g)}) = \mathrm{kernel}(\delta^q_{\mathrm{Ad}(g)}) / \mathrm{image}(\delta^{q-1}_{\mathrm{Ad}(g)}).$$

2.5. Definition. Let X be a compactified complex manifold.

An $\mathscr{A}$*-vector bundle* over X is a continuous vector bundle over X together with a finite family $\{(U_i, h_i)\}_{i\in I}$, where $\mathscr{U} = \{U_i\}_{i\in I}$ is an open covering of X and $h_i : E|U_i \to U_i \times \mathbb{C}^r$ are continuous vector bundle isomorphisms such that the corresponding cocycle of transition functions belongs to $Z^1(\mathscr{U}, \bar{\mathscr{A}}^{\mathrm{GL}(r)})$. Observe that then the restriction $E|(X\setminus\partial X)$ is a holomorphic vector bundle.

Let E, F be $\mathscr{A}$-vector bundles over X, and let $U \subseteq X$ be an open set. A continuous vector bundle homomorphism (isomorphism) from $E|U$ to $F|U$ is called an $\mathscr{A}$*-homomorphism* ($\mathscr{A}$*-isomorphism*) over U iff it is holomorphic over $U\setminus\partial X$.

Let $\mathscr{U} = \{U_i\}_{i\in I}$ be a finite open covering of X, and let E be an $\mathscr{A}$ vector bundle over X. We say the cocycle $g \in Z^1(\mathscr{U}, \bar{\mathscr{A}}^{\mathrm{GL}(r)})$ is $\mathscr{A}$*-associated* with E iff there exists a family of $\mathscr{A}$-isomorphisms $h_i : E|U_i \to U_i \times \mathbb{C}^r$ such that g is the corresponding cocycle of transition functions.

Let E be an $\mathscr{A}$ vector bundle over X and let $U \subseteq X$ be an open set. Then we denote by $\mathscr{A}^E(U)$ the space of all continuous sections $f: U \to E$ that are holomorphic in $U\setminus\partial X$, and by $\bar{\mathscr{A}}^E(U)$ we denote the subspace of all sections in $\mathscr{A}^E(U)$ that admit a continuous extension to $\bar{U}$. Observe again that $\mathscr{A}^E$ is a sheaf whereas $\bar{\mathscr{A}}^E$ is only a presheaf.

2.6. Remark. Let E be an $\mathscr{A}$ vector bundle over a compactified complex manifold X, $\mathscr{U}$ a finite open covering of X, and $g \in Z^1(\mathscr{U}, \bar{\mathscr{A}}^{\mathrm{GL}(r)})$ a cocycle which is $\mathscr{A}$-associated with E. We denote by $\mathrm{Ad}(E)$ the endomorphism bundle of E. Then it is easy to see that the group $H^q(\mathscr{U}, \bar{\mathscr{A}}^{\mathrm{Ad}(g)})$ can be identified with the Čech cohomology group $H^q(\mathscr{U}, \bar{\mathscr{A}}^{\mathrm{Ad}(E)})$ (cf. [14] for this identification).

2.7. Theorem. *Let X be a compactified complex manifold, $\mathscr{U}$ a finite open covering of X, and $g \in Z^1(\mathscr{U}, \bar{\mathscr{A}}^{\mathrm{GL}(r)})$ such that:*

(i) $H^1(\mathscr{U}, \bar{\mathscr{A}}^{\mathrm{Ad}(g)}) = 0$;

(ii) $\mathrm{image}(\delta^1_{\mathrm{Ad}(g)})$ *is topologically closed as a subspace of the Banach space* $C^2(\mathscr{U}, \bar{\mathscr{A}}^{\mathrm{L}(r)})$.

Then there exist $\varepsilon > 0$ and a continuous map

$$R: \{f \in Z^1(\mathscr{U}, \bar{\mathscr{A}}^{\mathrm{GL}(r)}) : \|f - g\| < \varepsilon\} \to C^0(\mathscr{U}, \bar{\mathscr{A}}^{\mathrm{GL}(r)})$$

such that $(Rf) \square f = g$ for all $f \in Z^1(\mathscr{U}, \bar{\mathscr{A}}^{\mathrm{GL}(r)})$ with $\|f - g\| < \varepsilon$.

2.8. Remark. If

$$\dim H^2(\mathscr{U}, \bar{\mathscr{A}}^{\mathrm{Ad}(g)}) < \infty, \tag{2.1}$$

then condition (ii) in Theorem 2.7 is fulfilled. This is a consequence of the following corollary of Banach's open mapping theorem: if the image of a bounded linear operator between Banach spaces is algebraically finitely codimensional, then this image is topologically closed.

We shall see that Theorem 2.7 is a special case of the following Implicit Function Theorem in Banach spaces:

2.9. Theorem (see [46]). *Let E, F, G be real Banach spaces and $U \subseteq E$, $V \subseteq F$ be open sets. Let $T: U \to V$, $S: V \to G$ be C^1-maps and $z \in U$ such that the following conditions are fulfilled:* $T(U) \subseteq S^{-1}(0)$, image(T'_z) = kernel(S'_{Tz}) *and* image(S'_{Tz}) *is topologically closed in* G. (Here T'_z and S'_{Tz} denote the derivatives of T and S at z and Tz.) *Then there exist $\varepsilon > 0$ and a continuous map*

$$R: \{f \in S^{-1}(0): \|f - Tz\| < \varepsilon\} \to U$$

such that $TRf = f$ for all $f \in S^{-1}(0)$ with $\|f - Tz\| < \varepsilon$.

Proof of Theorem 2.7. Set $E = C^0(\mathcal{U}, \bar{\mathcal{A}}^{L(r)})$, $F = C^1(\mathcal{U}, \bar{\mathcal{A}}^{L(r)})$, $G = C^2(\mathcal{U}, \bar{\mathcal{A}}^{L(r)})$, $U = \{y \in C^0(\mathcal{U}, \bar{\mathcal{A}}^{L(r)}): \|y\| < 1\}$, $V = C^1(\mathcal{U}, \bar{\mathcal{A}}^{\mathrm{GL}(r)})$,

$$Ty = \{(1 + y_i)^{-1} g_{ij}(1 + y_j)\}_{i,j \in I}, \quad y \in U,$$

$$Sf = \{f_{ij} f_{jk} - f_{ik}\}_{i,j,k \in I}, \quad f \in V.$$

Then $T(0) = g$, $T(U) \subseteq S^{-1}(0)$, $T'_0 = \delta^0_{\mathrm{Ad}(g)}$, $S'_g = \delta^1_{\mathrm{Ad}(g)}$. Therefore by (i) and (ii) the hypotheses of Theorem 2.9 are fulfilled (for $z = 0$).

2.10. Corollary. *Let X be a compact complex manifold* (*without boundary*) *and let E be a holomorphic vector bundle on X with*

$$H^1(X, \mathcal{O}^{\mathrm{Ad}(E)}) = 0,^1 \tag{2.2}$$

where Ad(E) *is the endomorphism bundle of E. Then all holomorphic vector bundles on X which are sufficiently "close" to E are holomorphically isomorphic to E.* More precisely, we have the following statement: *Let $\mathcal{U}$ be a finite open covering of X, and let $g \in Z^1(\mathcal{U}, \bar{\mathcal{A}}^{\mathrm{GL}(r)})$ be $\mathcal{A}$-associated with E. Then there exist $\varepsilon > 0$ and a continuous map*

$$R: \{f \in Z^1(\mathcal{U}, \bar{\mathcal{A}}^{\mathrm{GL}(r)}): \|f - g\| < \varepsilon\} \to C^0(\mathcal{U}, \bar{\mathcal{A}}^{\mathrm{GL}(r)})$$

such that $(Rf) \square f = g$ for all $f \in Z^1(\mathcal{U}, \bar{\mathcal{A}}^{\mathrm{GL}(r)})$ with $\|f - g\| < \varepsilon$.

Proof. After passing to a refinement we can assume that $\mathcal{U} = \{U_i\}$, where each $\bar{U}_i$ is a ball with respect to some local holomorphic coordinates. It suffices to show (i) and (ii) in Theorem 2.7. Condition (i) follows from (2.2). To prove (ii) we recall that for each $\bar{\partial}$-closed continuous (0, 1)-form φ on $\bar{U}_i$ there is a continuous solution of $\bar{\partial}\psi = \varphi$ on $\bar{U}_i$ (see, for instance, [34] and [34a]). By standard arguments this gives $H^2(\mathcal{U}, \bar{\mathcal{A}}^{\mathrm{Ad}(E)}) \cong H^2(X, \mathcal{O}^{\mathrm{Ad}(E)})$. Since X is compact, this implies (2.1).

2.11. Remark. This corollary is not valid for arbitrary non-compact complex manifolds. An example will be given in 12.7. The Stein manifolds form another class of manifolds where Theorem 2.7 can be applied. First we consider the special case of strictly pseudoconvex subsets of Stein manifolds [29, 34, 34a, 36].

---

[1] For the meaning of the number dim $H^1(X, \mathcal{O}^{\mathrm{Ad}(E)})$ if $\neq 0$, see, for instance, [14] or Article III in this volume.

2.12. Theorem. *Let M be a Stein manifold, and let $D \subset\subset M$ be a strictly pseudoconvex domain with C^2-boundary. Suppose $\mathscr{U} = \{U_i\}_{i \in I}$ is a finite covering of $\bar{D}$ by relatively open sets $U_i \subseteq \bar{D}$, and $g : [0, 1] \to Z^1(\mathscr{U}, \mathscr{A}^{GL(r)})$ is a continuous map. Then there exists a continuous map $H : [0, 1] \to C^0(\mathscr{U}, \mathscr{A}^{GL(r)})$ such that $h(0) = 1$ and $h(t) \,\square\, g(0) = g(t)$ for all $0 \leqslant t \leqslant 1$.*

In the proof of this theorem we use the following "linear"

2.13. Lemma (see, for instance [34, 34a]). *Let M be a Stein manifold, let $D \subset\subset M$ be a strictly pseudoconvex open set with C^2-boundary, and let E be an $\mathscr{A}$ vector bundle over $\bar{D}$. Then, for each continuous $\bar{\partial}$-closed $(0, q)$-form φ on $\bar{D}$ with values in E, the equation $\bar{\partial}\psi = \varphi$ can be solved with continuous ψ on $\bar{D}$.*

Proof of Theorem 2.12. After passing to a refinement we can assume that all $\bar{U}_i$ are strictly convex with respect to some local holomorphic coordinates (see, for instance, [34]). Then it follows by standard arguments from Lemma 2.13 that, for each $\mathscr{A}$ vector bundle E on $\bar{D}$, $H^1(\mathscr{U}, \mathscr{A}^E) = 0$ and $H^2(\mathscr{U}, \mathscr{A}^E) = 0$. In particular (cf. Remark 2.6), $H^1(\mathscr{U}, \mathscr{A}^{\mathrm{Ad}(g(t))}) = 0$ and $H^2(\mathscr{U}, \mathscr{A}^{\mathrm{Ad}(g(t))}) = 0$ for all $0 \leqslant t \leqslant 1$. Now the assertion follows from Theorem 2.7.

2.14. Definition. Let M be a complex manifold. A *continuous* $[0, 1]$-*family of holomorphic vector bundles* over M is a continuous vector bundle V over $M \times [0, 1]$ together with an open covering $\mathscr{U}$ of $M \times [0, 1]$ and a $\mathscr{C}$-associated cocycle $v \in Z^1(\mathscr{U}, \mathscr{C}^{GL(r)})$ such that $v|(M \times t)$ is holomorphic for all fixed $0 \leqslant t \leqslant 1$. We shall say that the holomorphic vector bundle E over M can be *deformed* into the holomorphic vector bundle F over M iff there exists a continuous $[0, 1]$-family V of holomorphic vector bundles over M such that $V|(M \times 0)$ (resp. $V|(M \times 1)$) is holomorphically isomorphic to E (resp. F).

2.15. Theorem. *Let E, F be holomorphic vector bundles over a Stein manifold M. If E can be deformed into F, then E is holomorphically isomorphic to F.*

This theorem is an elementary special case of Grauert's theorem (see § 3). Here we give a direct proof. The ideas of this proof will be used also in the proof of Grauert's theorem in the general case (see §§ 4–6). We need the following

2.16. Lemma. *Let E be a holomorphic vector bundle over a Stein manifold M.* For each open $U \subseteq M$ we denote by $\mathcal{O}^{\mathrm{Ad}(E)}(U)$ the Fréchet algebra of holomorphic endomorphisms of $E|U$ with the topology of uniform convergence on compact sets. Let $\mathcal{O}^{\mathrm{Aut}(E)}(U)$ be the group of automorphisms in $\mathcal{O}^{\mathrm{Ad}(E)}(U)$, and let $\mathcal{O}^{\mathrm{Aut}(E)}_{\mathrm{id}}(U)$ be the connected component of $\mathcal{O}^{\mathrm{Aut}(E)}(U)$ which contains the identity automorphism id of E. Then, *for each $\mathcal{O}_M$-convex compact set $K \subset\subset M$ and each neighborhood U of K, all automorphisms in $\mathcal{O}^{\mathrm{Aut}(E)}_{\mathrm{id}}(U)$ can be approximated uniformly on K by automorphisms in $\mathcal{O}^{\mathrm{Aut}(E)}_{\mathrm{id}}(M)$.*

Proof. Let $H \in \mathcal{O}^{\mathrm{Aut}(E)}_{\mathrm{id}}(U)$. Then there is a continuous map $h : [0, 1] \to \mathcal{O}^{\mathrm{Aut}(E)}_{\mathrm{id}}(U)$ such that $h(0) = \mathrm{id}$ and $h(1) = H$. Choose a neighborhood $V \subset\subset U$ of K and points $0 = t_0 < t_1 < \cdots < t_m = 1$ such that $h(t_j)h(t_{j-1})^{-1} = \mathrm{id} - X_j$, where all X_j

are so small over V that $\ln(\mathrm{id} - X_j) := -\sum_1^\infty (1/k)X_j^k$ converges uniformly on V. Then $H = \exp(\ln(\mathrm{id} - X_m))\dots\exp(\ln(\mathrm{id} - X_1))$ on V. Since K is $\mathcal{O}_M$-convex, the endomorphisms $\ln(\mathrm{id} - X_j)$ can be approximated uniformly on K by endomorphisms $A_j \in \mathcal{O}^{\mathrm{Ad}(E)}(M)$. Then $\exp(A_m)\dots\exp(A_1)$ is the required approximation.

Proof of Theorem 2.15. Choose a sequence $D_k \subset\subset M$ $(k = 1, 2, \dots)$ of strictly pseudoconvex open sets with C^2-boundary such that each $\bar{D}_k$ is $\mathcal{O}_M$-convex, $D_k \subset\subset D_{k+1}$, and each compact set $K \subset\subset M$ is contained in some D_k. By hypothesis there is a continuous $[0, 1]$-family V of holomorphic vector bundles over M such that $V|(M \times 0) \cong E$ and $V|(M \times 1) \cong F$. Denote by $\tilde{E}$ the continuous $[0, 1]$-family of holomorphic vector bundles over M which is obtained by lifting $V|(M \times 0)$ with respect to the projection $M \times [0, 1] \to M \times 0$. Then from Theorem 2.12 one obtains continuous vector bundle isomorphisms H_k from $\tilde{E}|(\bar{D}_k \times [0, 1])$ onto $V|(\bar{D}_k \times [0, 1])$ whose restrictions to each $\bar{D}_k \times t$, $0 \leqslant t \leqslant 1$, are $\mathscr{A}$-isomorphisms, and such that $H_k|(\bar{D}_k \times 0)$ is the identical automorphism of $V|(\bar{D}_k \times 0)$. Fix some isomorphism $T: E \to V|(M \times 0)$ and let $K(t): V|(M \times 0) \to \tilde{E}|(M \times t)$ be the canonical isomorphism. Then $H_k(t) := H_k \circ K(t) \circ T$ are isomorphisms from $E|D_k$ onto $V|(D_k \times t)$. In particular, so we obtain holomorphic isomorphisms $H_k(1)$ from $E|D_k$ onto $V|(D_k \times 1)$. The families $H_k(t)H_{k+1}(t)^{-1}$, $0 \leqslant t \leqslant 1$, form continuous curves in the groups $\mathcal{O}^{\mathrm{Aut}(E)}(D_k)$ connecting $H_k(1)H_{k+1}(1)^{-1}$ with id. Therefore it follows from Lemma 2.16 that the automorphisms $H_k(1)[H_{k+1}(1)]^{-1}$ can be approximated uniformly on D_{k-1} by automorphisms A_k of E over M such that the sequence $H_k(1)A_{k-1}\dots A_1$ $(k = 2, 3, \dots)$ converges uniformly on the compact sets in M to some holomorphic automorphism from E onto $V|(M \times 1) \cong F$.

2.17. Corollary of Theorem 2.15. *Let $D \subseteq \mathbb{C}^n$ be a domain of holomorphy which is starlike* (this means, there exists $z_0 \in D$ with $z_0 + t(z - z_0) \in D$ for all $z \in D$ and $0 \leqslant t \leqslant 1$). *Then every holomorphic vector bundle over D is holomorphically trivial.*

§ 3. Grauert's Theorem

In this § we give the formulation of Grauert's theorem on Oka's principle for bundles over Stein analytic spaces.

3.1. Notation. Let X be an analytic space, and let G be a complex Lie group. We say E is a *holomorphic fibre bundle with characteristic fibre G over X* (see [6]) iff E is obtained in the following way: Let $\mathscr{U} = \{U_i\}_{i \in I}$ be an open covering of X, and $f_{ij}: (U_i \cap U_j) \times G \to G$ a family of holomorphic maps such that the following conditions are fulfilled: (i) $f_{ij}(z, f_{jk}(z, y)) = f_{ik}(z, y)$ for all $z \in U_i \cap U_j \cap U_k$ and $y \in G$; (ii) for each $z \in U_i \cap U_j$, $f_{ij}(z, \cdot)$ is an automorphism of G. Then E is the factor structure of the disjoint union of all $U_i \times G$ with respect to the equivalence relation: $U_j \times G \ni (z, y) \sim (z, f_{ij}(z, y)) \in U_i \times G$.

Denote by $\mathcal{O}^E$ ($\mathcal{C}^E$) the sheaf of germs of local holomorphic (continuous) sections of E. That is, for each open $U \subseteq X$, $\mathcal{O}^E(U)$ ($\mathcal{C}^E(U)$) is the group of holomorphic (continuous) sections $f: U \to E$. $\mathcal{O}^E(U)$ and $\mathcal{C}^E(U)$ will be considered as topological groups with the topology of uniform convergence on the compact subsets of U.

Let M be a topological group, and N a subgroup of M. Let I^q ($q = 1, 2, \ldots$) be the closed unit cube in $\mathbb{R}^q$, and ∂I^q the boundary of I^q. Set

$$J_{q-1} = \{(t_1, \ldots, t_q) \in \partial I^q : t_q \neq 0\}.$$

Denote by $\rho_q(M, N)$ ($q = 1, 2, \ldots$) the topological group of continuous maps $f: I^q \to M$ such that $f(\partial I^q) \subseteq N$ and $f(J_{q-1}) = 1$. Set $\rho_0(M, N) = M$. The group of connected components of $\rho_q(M, N)$ will be denoted by $\pi_q(M, N)$. Set $\rho_q(M) = \rho_q(M, 1)$ and $\pi_q(M) = \pi_q(M, 1)$.

3.2. Theorem (*Grauert's theorem*). *Let X be a Stein analytic space, G a complex Lie group, and E a holomorphic fibre bundle with characteristic fibre G over X. Then*:

(i) *Let* $\mathcal{U}$ *be an open covering of* X, *and* $f, g \in Z^1(\mathcal{U}, \mathcal{O}^E)$ (cf. 1.7). *If* f *and* g *are* $\mathcal{C}^E$*-equivalent, then they are* $\mathcal{O}^E$*-equivalent. Moreover, if* $f = c \,\square\, g$ *for some* $c \in C^0(\mathcal{U}, \mathcal{C}^E)$, *then there exists a continuous map* $H: [0, 1] \to C^0(\mathcal{U}, \mathcal{C}^E)$ *such that* $H(1) = c$, $H(0) \in C^0(\mathcal{U}, \mathcal{O}^E)$ *and* $f = H(t) \,\square\, g$ *for all* $0 \leqslant t \leqslant 1$.

(ii) *Let* $\mathcal{U} = \{U_i\}_{i \in I}$ *be an open covering of* X *such that each* U_i *is Stein. Then for each* $f \in Z^1(\mathcal{U}, \mathcal{C}^E)$ *there exists* $c \in C^0(\mathcal{U}, \mathcal{C}^E)$ *with* $c \,\square\, f \in C^0(\mathcal{U}, \mathcal{O}^E)$.

(iii) *Let* Y *be an* $\mathcal{O}_X$*-convex open subset of* X. *If a section* $f \in \mathcal{O}^E(Y)$ *can be approximated uniformly on the compact subsets of* Y *by sections in* $\mathcal{C}^E(X)$, *then* f *can be approximated uniformly on the compact subsets of* Y *by sections in* $\mathcal{O}^E(E)$.

(iv) *For* $q = 0, 1, 2, \ldots$ *the natural map* $\pi_q(\mathcal{O}^E(X)) \to \pi_q(\mathcal{C}^E(X))$ *is bijective.*

For $E = X \times \mathrm{GL}(r)$, where $\mathrm{GL}(r)$ is the group of invertible complex $r \times r$-matrices, one obtains the following

3.3. Corollary. *Let X be a Stein analytic space. Then*:

(i) *If E and F are holomorphic vector bundles over X which are continuously isomorphic, then E and F are also holomorphically isomorphic.*

(ii) *Each continuous vector bundle over X carries a uniquely determined structure of a holomorphic vector bundle.*

Corollary 3.3(i) contains Theorem 2.15 as special case. Further it implies the following deep strengthening of the elementary Corollary 2.17:

3.4. Corollary. *Let X be a Stein analytic space which is contractible as a topological space. Then all holomorphic vector bundles over X are holomorphically trivial.*

Notice also

3.5. Theorem. *Every holomorphic vector bundle over an open Riemann surface is holomorphically trivial.*

3.6. Notes. Theorem 3.2 is due to H. Grauert [21, 22, 23] (1957/58). The proof of this theorem will be given in §§ 4–6. In this proof we follow H. Cartan's [6] (1958) presentation of Grauert's theorem as well as the paper of O. Forster and K. J. Ramspott [15] (1966), where an important generalization is given (see § 7). Only the so-called "fundamental problem" (see § 4) will be solved in another way (by the Implicit Funcion Theorem in Banach spaces).

We point out the paper of M. Cornalba and Ph. Griffiths [7] (1975), which contains important new ideas concerning Grauert's theorem and its applications. In particular, there can be found a new proof of Grauert's theorem in the case of Corollary 3.3.

Notice that some special cases of Grauert's theorem were obtained earlier by J-P. Serre [58] and J. Frenkel [19].

Theorem 3.5 was obtained by H. Röhrl [53] (1957). H. Röhrl gives a direct proof as well as topological arguments which reduce Theorem 3.5 to Corollary 3.3(i). For elegant direct proofs of Theorem 3.5 we refer also to [13, 26].

Remark. All proofs of Grauert's theorem mentioned above use induction with respect to the dimension of the base space. Only recently, for smooth base spaces, a proof without such an induction was obtained in [34b]. Instead in [34b] an induction with respect to the levels of a strictly plurisubharmonic exhaustion function is used. It seems that this new approach has some advantages. For instance, in this way, one can prove certain "relative" versions of Grauert's theorem for appropriate pairs of complex manifolds $D \subseteq X$. To explain what this means, we now formulate one of the results obtained in [34b].

Let X be a complex manifold endowed with a C^2 function $\varrho : X \to \mathbb{R}$ such that the following two conditions are fulfilled: (i) For all $t \in \mathbb{R}$, the set $D_t := \{x \in X : \varrho(z) < t\}$ is relatively compact. (ii) ϱ is strictly plurisubharmonic on $X \setminus D_0$. (Recall that complex manifolds X which admit such a function ϱ are called *pseudoconvex*, and X is Stein if and only if ϱ can be chosen to be strictly plurisubharmonic over all of X—the solution of the Levi problem.)

Theorem ([34b]). *If E is a continuous vector bundle over X which admits a holomorphic structure over D_0, then this structure exends uniquely to a holomorphic structure of E over X.*

In other words, the following two statements hold true:

(a) *If E is a continuous vector bundle over X and if there exists a holomorphic vector bundle over D_0 which is continuously isomorphic to $E|D_0$, then there exists also a holomorphic vector bundle over X which is continuously isomorphic to E.*

(b) *Let E und F be two holomorphic vector bundles over X. If E and F are continuously isomorphic over X and if, moreover, the restrictions of E and F to D_0*

are holomorphically isomorphic, then E und F are holomorphically isomorphic over X.

Remark. This theorem contains the following special result: *If E is a continuous vector bundle over X which admits over D_0 the structure of the trivial holomorphic vector bundle, then there exists a uniquely determined holomorphic structure of E over X whose restriction to D_0 is trivial.*

This result is equivalent to Corollary 3.3 in the case when X is a Stein analytic space which has not more than a finite number of singular points, because of the following well-known fact (cf., for instance, Theorem C4 in Chapter IX of [29]): There is a holomorphic map p from X onto some Stein analytic space $\tilde{X}$ such that the following conditions are fulfilled: (1) The set S of singular points of $\tilde{X}$ is finite. (2) $p^{-1}(S)$ is compact. (3) p is biholomorphic from $X \setminus p^{-1}(S)$ onto $\tilde{X} \setminus S$.

§4. Proof of Grauert's Theorem. I. The Fundamental Problem

4.1. Notation. Let $R = R^1 \times \cdots \times R^n \subseteq \mathbb{C}^n$, where $R^1, \ldots, R^n$ are open rectangles in $\mathbb{C}^1$. Let $R^1 = R^1_1 \cup R^1_2$, where R^1_1 and R^1_2 are also open rectangles. Set $R_j = R^1_j \times R^2 \times \cdots \times R^n$. Let $D \subseteq \mathbb{C}^n$ be a Stein neighborhood of $\bar{R}$, and X a closed analytic subset of D.

Let G be a complex Lie group, and let E be a holomorphic fibre bundle with the characteristic fibre G over X (see 3.1). Denote by $A(G)$ the Lie algebra of G, and denote by $A(E)$ the associated holomorphic vector bundle with the characteristic fiber $A(G)$ defined as follows (cf. [6]): If $\{U_i\}$ and $\{f_{ij}\}$ are as in 3.1, then $A(E)$ is defined by the cocycle $\{f'_{ij}\}$, where $f'_{ij}(z)$ is the derivative of the automorphism $f_{ij}(z, \cdot)$ at the unit element in G.

The exponential map $\exp : A(G) \to G$ defines a holomorphic map $\exp : A(E) \to E$. There are a neighborhood $\Theta_{A(E)}$ of the zero section in $A(E)$ and a neighborhood Θ_E of the unit section in E such that exp is biholomorphic from $\Theta_{A(E)}$ onto Θ_E. Set $\log = \exp^{-1}$ on Θ_E.

4.2. Theorem (cf. "le problème fondamental" in [6]). *Let $U \subseteq X$ be a neighborhood of $\overline{(R_1 \cap R_2)} \cap X$, and let $U_j := R_j \cap X$. Then there exist a neighborhood $\Sigma \subseteq \mathcal{O}^E(U)$ of the unit element and continuous maps $\sigma_j : \Sigma \to \mathcal{O}^E(U_j)$ such that $\sigma_j(1) = 1$ and*

$$f = \sigma_1(f)\sigma_2(f) \quad \text{on } U_1 \cap U_2 \quad \text{for all } f \in \Sigma.$$

Proof. Let $\tilde{D}$ be a Stein domain with $R \subset\subset \tilde{D} \subset\subset D$. Set $\tilde{X} = X \cap \tilde{D}$, and let M be the Banach space of bounded sections in $\mathcal{O}^{A(E)}(\tilde{X})$. Denote by B_1, B_2 and B_{12} the Banach spaces of bounded holomorphic M-valued maps defined on R_1,

R_2 and $R_1 \cap R_2$. Then we can find neighborhoods S_j of $0 \in B_j$ such that for $f_j \in S_j$ the map

$$\Phi(f_1, f_2) = \log((\exp f_1)(\exp f_2))$$

is defined. This map acts holomorphically from $S_1 \times S_2$ into B_{12}, and the differential of Φ at $(0,0)$ is the map: $(f_1, f_2) \to f_1 + f_2$. By the same arguments as in the proof of Theorem 2.2, we see that this differential has a bounded linear right inverse. Therefore the Implicit Function Theorem in Banach spaces gives a neighborhood S_{12} of $0 \in B_{12}$ and continuous maps $a_j : S_{12} \to S_j$ such that $a_j(0) = 0$ and

$$\Phi(a_1(g), a_2(g)) = g \quad \text{for all } g \in S_{12}.$$

We can assume that $U = \tilde{X} \cap V$, where $V \subseteq \tilde{D}$ is a Stein domain such that $R_{12} \subset\subset V$. By Cartan's Theorem A we can find a finite number of sections $s_1, \ldots, s_m \in M$ which generate $\mathcal{O}^{A(E)}$ over $\tilde{X}$. By Cartan's Theorem B and by the theorem that every continuous linear epimorphism between Fréchet spaces admits a continuous (in general, non-linear) right inverse, one obtains continuous maps $b_i : \mathcal{O}^{A(E)}(U) \to \mathcal{O}(V)$ (with respect to the topology of uniform convergence on the compact subsets of U) such that

$$f = \sum_{i=1}^{m} (b_i(f) | U) s_i \quad \text{for all } f \in \mathcal{O}^{A(E)}(U).$$

For $f \in \mathcal{O}^{A(E)}(U)$ and $\zeta \in R_1 \cap R_2$ we define by

$$c(f)(\zeta) = \sum_{i=1}^{m} b_i(f)(\zeta) s_i$$

a continuous map $c : \mathcal{O}^{A(E)}(U) \to B_{12}$. Choose a neighborhood Σ of the unit element in $\mathcal{O}^E(U)$ so small that $c(\log \Sigma) \subseteq S_{12}$. Setting

$$\sigma_i(f)(\zeta) = ((a_i(c(\log f)))(\zeta))(\zeta) \quad \text{for } f \in \Sigma, \zeta \in U_i,$$

we conclude the proof.

From Theorem 4.2 one easily obtains

4.3. Corollary. *Under the hypotheses of Theorem 4.2 and for $q = 1, 2, \ldots$, there exist a neighborhood $\Gamma \subseteq \varrho_q(\mathscr{C}^E(U), \mathcal{O}^E(U))$ of the unit element and continuous maps $\gamma_i : \Gamma \to \varrho_q(\mathscr{C}^E(U_i), \mathcal{O}^E(U_i))$ such that $\gamma_i(1) = 1$ and*

$$f = \gamma_1(f)\gamma_2(f) \quad \text{on } U_1 \cap U_2 \quad \text{for all } f \in \Gamma.$$

§ 5. Proof of Grauert's Theorem. II. The Principal Theorem

In this § we use the notations from § 3, Section 3.1 and suppose that the hypotheses of Theorem 3.2 are fulfilled. For $q = 1, 2, \ldots$ and each open set $U \subseteq X$

we set

$$\varrho_q^E(U) = \varrho_q(\mathscr{C}^E(U), \mathcal{O}^E(U)) \quad \text{and} \quad \pi_q^E(U) = \pi_q(\mathscr{C}^E(U), \mathcal{O}^E(U)).$$

In this way sheaves ϱ_q^E and π_q^E over X are defined. The aim of this § is the proof of the following auxiliary

5.1. Theorem (cf. "le Théorème principal" in [6]). *For each* $q = 1, 2, \ldots$ *we have*:

(1) $\pi_q^E(X) = 1$;

(2) $H^1(X, \varrho_q^E) = 1$ (for the definition of $H^1(X, \varrho_q^E)$, see, for instance, [35] and article I in this volume);

(3) *for each* $\mathcal{O}_X$*-convex open set* $Y \subseteq X$, *the restriction map* $\varrho_q^E(X) \to \varrho_q^E(Y)$ *has dense image in* $\varrho_q^E(Y)$.

5.2. Proof of Theorem 5.1 on compact sets. Let $y_j(z)$, $z \in \mathbb{C}^n$, be the real coordinates in $\mathbb{C}^n$ such that $z_j = y_j(z) + iy_{j+n}(z)$. Let $Q = \{z \in \mathbb{C}^n : 0 < y_1(z), \ldots, y_{2n}(z) < 1\}$, and let $R \subset\subset Q$ be a closed cube whose sides are parallel to the sides of Q, where we admit that some of the sides of R consist only of one point. For $0 \leqslant t \leqslant s \leqslant 1$ and $1 \leqslant j \leqslant 2n$ we set $R(j, t, s) = \{z \in R : t \leqslant y_j(z) \leqslant s\}$. In this Section 5.2 we assume that X is a closed analytic subset of Q. Set $K = X \cap R$ and $K(j, t, s) = X \cap R(j, t, s)$.

5.2.1. Lemma. *For* $q = 1, 2, \ldots$ *we have*:

(1′) *Let* $U \subseteq X$ *be a neighborhood of* K *and* $f \in \varrho_q^E(U)$. *Then there exist a neighborhood* $U' \subseteq U$ *of* K *and a continuous map* $F : [0, 1] \to \varrho_q^E(U')$ *such that* $F(1) = 1$ *and* $F(0) = f | U'$.

(2′) *Let* $0 < t < 1$, *let* $U \subseteq X$ *be a neighborhood of* $K(j, t, t)$, *and let* $f \in \varrho_q^E(U)$. *Then there exist neighborhoods* $U_1 \subseteq X$ *of* $K(j, 0, t)$, $U_2 \subseteq X$ *of* $K(j, t, 1)$ *and sections* $f_i \in \varrho_q^E(U_i)$ *such that* $f = f_1 f_2^{-1}$ *on* $U_1 \cap U_2 \cap U$.

(3′) *Let* $U \subseteq X$ *be a neighborhood of* K, *and let* $f \in \varrho_q^E(U)$. *Then there exist a neighborhood* $U' \subseteq U$ *of* K *and a sequence* $f_i \in \varrho_q^E(X)$ *such that* $f_i | U'$ *converges to* f *in the topology of* $\varrho_q^E(U')$.

Proof of Lemma 5.2.1. We denote by $(m')_k$ ($m = 1, 2, 3$) the statement (m') under the additional assumption that R has real dimension k. It is easy to see that $(1')_0$ holds true. Hence, it suffices to prove the implications $(1')_k \Rightarrow (3')_k$, $(3')_k \Rightarrow (2')_k$ and $(1')_k, (2')_{k+1} \Rightarrow (1')_{k+1}$.

Proof of $(1')_k \Rightarrow (3')_k$. Suppose that the real dimension of K is k. Let $U \subseteq X$ be a neighborhood of K, and let $f \in \varrho_q^E(U)$. By $(1')_k$, after shrinking U, f can be written $f = f_1 \ldots f_m$, where $f_1, \ldots, f_m \in \varrho_q^E(U)$ and all values of $f_1, \ldots, f_m$ belong to Θ_E (see 3.1). Choose an $\mathcal{O}_X$-convex neighborhood $U' \subset\subset U$ of K and fix $1 \leqslant i \leqslant m$. Now it suffices to prove that f_i can be approximated uniformly on U' by sections in $\varrho_q^E(X)$.

To do this we choose continuous functions $\varphi_0, \varphi_1, \ldots, \varphi_\nu, \varphi_{\nu+1}, \ldots, \varphi_\tau$ on I^q such that:

(I) $\sum_{r=0}^{\tau} \varphi_r = 1$ on I^q;
(II) the supports $\operatorname{supp} \varphi_r$ of φ_r are sufficiently small;
(III) $\operatorname{supp} \varphi_0 \subseteq I^q \backslash \partial I^q$;
(IV) $\operatorname{supp} \varphi_r \cap \partial I^q \neq \emptyset$ but $\operatorname{supp} \varphi_r \cap J_{q-1} = \emptyset$ for $r = 1, \ldots, v$.
(V) $\operatorname{supp} \varphi_r \cap J_{q-1} \neq \emptyset$ for $r = v + 1, \ldots, \tau$.

For each $r = 1, \ldots, v$ we fix some point $x_r \in (\operatorname{supp} \varphi_r) \cap \partial I^q$. Since U' is $\mathcal{O}_X$-convex, for $r = 1, \ldots, v$, we can find sections $g_r \in \mathcal{O}^{A(E)}(X)$ which are sufficiently close to $\log f_i(x_r)$ uniformly over U'. Further we choose a continuous function $\psi : X \to [0, 1]$ such that $\psi = 1$ on U' and $\psi = 0$ on $X \backslash U$. Then

$$\exp(\varphi_0(x)\psi \log F_i(x) + \sum_{r=1}^{v} \varphi_r(x) g_r), \quad x \in I^q,$$

is the required approximation for f_i.

Proof of $(3')_k \Rightarrow (2')_{k+1}$. Let $0 < t < 1$, let $U \subseteq X$ be a neighborhood of $K(j, t, t)$, and let $f \in \varrho_q^E(U)$. We can assume that $R(j, t, t)$ has real dimension k (otherwise $K(j, t, t) = K$ and the assertion is trivial). Then by $(3')_k$ after shrinking U we can find $g \in \varrho_q^E(X)$ such that fg is sufficiently close to the unit element in $\varrho_q^E(U)$, and Corollary 4.3 gives neighborhoods $U_1 \subseteq X$ of $K(j, 0, t)$, $U_2 \subseteq X$ of $K(j, t, 1)$ and sections $h_i \in \varrho_q^E(U_i)$ such that $fg = h_1 h_2$ on $U_1 \cap U_2 \cap U$. It remains to set $f_1 = h_1$ and $f_2 = g h_2^{-1}$.

Proof of $(1')_k$, $(2')_{k+1} \Rightarrow (1')_{k+1}$. Suppose K has real dimension $k + 1$. Let $U \subseteq X$ be a neighborhood of K and $f \in \varrho_q^E(U)$. Choose $1 \leqslant j \leqslant 2n$ such that, for all $0 \leqslant t \leqslant 1$, $R(j, t, t)$ has real dimension k. Then by $(1')_k$ we can find numbers $0 \leqslant t_0 < t_1 < \cdots < t_r = 1$, neighborhoods $U_i \subseteq U$ of $K(j, t_{i-1}, t_i)$ and continuous maps $F_i : [0, 1] \to \varrho_q^E(U_i)$ such that $F_i(1) = 1$ and $F_i(0) = f | U_i$. The maps $F_i^{-1} F_{i+1} | (U_i \cap U_{i+1})$ can be considered as elements in $\varrho_{q+1}^E(U_i \cap U_{i+1})$. Applying $r - 1$ times $(2')_{k+1}$, we obtain $h_i \in \varrho_{q+1}^E(U_i)$ such that $F_i^{-1} F_{i+1} = h_i h_{i+1}^{-1}$. Setting $F = F_i h_i$ on U_i we obtain a continuous map $F : [0, 1] \to \varrho_q^E(U_1 \cup \cdots \cup U_r)$ such that $F(1) = 1$ and $F(0) = f | (U_1 \cup \cdots \cup U_r)$.

5.3. End of Proof of Theorem 5.1

5.3.1. Proof of (3). Let K be a compact subset of Y, and let $f \in \varrho_q^E(Y)$. We have to prove that f can be approximated uniformly on K by sections in $\varrho_q^E(X)$. Since Y is $\mathcal{O}_X$-convex, after enlarging K, we can find a neighborhood $U \subset\subset Y$ of K, a closed cube $R \subset\subset Q$ (see 5.2), whose sides are parallel to the sides of Q, and a biholomorphic map h from U onto a closed analytic subset of Q such that $h(K) = h(U) \cap R$ (see, for instance, [29], VII A). Then by Lemma 5.2.1(1'), after shrinking U, $f | U$ can be written $f | U = f_1 \ldots f_m$, where $f_1, \ldots, f_m \in \varrho_q^E(U)$ and all values of $f_1, \ldots, f_m$ belong to Θ_E (see 3.1). Now we proceed as in the proof of the implication $(1')_k \Rightarrow (3')_k$ above.

5.3.2. Proof of (2). Choose a sequence of $\mathcal{O}_X$-convex open sets $W_m \subseteq X$ such that (see, for instance, [22], VII A): $W_m \subset\subset W_{m+1}$; each compact subset of X is contained in some W_m; there exist closed cubes $R_m \subset\subset Q$ (see 5.2 for some n depending on m), whose sides are parallel, to the sides of Q, and biholomorphic maps h_m from W_m onto closed analytic subsets of Q such that $h_m(\bar{W}_{m-1}) = h_m(W_m) \cap R_m$. Then it follows from Lemma 5.2.1(2′) that

$$H^1(\bar{W}_m, \varrho_q^E) = 1 \quad \text{for all } m. \tag{5.1}$$

Now let $\mathscr{U}$ be an arbitrary open covering of X, and let $f \in Z^1(\mathscr{U}, \varrho_q^E)$ (see 1.7). Then by (5.1), we can find a sequence $c_m \in C^0(\mathscr{U} \cap W_m, \varrho_q^E)$ such that $f = c_m \square 1$ on W_m. In view of (3) from this sequence we can obtain a new sequence $\tilde{c}_m \in C^0(\mathscr{U} \cap W_{m-1}, \varrho_q^E)$ such that also $f = \tilde{c}_m \square 1$ on W_{m-1} and, moreover, $c := \lim \tilde{c}_m$ exists uniformly on the compact subsets of X (then $f = c \square 1$ on X). Actually, assume that $\tilde{c}_1, \ldots, \tilde{c}_m$ are already constructed. Then $\tilde{c}_m^{-1} c_{m+1} \in \varrho_q^E(W_{m-1})$ and by (3) there is a section $b \in \varrho_q^E(X)$ such that $\tilde{c}_m^{-1} c_{m+1} b$ is sufficiently close to 1 uniformly on W_{m-2}. Set $\tilde{c}_{m+1} = c_{m+1} b$.

5.3.3. Proof of (1). Let $f \in \varrho_q^E(X)$. We have to show that there is a continuous map $F: [0,1] \to \varrho_q^E(X)$ such that $F(1) = 1$ and $F(0) = f$. By Lemma 5.2.1(1′) we can find an open covering $\mathscr{U} = \{U_i\}$ of X and continuous maps $F_i: [0,1] \to \varrho_q^E(U_i)$ such that $F_i(1) = 1$ and $F_i(0) = f \mid U_i$. Then $\{F_i^{-1} F_j\} \in Z^1(\mathscr{U}, \varrho_{q+1}^E)$ and by (2) there exist $\{g_i\} \in C^0(\mathscr{U}, \varrho_{q+1}^E)$ with $F_i^{-1} F_j = g_i g_j^{-1}$. Setting $F = F_i g_i$ on U_i we obtain the required map.

§ 6. Proof of Grauert's Theorem. III. Deduction from the Principal Theorem

6.1. Twisting of Bundles by Cocycles. We use the notations from 3.1. Let $\mathscr{V} = \{V_j\}$ be an open covering of X, let $v \in Z^1(\mathscr{V}, \mathscr{C}^E)$ (see 1.7), and let $p: E \to X$ be the projection of E. Then we denote by E^v the fibre bundle with characteristic fibre G over X which is defined as the factor structure of the disjoint union of all $p^{-1}(V_j)$ with respect to the following equivalence relation: $p^{-1}(V_j) \ni x \sim y \in p^{-1}(V_i)$ iff $z := p(x) = p(y)$ and $x = V_{ji}(z) y V_{ij}(z)$.

Let $d = \{d_i\} \in C^0(\mathscr{V}, \mathscr{C}^E)$. Denote by d_i^v the section in $\mathscr{C}^{E^v}(V_i)$ defined in $p^{-1}(V_i)$ by d_i, and set $d^v = \{d_i^v\}$.

Let $h = \{h_{ij}\} \in Z^1(\mathscr{V}, \mathscr{C}^E)$. Denote by h_{ij}^v the section in $\mathscr{C}^{E^v}(V_i \cap V_j)$ defined in $p^{-1}(V_k)$ by $v_{ki} h_{ij} v_{jk}$. Then $h^v := \{h_{ij}^v\} \in Z^1(\mathscr{V}, \mathscr{C}^{E^v})$.

It is easy to see that $v^v = 1$ and, for $d \in C^0(\mathscr{V}, \mathscr{C}^E)$, $h \in Z^1(\mathscr{V}, \mathscr{C}^E)$, $(d \square h)^v = d^v \square h^v$. In particular, $v = d \square h$ is equivalent with $1 = d^v \square h^v$. Moreover, it is easy to see that the map $d \to d^v$ is a continuous linear isomorphism between $C^0(\mathscr{V}, \mathscr{C}^E)$ and $C^0(\mathscr{V}, \mathscr{C}^{E^v})$.

6.2. Proof of Theorem 3.2(i). By 6.1 we can assume that $f = 1$. After passing to a refinement of $\mathscr{U} = \{U_i\}$ we can find continuous maps $C_i: [0,1] \to \mathscr{C}^E(U_i)$

such that $C_i(1) = c_i$ and $C_i(0) \in \mathcal{O}^E(U_i)$. Then $\{C_i^{-1} g_{ij} C_j\} \in Z^1(\mathcal{U}, \varrho_1^E)$ and by Theorem 5.1(2) there are $G_i \in \varrho_1^E(U_i)$ with $C_i^{-1} g_{ij} C_j = G_i G_j^{-1}$. Setting $H_i = C_i G_i$ we conclude the proof.

6.3. Proof of Theorem 3.2(iv). It follows from Theorem 5.1(1) and the exact homotopy sequence

$$\pi_q(\mathcal{O}^E(X)) \to \pi_q(\mathscr{C}^E(X)) \to \pi_q^E(X) \to \pi_{q-1}(\mathcal{O}^E(X)) \to \pi_{q-1}(\mathscr{C}^E(X))$$

that the map $\pi_q(\mathcal{O}^E(X)) \to \pi_q(\mathscr{C}^E(X))$ is bijective for $q \geqslant 1$ and injective for $q = 0$. It remains to prove that for each $f \in \mathscr{C}^E(X)$ there is a continuous map $F: [0, 1] \to \mathscr{C}^E(X)$ such that $F(0)$ is holomorphic and $F(1) = f$. Choose a sufficiently fine open covering $\mathcal{U} = \{U_i\}$ of X such that there are continuous maps $F_i: [0, 1] \to \mathscr{C}^E(U_i)$ with $F(0) \in \mathcal{O}^E(U_i)$ and $F(1) = f \mid U_i$. Then $\{F_i^{-1} F_j\} \in Z^1(\mathcal{U}, \varrho_1^E)$ and from Theorem 5.1(2) we obtain $g_i \in \varrho_1^E(U_i)$ such that $F_i^{-1} F_j = g_i g_j^{-1}$. Setting $F = F_i g_i$ on U_i, we obtain the required map.

6.4. Proof of Theorem 3.2(iii). Let $f \in \mathcal{O}^E(Y)$ which can be approximated uniformly on the compact subsets of Y by sections in $\mathscr{C}^E(X)$. Let $K \subseteq Y$ be a compact set. Since Y is $\mathcal{O}_X$-convex, we can find an $\mathcal{O}_X$-convex neighborhood $U \subset\subset Y$ of K. Choose a section $g \in \mathscr{C}^E(X)$ such that $(fg^{-1})(z) \in \Theta_E$ (see 4.1) for all $z \in \bar{U}$. Then $f \mid U$ and $g \mid U$ belong to the same connected component of $\mathscr{C}^E(U)$. On the other hand, by (iv), the connected component of g in $\mathscr{C}^E(X)$ contains some section $h \in \mathcal{O}^E(X)$. Hence, there is a continuous map $F: [0, 1] \to \mathscr{C}^E(U)$ such that $F(1) = 1$ and $F(0) = fh^{-1} \mid U$. Then $F \in \varrho_1^E(U)$ and, by Theorem 5.1(3), F can be approximated uniformly on K by a sequence $F_m \in \varrho_1^E(X)$. Then the sequence $F_m(0)h \in \mathcal{O}^E(X)$ converges uniformly on K to f.

6.5. Proof of Theorem 3.2(ii). Let $\mathcal{U} = \{U_i\}_{i \in I}$ be an open covering of X by Stein sets, and let $f \in Z^1(\mathcal{U}, \mathscr{C}^E)$.

6.5.1. Lemma. *Let $\mathscr{V} = \{V_\nu\}_{\nu \in N}$ be an open covering of X, which is a refinement of $\mathcal{U}$, and let $a: N \to I$ be a map with $V_\nu \subseteq U_{a(\nu)}$ for all $\nu \in N$. Let $(a(f))_{\nu\mu} := f_{a(\nu)a(\mu)} \mid V_\nu \cap V_\mu$ (cf. 1.9). If there exists $v \in C^0(\mathscr{V}, \mathscr{C}^E)$ such that $v \,\square\, a(f)$ is holomorphic, then there exists also $u \in C^0(\mathcal{U}, \mathscr{C}^E)$ such that $u \,\square\, f$ is holomorphic.*

Proof. Since the restriction of $v \,\square\, a(f)$ to each U_j is $\mathscr{C}^E$-trivial and each U_j is Stein, we obtain from Theorem 3.2(i) sections $h_{j\nu} \in \mathcal{O}^E(U_j \cap V_\nu)$ such that

$$h_{j\nu}^{-1} v_\nu^{-1} f_{a(\nu)a(\mu)} v_\mu h_{j\mu} = 1 \quad \text{on } V_\nu \cap V_\mu \cap U_j.$$

Hence, by $u_j := f_{ja(\nu)} v_\nu h_{j\nu}$ on $U_j \cap V_\nu$, sections $u_j \in \mathscr{C}^E(U_j)$ are defined. Since $u_i^{-1} f_{ij} u_j = h_{i\nu}^{-1} h_{j\nu}$ on $U_i \cap U_j \cap V_\nu$, all $u_i^{-1} f_{ij} u_j$ are holomorphic.

6.5.2. Lemma. *Let $K_1, K_2 \subseteq X$ be compact sets such that K_1, K_2 and $K_1 \cup K_2$ are $\mathcal{O}_X$-convex. Suppose there exist neighborhoods V_k of K_k and $c_k \in C^0(\mathcal{U} \cap V_k, \mathscr{C}^E)$ such that*

$$g_k := c_k \,\square\, (f \mid V_k) \in Z^1(\mathcal{U} \cap V_k, \mathcal{O}^E). \tag{6.1}$$

Then there exist also a neighborhood V of $K_1 \cup K_2$ and $c \in C^0(\mathcal{U} \cap V, \mathscr{C}^E)$ such that $c \,\square\, (f|V)$ is holomorphic.

Proof. We can assume that V_1 and V_2 are Stein. Since $g_1 = c_1 c_2^{-1} \,\square\, g_2$ on $V_1 \cap V_2$, then by Theorem 3.2(i) there is a continuous map $H:[0,1] \to C^0(\mathcal{U} \cap V_1 \cap V_2, \mathscr{C}^E)$ such that $H(1) = c_1 c_2^{-1}$, $H(0)$ is holomorphic and $g_1 = H(t) \,\square\, g_2$ for all $0 \leqslant t \leqslant 1$. It follows (see 6.1) that $1 = (H(t)c_2 c_1^{-1})^{g_1} \,\square\, 1$ for all $0 \leqslant t \leqslant 1$. This means that, for each $0 \leqslant t \leqslant 1$, $(H(t)c_2 c_1^{-1})^{g_1}$ defines a section in $\mathscr{C}^{E^{g_1}}(V_1 \cap V_2)$. Since $(H(1)c_2 c_1^{-1})^{g_1} = 1$, after shrinking V_1 and V_2, we can find $f_1, \ldots, f_m \in \mathscr{C}^{E^{g_1}}(V_1 \cap V_2)$ whose values belong to Θ_E (see 4.1) such that $(H(0)c_2 c_1^{-1})^{g_1} = f_1 \ldots f_m$. Choose neighborhoods $W_k \subset\subset V_k$ of K_k and a continuous function $\varphi: X \to [0,1]$ such that $\varphi = 1$ on $W_1 \cap W_2$ and $\varphi = 0$ on $X \setminus (V_1 \cap V_2)$. Set $v = \exp(\varphi \log f_1) \ldots \exp(\varphi \log f_m)$. Then $v \in \mathscr{C}^{E^{g_1}}(X)$ and $v = (H(0)c_2 c_1^{-1})^{g_1}$ on $W_1 \cap W_2$. Consider v as an element in $C^0(\mathcal{U} \cap W_1, \mathscr{C}^{E^{g_1}})$ and denote by w_1 the element in $C^0(\mathcal{U} \cap W_1, \mathscr{C}^E)$ with $w_1^{g_1} = v$. Then

$$w_1 \,\square\, g_1 = g_1 \qquad \text{on } W_1 \tag{6.2}$$

and $w_1 c_1 c_2^{-1} = H(0)$. After shrinking W_1 and W_2 we can assume that $W_1 \cup W_2$ is Stein. Since also all U_j are Stein, we can apply Theorem 3.2(i) to the coverings $\{U_j \cap W_1, U_j \cap W_2\}$ of $U_j \cap (W_1 \cup W_2)$ and obtain $d_k \in C^0(\mathcal{U} \cap V_k, \mathcal{O}^E)$ such that $w_1 c_1 c_2^{-1} = d_1^{-1} d_2$, i.e. $d_1 w_1 c_1 = d_2 c_2$ on $W_1 \cap W_2$. Therefore, by $c = d_1 w_1 c_1$ on W_1 and $c = d_2 c_2$ on W_2 an element $c \in C^0(\mathcal{U} \cap (W_1 \cup W_2), \mathscr{C}^E)$ is defined. It follows from (6.1) and (6.2) that $c \,\square\, (f|(W_1 \cup W_2))$ is holomorphic.

6.5.3. End of Proof. Choose a sequence W_m as in 5.3.2. Then by Lemmas 6.5.1 and 6.5.2 it is easy to find a sequence $c_m \in C^0(\mathcal{U} \cap W_m, \mathscr{C}^E)$ such that all $c_m \,\square\, (f|W_m)$ are holomorphic. Now we construct a sequence $\tilde{c}_m \in C^0(\mathcal{U} \cap W_m, \mathscr{C}^E)$ such that also all $\tilde{c}_m \,\square\, (f|W_m)$ are holomorphic and, moreover, $\tilde{c}_{m+1} = \tilde{c}_m$ on W_{m-3}. Then $c := \lim \tilde{c}_m$ is an element in $C^0(\mathcal{U}, \mathscr{C}^E)$ such that $c \,\square\, f$ is holomorphic.

Assume that $\tilde{c}_1, \ldots, \tilde{c}_m$ are already constructed. Set $h = \tilde{c}_m \,\square\, (f|W_m)$. Then $(\tilde{c}_m c_{m+1}^{-1}) \,\square\, (c_{m+1} \,\square\, f) = h$ and by Theorem 3.2(i) we obtain a continuous map $H:[0,1] \to C^0(\mathcal{U} \cap W_m, \mathscr{C}^E)$ such that $H(1) = \tilde{c}_m c_{m+1}^{-1}$, $H(0)$ is holomorphic, and $H(t) \,\square\, (c_{m+1} \,\square\, f) = h$ for all $0 \leqslant t \leqslant 1$. It follows $H(t) c_{m+1} \tilde{c}_m^{-1} \,\square\, h = h$, i.e. $(H(t) c_{m+1} \tilde{c}_m^{-1})^h \,\square\, 1 = 1$ (see 6.1). By the same argument as in the proof of Lemma 6.5.2, we find $v \in \mathscr{C}^{E^h}(X)$ such that

$$v = (H(0) c_{m+1} \tilde{c}_m^{-1})^h \text{ on } W_{m-1} \setminus W_{m-2} \qquad \text{and} \qquad v = 1 \text{ on } W_{m-3}. \tag{6.3}$$

Consider v as element in $C^0(\mathcal{U} \cap W_m, \mathscr{C}^{E^h})$ and let w be the element in $C^0(\mathcal{U} \cap W_m, \mathscr{C}^E)$ with $w^h = v$. Define $\tilde{c}_{m+1,i} = w_i \tilde{c}_{m,i}$ if $U_i \cap W_{m-2} \neq \varnothing$ and $\tilde{c}_{m+1,i} = c_{m+1,i}$ if $U_i \cap W_{m-2} = \varnothing$. By (6.3) then $\tilde{c}_{m+1} = \tilde{c}_m$ on W_{m-3}. It remains to prove that all $\tilde{c}_{m+1,i} f_{ij} \tilde{c}_{m+1,j}^{-1}$ are holomorphic on $U_i \cap U_j \cap W_{m+1}$. If $U_i \cap W_{m-2} \neq \varnothing$ and $U_j \cap W_{m-2} \neq \varnothing$, this follows from $w \,\square\, h = h$. If $U_i \cap W_{m-2} = U_j \cap W_{m-2} = \varnothing$, this follows, because $c_{m+1} \,\square\, (f|W_{m+1})$ is holomorphic. If $U_i \cap W_{m-2} \neq \varnothing$ but $U_j \cap W_{m-2} = \varnothing$, this follows from (6.3).

§ 7. Grauert's Theorem for Oka Pairs (after Forster and Ramspott)

In 1966 0. Forster and K.J. Ramspott [15] obtained that in Theorem 3.2 the sheaves $\mathcal{O}^E$ and $\mathcal{C}^E$ can be replaced by appropriate pairs (*Oka pairs*) of subsheaves $\mathcal{F} \subseteq \mathcal{O}^E$ and $\mathcal{G} \subseteq \mathcal{C}^E$. This generalization of Theorem 3.2 was proved by Forster and Ramspott with the object of interesting applications. Some of these application will be given in the next §. In the present § we give the formulation of this theorem. For the proof, which is a generalization of H. Cartan's presentation of Grauert's proof given in §§ 4–6, we refer to [15] (for the solution of the "fundamental problem", see also [42]).

7.1. Notation. Let X be an analytic space, let G be a complex Lie group with the Lie algebra $A(G)$, E a holomorphic fibre bundle over X with characteristic fibre G (see 3.1), and $A(E)$ the associated holomorphic vector bundle with characteristic fibre $A(G)$ (see 4.1).

Let $\mathcal{G}$ be a subsheaf of groups of $\mathcal{C}^E$. For each open $U \subseteq X$ we denote by $A(\mathcal{G})(U)$ the set of all $f \in \mathcal{C}^{A(E)}(U)$ such that $\exp(\lambda f) \in \mathcal{G}(U)$ for all complex numbers λ. If $z \in X$, then we denote by $\mathcal{G}(z)$ the group of all x in the fibre E_z such that for some neighborhood U of z there exists $f \in \mathcal{G}(U)$ with $f(z) = x$. $\mathcal{G}(z)$ will be considered as topological subgroup of E_z.

An *Oka pair* in E is by definition a pair $(\mathcal{F}, \mathcal{G})$, where $\mathcal{F}$ is a subsheaf of groups of $\mathcal{O}^E$, $\mathcal{G}$ is a subsheaf of groups of $\mathcal{C}^E$ such that $\mathcal{F} \subseteq \mathcal{G}$ and the following conditions are fulfilled:

(i) There is a neighborhood $\Theta \subseteq \Theta_E$ (see 4.1) of the unit section in E such that if $f \in \mathcal{F}(U)$ and $g \in \mathcal{G}(U)$, $U \subseteq X$ open, with $f(z), g(z) \in \Theta$ for all $z \in U$, then $\log f \in A(\mathcal{F})(U)$ and $\log g \in A(\mathcal{G})(U)$.

(ii) $A(\mathcal{G})$ is a subsheaf of $\mathcal{C}$-modules of $\mathcal{C}^{A(E)}$ and, for each open $U \subseteq X$, $A(\mathcal{G})(U)$ is a closed subspace of the Fréchet space $\mathcal{C}^{A(E)}(U)$.

(iii) $A(\mathcal{F})$ is a coherent subsheaf of $\mathcal{O}$-modules of $\mathcal{O}^{A(E)}$.

(iv) For each $z \in X$ and $q = 0, 1, 2, \ldots$, the natural map $\pi_q(\mathcal{F}(z)) \to \pi_q(\mathcal{G}(z))$ (see 3.1) is bijective.

Observe that, clearly, $(\mathcal{O}^E, \mathcal{C}^E)$ is an Oka pair in E.

7.2. Example. Let $\mathcal{F}$ be a subsheaf of groups of $\mathcal{O}^E$ satisfying conditions (i) and (iii) in the definition of an Oka pair. For each open $U \subseteq X$ we denote by $\mathcal{F}^c(U)$ the group of all $f \in \mathcal{C}^E(U)$ such that $f(z) \in \mathcal{F}(z)$ for all $z \in U$. Then it is easy to see that $(\mathcal{F}, \mathcal{F}^c)$ is an Oka pair in E. Notice that $\mathcal{F}(z) = \mathcal{F}^c(z)$.

7.3. Theorem [15]. *Let X be a Stein analytic space, G a complex Lie group, and E a holomorphic fibre bundle with characteristic fibre G over X. Then, for each Oka pair $(\mathcal{F}, \mathcal{G})$ in E, assertions* (i)–(iv) *in Theorem* 3.2 *remain valid if we replace $\mathcal{O}^E$ by $\mathcal{F}$ and $\mathcal{C}^E$ by $\mathcal{G}$.*

7.4. Notes. Statements on Oka's principle for homotopy and approximation (as in parts (iii) and (iv) of Theorem 3.2) were obtained also for sections in bundles of homogeneous spaces on which a bundle of complex Lie groups acts holomorphically and transitively (see K.J. Ramspott [52] and H. Grauert/H. Kerner [24]). S. Hayes [31, 32, 33] generalized this to Oka pairs of homogeneous spaces [31]; these are certain pairs of topological spaces on which an Oka pair of groups acts transitively.

§ 8. Applications

Here we give some examples for results which are obtained by means of Grauert's theorem or by its generalization to Oka pairs of Forster and Ramspott.

Notation. If $t > 0$ is a real number, then we denote by $[t]$ the integral part of t. $\mathcal{O}^k$ denotes the trivial holomorphic vector bundle of rank k.

8.1. Theorem. *Let X be a pure n-dimensional Stein analytic space, and E a holomorphic vector bundle on X of rank $r \geqslant [n/2]$. Then:*

(i) *There exists a holomorphic vector bundle F over X such that holomorphically $E \cong F \oplus \mathcal{O}^{r-[n/2]}$.*

(ii) *If n is even, then there exists a holomorphic vector bundle F over X such that holomorphically $E \cong F \oplus \mathcal{O}^{r-(n/2)+1}$ precisely when the Chern class $c_{n/2}(E)$ vanishes.*

8.2. Theorem. *Let X be a pure n-dimensional Stein analytic space, and let E be a holomorphic vector bundle over X of rank r. Then:*

(i) *There exists a holomorphic vector bundle F over X such that holomorphically $E \oplus F \cong \mathcal{O}^{r+[n/2]}$.*

(ii) *If n is even, then there exists a holomorphic vector bundle F over X such that holomorphically $E \oplus F \cong \mathcal{O}^{r+(n/2)-1}$ precisely when the Segre class $s_{n/2}(E)$ vanishes.*

For the case of a smooth Stein manifold X, Theorems 8.1(i) and 8.2(i) were proved in 1968 by O. Forster and K.J. Ramspott [16, 18, 12] by means of topological facts and Oka's principle for fibre spaces with homogeneous fibre [52]. In the formulation given here Theorems 8.1 and 8.2 were obtained in 1982 by M. Schneider [55, 56] using the following recent result of H. Hamm [30]: Every n-dimensional Stein analytic space has the homotopy type of an n-dimensional CW-complex. M. Schneider [55, 56] proved that for n-dimensional CW-complexes the continuous versions of Theorems 8.1 and 8.2 hold true. By Corollary 3.3 this implies Theorems 8.1 and 8.2.

By H. Cartan's Theorems A and B, part (i) in Theorem 8.2 can be formulated also as follows: The $\mathcal{O}(X)$-module $\mathcal{O}^E(X)$ is generated by $r + [n/2]$ elements. In this form Theorem 8.2(i) admits a generalization to arbitrary (not necessarily locally free) coherent analytic sheaves:

8.3. Theorem. *Let X be a Stein analytic space, and let $\mathscr{F}$ be a coherent analytic sheaf over X. For each integer $k \geqslant 1$, set $S_k(\mathscr{F}) = \{z \in X : \dim(\mathscr{F}_z/m_z\mathcal{O}_z) \geqslant k\}$* (here m_z is the maximal ideal in $\mathcal{O}_z$) *and $n_k = \dim S_k(\mathscr{F})$. Set*

$$r = \sup_{k \geqslant 1, S_k \setminus S_{k+1} \neq \varnothing} \left[\frac{n_k}{2}\right] + k.$$

Then the $\mathcal{O}(X)$-module $\mathscr{F}(X)$ is finitely generated precisely when $r < \infty$. In this case $\mathscr{F}(X)$ is generated by r elements.

This theorem was proved by O. Forster and K.J. Ramspott [11, 12, 15, 16, 17] with the weaker estimate $n_k + 1$ instead of n_k by means of Theorem 7.3. M. Schneider [55] observed that, by H. Hamm's result [30] cited above, the same proof gives also the estimate stated in Theorem 8.3.

8.4. Complete intersections. Let Y be a pure r-codimensional analytic subspace of a pure n-dimensional Stein analytic space X, and let $\mathscr{J}_Y$ be the ideal sheaf of Y.

Y is called a *locally complete intersection* in X iff for all $y \in Y$ the $\mathcal{O}_{X,Y}$-module $\mathscr{J}_{Y,y}$ is generated by r elements.

Y is called an (ideal theoretic) *complete intersection* in X iff the $\mathcal{O}(X)$-module $\mathscr{J}_Y(X)$ is generated by r functions.

If Y is a locally complete intersection in X, then the sheaf $J_Y/\mathscr{J}_Y^2 | Y$ is locally free of rank r, i.e. a holomorphic vector bundle, which is called the *conormal bundle* of Y.

Observe that if X and Y are smooth, then Y is always a locally complete intersection in X and the conormal bundle of Y is the usual conormal bundle.

If Y is a locally complete intersection in X, then a simple necessary condition for Y to be a complete intersection is the triviality of the conormal bundle of Y. For, if $f_1, \ldots, f_r \in \mathscr{J}_Y(X)$ generate $\mathscr{J}_Y(X)$, then $f_1, \ldots, f_r \bmod \mathscr{J}_Y^2$ form a global holomorphic basis in the conormal bundle of Y.

8.5. Theorem. *Let Y be a pure k-dimensional analytic subspace of a pure n-dimensional Stein analytic space. Suppose Y is a locally complete intersection with trivial conormal bundle. Then each of the following conditions is sufficient for Y to be a complete intersection:*

(i) $k < n/2$;

(ii) *X is contractible (for instance, $X = \mathbb{C}^n$) and $k \leqslant 2(n-1)/3$.*

(iii) *$k = n/2$ and the dual class $d(Y) \in H^{2k}(X, \mathbb{Z})$ vanishes.*

If X and Y are smooth, part (i) of this theorem was proved by O. Forster and K.J. Ramspott [16] (1966) using Theorem 7.3. In the formulation given here Theorems 8.5(i) and (ii) were obtained in 1982 by C. Banica and O. Forster [2] and M. Schneider [55, 56] (again by means of Hamm's result [30]). Theorem 8.5(iii) is due to M. Schneider (see [54] if X and Y are smooth, and [55] for the general case). The reader who is interested in more information about complete intersections (in Stein or non-Stein spaces) is refered to [2, 55, 56, 50].

Notice also

8.6. Theorem. *Let E be a holomorphic vector bundle over the Stein analytic space X. If $f : X \to E$ is a continuous section which is holomorphic in a neighborhood of $\{z \in X : f(z) = 0\}$, then there is a holomorphic section $h : X \to E$ with the following properties:*

(i) *$h(z) = 0$ precisely when $f(z) = 0$, $z \in X$.*

(ii) *If $f(z) = 0$, then there exist a neighborhood U of z and a holomorphic automorphism A of $E|U$ such that $h = Af$ on U.*

This theorem can be proved by means of Theorem 7.3. If E is trivial and $\dim X < \infty$, then it is contained in Satz 8 in [16].

Finally we point out the following two results of M. Cornalba and Ph. Griffiths [7], the proof of which are based on Grauert's theorem (among other things).

8.7. Theorem. *Let E be a continuous vector bundle of rank r over a Stein manifold X. Then there exist closed analytic subsets Y_k of X such that $c_k(E) = d(Y_k)$ for $k = 1, \ldots, r$. (Here $c_k(E)$* is the k-th Chern class of E and $d(Y_k) \in H^{2k}(X, \mathbb{Z})$ is the dual class of Y_k.)

8.8. Theorem. *Let X be a Stein manifold, and let $\mathbb{Q}$ be the field of rational numbers. Then each class in $H^{2k}(X, \mathbb{Q})$ is a rational multiple of $d(Y)$ for some closed analytic subset Y of X.*

§9. Infinite Dimensional Generalizations of Grauert's Theorem

Grauert's theorem as well as its generalization to Oka pairs of Forster and Ramspott admit generalizations to bundles whose fibres are Banach spaces or Banach Lie groups, respectively. Here we give the formulations of these generalizations.

9.1. Notation. Let X be an analytic space, and G a complex Banach Lie group (for instance, the group of invertible elements of a complex Banach algebra with unit). Then the notion of a holomorphic fibre bundle with characteristic fibre G on X can be defined as in the case of a finite dimensional Lie group (see 3.1).

Let E be a holomorphic fibre bundle with characteristic fibre G over X. Denote by $\mathcal{O}^E$ (resp. $\mathcal{C}^E$) the sheaf of germs of local holomorphic (resp. continuous) sections of E. That is, if $U \subseteq X$ is open, then $\mathcal{O}^E(U)$ (resp. $\mathcal{C}^E(U)$) is the group of holomorphic (resp. continuous) sections $f : U \to E$. $\mathcal{O}^E(U)$ and $\mathcal{C}^E(U)$ will be considered as topological groups with the topology of uniform convergence on the compact subsets of U.

In 1968 L. Bungart [4] proved the following generalization of Theorem 3.2:

9.2. Theorem. *Let X be a Stein analytic space, let G be a complex Banach Lie group, and let E be a holomorphic fibre bundle with characteristic fibre G on X. Then assertions* (i)–(iv) *in Theorem* 3.2 *hold true.*

9.3. Banach Vector Bundles. Let A, B be complex Banach spaces. We denote by $L(A, B)$ the Banach space of bounded linear operators from A into B. Set $L(A) = L(A, A)$. Denote by $\mathrm{GL}(A)$ the group of invertible operators in $L(A)$. Let X be an analytic space.

E is called a *holomorphic (resp. continuous) Banach vector bundle* with characteristic fibre A over X iff E is obtained in the following way: There are given an open covering $\{U_i\}$ of X and a family $f_{ij}: U_i \cap U_j \to \mathrm{GL}(A)$ of holomorphic (resp. continuous) maps with $f_{ij} f_{jk} = f_{ik}$ on $U_i \cap U_j \cap U_k$ such that E is the factor structure of the disjoint union of all $U_i \times A$ with respect to the equivalence relation: $U_j \times A \ni (z, a) \sim (z, f_{ij}(z)a) \in U_i \times A$.

Let E, F be continuous (resp. holomorphic) Banach vector bundles on an analytic space X. A continuous (resp. holomorphic) fibre preserving map $H: E \to F$ is called a *continuous (resp. holomorphic) homomorphism* iff H induces bounded linear operators between the fibres of E and F. H is called a *continuous (resp. holomorphic) isomorphism* iff H induces bounded linear isomorphisms between the fibres of E and F.

For $E = X \times \mathrm{GL}(A)$, Theorem 9.2 implies the following generalization of Corollary 3.3:

9.4. Corollary. *Let X be a Stein analytic space. Then*:

(i) *If E and F are holomorphic Banach vector bundles over X which are continuously isomorphic, then E and F are holomorphically isomorphic.*

(ii) *Each continuous Banach vector bundle over X carries a uniquely defined structure of a holomorphic Banach vector bundle.*

It follows from Corollary 9.4(i) that Corollary 3.4 remains valid also for holomorphic Banach vector bundles. However, in general, Theorem 3.5 does not hold for Banach vector bundles, for there are Banach spaces A with non-connected group $\mathrm{GL}(A)$ [8]. On the other hand, for each infinite dimensional Hilbert space H, $\mathrm{GL}(H)$ is contractible [40]. Therefore, Corollary 9.4(i) implies

9.5. Theorem. *Let X be a Stein analytic space. Then each holomorphic Banach vector bundle over X with an infinite dimensional Hilbert space as characteristic fibre is holomorphically trivial.*

9.6. Infinite Dimensional Oka Pairs (see [42]). First we give an infinite dimensional generalization of the notion of a coherent analytic sheaf. This will be used to define infinite dimensional Oka pairs. Let X be an analytic space.

An analytic sheaf $\mathscr{F}$ on X is called a *Fréchet analytic sheaf* iff, for each open $U \subseteq X$, $\mathscr{F}(U)$ is a Fréchet space such that the multiplication map $\mathcal{O}(U) \times \mathscr{F}(U) \to \mathscr{F}(U)$ is continuous (here $\mathcal{O}(U)$ carries the topology of uniform convergence on the compact subsets of U) and, for each open $V \subseteq U$, the restriction map $\mathscr{F}(U) \to \mathscr{F}(V)$ is continuous.

Let $\mathscr{F}$, $\mathscr{G}$ be Fréchet analytic sheaves on X, and let $f:\mathscr{F}\to\mathscr{G}$ be a homomorphism of $\mathcal{O}$-module sheaves. Then f is called a *holomorphic Fréchet homomorphism* iff, for each open $U\subseteq X$, the induced map $\mathscr{F}(U)\to\mathscr{G}(U)$ is continuous.

Let E be a holomorphic Banach vector bundle over X. Denote by $\mathcal{O}^E$ the Fréchet analytic sheaf of germs of local holomorphic sections of E. That is, for each open $U\subseteq X$, $\mathcal{O}^E(U)$ is the Fréchet space of holomorphic sections $f:U\to E$ with the topology of uniform convergence on the compact subsets of U. If A is a complex Banach space, then we write $\mathcal{O}^A=\mathcal{O}^{X\times A}$.

A Fréchet analytic sheaf $\mathscr{F}$ over X is called a *Banach coherent* analytic sheaf iff for each $z\in X$ and all integers $p\geqslant 0$ there exists an exact sequence

$$\mathcal{O}^{B_n}|U\xrightarrow{h_n}\cdots\xrightarrow{h_1}\mathcal{O}^{B_0}|U\xrightarrow{h_0}\mathscr{F}|U,$$

where $B_0,\ldots,B_n$ are complex Banach spaces and $h_0,\ldots,h_n$ are holomorphic Fréchet homomorphisms.

It is clear that every coherent analytic sheaf is Banach coherent. Then the spaces $B_0,\ldots,B_n$ can be chosen to be finite dimensional. Notice the following infinite dimensional example: Let A, B be complex Banach spaces, and let $T:X\to L(A,B)$ be a holomorphic map, whose values are epimorphisms (with not necessarily complemented kernel). For each open $U\subseteq X$, we denote by $\mathscr{K}er\,T(U)$ the space of all holomorphic maps $f:U\to A$ such that $Tf=0$ on U. Then $\mathscr{K}er\,T$ is a Banach coherent analytic sheaf.

Let E be a holomorphic fibre bundle over X, whose characteristic fibre is a complex Banach Lie group G. Let $A(G)$ be the Banach Lie algebra of G. As in 4.1 and 7.1 we introduce the notations $A(E)$, exp, log, $\Theta_{A(E)}$, Θ_E and $A(\mathscr{G})$, where $\mathscr{G}$ is a subsheaf of groups of $\mathscr{C}^E$.

An *Oka pair* $(\mathscr{F},\mathscr{G})$ in E is defined in the same way as in 7.1 with the exception that condition (iii) is replaced by

(iii)′ $A(\mathscr{F})$ is a Banach coherent analytic subsheaf of $\mathcal{O}^{A(E)}$.

9.7. Theorem (see [42]). *Let X be a Stein analytic space, and let E be a holomorphic fibre bundle over X, whose characteristic fibre is a complex Banach Lie group. Then, for each Oka pair $(\mathscr{F},\mathscr{G})$ in E, assertions* (i)–(iv) *in Theorem* 3.2 *remain valid if we replace $\mathcal{O}^E$ by $\mathscr{F}$ and $\mathscr{C}^E$ by $\mathscr{G}$.*

By means of this theorem one can prove that Theorem 8.6 holds true also for arbitrary holomorphic Banach vector bundles. Notice also the following application:

9.8. Theorem (see [42]). *Let X be a Stein analytic space, let A, B be complex Banach algebras with units, and let* GA *be the group of invertible elements in A. Suppose $T:X\to L(A,B)$ is a holomorphic map, whose values are multiplicative epimorphisms. If $c:X\to$* GA *is a continuous map such that Tc is holomorphic on X, then there exists a holomorphic map $h:X\to$* GA *with $Th=Tc$ on X.*

9.9. Note. In 1969 J. Kajiwara and H. Kazama [38] proved the following generalization of Bungart's Theorem 9.2: *Let X and E be as in Theorem* 9.2, *let*

$Y \subseteq X$ be a closed analytic set, and let $\mathcal{O}^{E,Y}$ ($\mathcal{C}^{E,Y}$) be the subsheaf of sections in $\mathcal{O}^E$ ($\mathcal{C}^E$) which are equal to 1 *on Y. Then assertions* (i)–(iv) *in Theorem* 3.2 *remain valid if we replace $\mathcal{O}^E$ by $\mathcal{O}^{E,Y}$ and $\mathcal{C}^E$ by $\mathcal{C}^{E,Y}$.* Observe that the result of Kajiwara and Kazama is a special case of Theorem 9.7 (which was obtained later).

§ 10. Grauert's Theorem for $\mathcal{A}$-Bundles

Let X be a Stein manifold, and let $D \subset\subset X$ be a strictly pseudoconvex domain with C^2-boundary. In view of the uniform estimates obtained for the $\bar{\partial}$-equation on such domains (see, for instance, [34], one can prove (by a modification of the proof given in §§ 4–6) Grauert's Theorem 3.2 as well as its infinite dimensional generalization Theorem 9.2 also for holomorphic bundles on D which admit a continuous continuation to $\bar{D}$ (see [41, 51]). For example, we have

10.1. Theorem. *If two $\mathcal{A}$ vector bundles over $\bar{D}$* (see 2.5) *are continuously isomorphic, then they are also $\mathcal{A}$-isomorphic.*

Observe also the following interesting result of D. Heunemann [34c] (1986): *Each $\mathcal{A}$ vector bundle over $\bar{D}$ can be "approximated" by holomorphic vector bundles over neighborhoods of $\bar{D}$.* Together with Theorem 2.12 this makes it possible to obtain Theorem 10.1 from Corollary 3.3(i).

§ 11. Characterization of Stein Domains by Oka's Principle

If G is a complex Lie group, then we denote by $\mathcal{O}^G$ ($\mathcal{C}^G$) the sheaf of germs of local holomorphic (continuous) G-valued maps.

11.1. Theorem. *Let X be a 2-dimensional Stein manifold, and let $D \subseteq X$ be an open set. Suppose there exists a complex Lie group such that the following Oka principle holds true:*

If $\mathcal{U}$ is an open covering of D and two cocycles $f, g \in Z^1(\mathcal{U}, \mathcal{O}^G)$ are $\mathcal{C}^G$-equivalent (see 1.7), *then f and g are also $\mathcal{O}^G$-equivalent.*

Then D is Stein.

This theorem was proved in 1978 by J. Kajiwara and M. Nishihara [39]. It is not true for $\dim X \geqslant 3$. For example, $\mathbb{C}^3 \backslash 0$ is not Stein, but $H^1(\mathbb{C}^3 \backslash 0, \mathcal{O}) = 0$ (see, for instance, [26]).

11.2. Theorem. *Let X be a Stein manifold* (of arbitrary dimension), *and let $D \subseteq X$ be a domain with continuous boundary. Suppose there exists a complex Lie group G such that for every polydisc $P \subseteq X$ the following Oka principle holds true:*

If $\mathcal{U}$ is an open covering of $D \cap P$ and $f \in Z^1(\mathcal{U}, \mathcal{O}^G)$ is $\mathcal{C}^G$-trivial, *then f is also* $\mathcal{O}^G$-*trivial.*

Then D is Stein.

This theorem was proved in 1978 by J. Kajiwara [37]. It is not valid for domains D with discontinuous boundary. A counter-example was given also in [37].

11.3. Theorem (see [43]). *Let D be an arbitrary open subset of a Stein manifold. If $H^1(D, \mathcal{O}) = 0$ and all continuously trivial holomorphic vector bundles over D are holomorphically trivial, then D is Stein.*

§ 12. Holomorphic Vector Bundles over the Riemann Sphere

12.1. Notation. Let $\mathbb{P}^1 = \mathbb{C}^1 \cup \{\infty\}$ be the Riemann sphere. Set $\mathbb{P}^1_+ = \mathbb{P}^1 \setminus \infty$ and $\mathbb{P}^1_- = \mathbb{P}^1 \setminus 0$. For each integer k we denote by $\mathcal{O}(k)$ a holomorphic vector bundle of rank 1 over $\mathbb{P}^1$ which is associated with the cocycle $f \in Z^1(\{\mathbb{P}^1_+, \mathscr{P}^1_-\}, \mathcal{O}^{\mathbb{C}\setminus 0})$ (see 1.7) defined by

$$f_{+-}(z) = z^k, \quad z \in \mathbb{P}^1_+ \cap \mathbb{P}^1_-.$$

12.2. Theorem. *For every holomorphic vector bundle E over $\mathbb{P}^1$ there exist uniquely determined integers $a_1 \geqslant \cdots \geqslant a_r$ such that holomorphically*

$$E \cong \mathcal{O}(a_1) \oplus \cdots \oplus \mathcal{O}(a_r).$$

We point out the contrast of this theorem with Oka's principle: The holomorphic structure of a holomorphic vector bundle on $\mathbb{P}^1$ is characterized by a collection of integers $a_1 \geqslant \cdots \geqslant a_r$ whereas, as it is well-known, the underlying continuous structure is defined by the sum $a_1 + \cdots + a_r$. In particular, the topologically trivial bundle of rank $r \geqslant 2$ admits infinitely many different holomorphic structures. An exception is the case $r = 1$: Theorem 12.2 shows that the topological and the holomorphic classifications of rank 1 vector bundles over the Riemann sphere coincide.

To give another formulation and the proof of Theorem 12.2 we introduce some notations:

12.3. Notation. Let $L(r)$ be the algebra of complex $r \times r$-matrices endowed with some norm $\|\cdot\|$ such that $L(r)$ becomes a normed algebra. Let $\mathrm{GL}(r)$ be the group of invertible elements in $L(r)$. For each open $U \subseteq \mathbb{P}^1$ we denote by $\bar{\mathscr{A}}^{L(r)}(U)$ the Banach algebra of all continuous maps $f: \bar{U} \to L(r)$ which are holomorphic in U with the norm

$$\|f\|_U = \max_{z \in \bar{U}} \|f(z)\|.$$

Let $\mathscr{A}^{\mathrm{GL}(r)}(U)$ be the group of invertible elements in $\mathscr{A}^{L(r)}(U)$.

Further, we fix numbers $0 < r < R < \infty$ and set $U_+ = \{z \in \mathbb{P}^1 : |z| < R\}$, $U_- = \{z \in \mathbb{P}^1 : |z| > r\}$.

Since every holomorphic vector bundle over the discs U_+ and U_- is trivial (this follows, for example, from Corollary 2.17 or Theorem 3.5), Theorem 12.2 is equivalent to the following

12.2′. Theorem. *For every $M \in \mathscr{A}^{\mathrm{GL}(r)}(U_+ \cap U_-)$ there exist $M_\pm \in \mathscr{A}^{\mathrm{GL}(r)}(U_\pm)$ and uniquely determined integers $a_1 \geqslant \cdots \geqslant a_r$ such that*

$$M(z) = M_+(z)D(z)M_-(z), \qquad z \in \overline{U_+ \cap U_-},$$

where $D(z)$ is the diagonal matrix with entries $z^{a_1}, \ldots, z^{a_r}$.

Theorem 12.2 is called the splitting theorem of Grothendieck [28] (1957). Theorem 12.2′ was known already at the beginning of the century (for historical remarks we refer to [50, 60] and article I in volume 1 of this Encyclopaedia). There are several proofs of Theorems 12.2 and 12.2′ (see, for example, [60, 26, 28, 50]. Below we give a proof of Theorem 12.2′ which is obtained by a new combination of old ideas from earlier proofs.

It is clear that Theorem 12.2′ is a consequence of the following Lemmas 12.4 and 12.5

12.4. Lemma. *Theorem 12.2′ holds true for all $M \in \mathscr{A}^{\mathrm{GL}(r)}(U_+ \cap U_-)$ of the form*

$$M = \sum_{j=-m}^{n} z^j A_j, \quad \textit{where } A_j \in L(r) \textit{ and } n, m < \infty.$$

12.5. Lemma. *Under the hypotheses of Theorem 12.2′ there exist $M_\pm \in \mathscr{A}^{\mathrm{GL}(r)}(U_\pm)$ such that*

$$M_+ M M_- = \sum_{j=-m}^{n} z^j A_j, \quad \textit{where } A_j \in L(r) \textit{ and } n, m < \infty.$$

In the proof of Lemma 12.5 we use

12.6. Lemma. *There exists $\varepsilon > 0$ such that for each $M \in \mathscr{A}^{\mathrm{GL}(r)}(U_+ \cap U_-)$ with $\|1 - M\|_{U_+ \cap U_-} < \varepsilon$ there exist $M_\pm \in \mathscr{A}^{\mathrm{GL}(r)}(U_\pm)$ with $M = M_+ M_-$.*

First Proof. Since $H^1(\mathbb{P}^1, \mathcal{O}) = 0$ (the Laurent decomposition), the assertion follows from Corollary 2.10.

Second Proof. The map Φ defined by $\Phi(X_+, X_-) = (1 + X_+)(1 + X_-)$ for $X_\pm \in \mathscr{A}^{L(r)}(U_\pm)$ is holomorphic from the Banach space $\mathscr{A}^{L(r)}(U_+) \oplus \mathscr{A}^{L(r)}(U_-)$ into the Banach space $\mathscr{A}^{L(r)}(U_+ \cap U_-)$ (it is a quadratic polynomial). Its derivative Φ_0^1 at the point $(0,0)$ is given by $\Phi^1(X_+, X_-) = X_+ + X_-$ and hence is surjective. Therefore it follows from the Implicit Function Theorem in

Banach spaces that for any neighborhod W of $(0,0)$ in $\bar{\mathscr{A}}^{L(r)}(U_+) \oplus \bar{\mathscr{A}}^{L(r)}(U_-)$, the image $\Phi(W)$ contains a neighborhood of $\Phi(0,0) = 1 \in \bar{\mathscr{A}}^{L(r)}(U_+ \cap U_-)$. It remains to choose W so small that $1 + X_\pm \in \bar{\mathscr{A}}^{GL(r)}(U_\pm)$ for all $(X_+, X_-) \in W$.

Third Proof. Set $\bar{\mathscr{A}}_0(U_-) = \{f \in \bar{\mathscr{A}}^{L(r)}(U_-) : f(\infty) = 0\}$. Then we have projections (the Laurent decomposition) P_+ and P_- from $\bar{\mathscr{A}}^{L(r)}(U_+ \cap U_-)$ onto $\bar{\mathscr{A}}^{L(r)}(U_+)$ resp. $\bar{\mathscr{A}}_0^{L(r)}(U_-)$ with $f = P_+ f + P_- f$ for all $f \in \bar{\mathscr{A}}^{L(r)}(U_+ \cap U_-)$. Set

$$\varepsilon = \tfrac{1}{4}\min\{\|P_+\|^{-1}, \|P_-\|^{-1}\}.$$

where $\|P_\pm\|$ are the norms of $P_\pm$ as operators between the Banach spaces $\bar{\mathscr{A}}^{L(r)}(U_+ \cap U_-)$ and $\bar{\mathscr{A}}^{L(r)}(U_\pm)$.

Now let $M = 1 - A \in \bar{\mathscr{A}}^{GL(r)}(U_+ \cap U_-)$ with $\|A\| \leqslant \varepsilon$. Set $W_1 = P_- A$ and $W_{n+1} = P_-(AW_n)$ for $n = 1, 2, \ldots$. Then the series

$$A_+ := P_+ A + \sum_{n=1}^{\infty} P_+(AW_n) \quad \text{and} \quad A_- := -\sum_{n=1}^{\infty} W_n$$

converge in $\bar{\mathscr{A}}^{L(r)}(U_\pm)$, where $\|A_+\|_{U_+} \leqslant 1/2$ and $\|A_-\|_{U_-} \leqslant 1/2$. Therefore $1 - A_\pm \in \bar{\mathscr{A}}^{GL(r)}(U_\pm)$. We have

$$M(1 - A_-) = 1 - A_- - \left(A + \sum_{n=1}^{\infty} AW_n\right),$$

where

$$A + \sum_{n=1}^{\infty} AW_n = P_+\left(A + \sum_{n=1}^{\infty} AW_n\right) + P_-\left(A + \sum_{n=1}^{\infty} AW_n\right) = A_+ - A_-.$$

Hence $M = (1 - A_+)(1 - A_-)^{-1}$.

Proof of Lemma 12.5. Using the Laurent expansion, M can be approximated uniformly on $\overline{U_+ \cap U_-}$ by maps of the form

$$\sum_{j=-m}^{n} z^j A_j, \quad \text{where } A_j \in L(r) \text{ and } m, n < \infty. \tag{12.1}$$

Therefore it follows from Lemma 12.6 that M can be written $M = AA_+A_-$, where $A_\pm \in \bar{\mathscr{A}}^{GL(r)}(U_\pm)$ and A is of the form (12.1). Applying the same argument to AA_+ we obtain $AA_+ = B_+B_-B$, where $B_\pm \in \bar{\mathscr{A}}^{GL(r)}(U_\pm)$ and B is of the form (12.1). Then $M = B_+(B_-B)A_-$. Since A and B are of the form (12.1) and since $B_-B = B_+^{-1}AA_+$, it follows from Liouville's theorem that B_-B is of the form (12.1).

Proof of Lemma 12.4.

I. *Existence.* If a is an integer and $x \in \mathbb{C}^r \backslash 0$, then we shall say the pair (φ^+, φ^-) is an *a-section of* x iff $\varphi_\pm : U_\pm \to \mathbb{C}^r$ are holomorphic maps such that $\varphi_-(\infty) = x$ and

$$z^a \varphi^+(z) = M(z)\varphi^-(z) \quad \text{for } z \in U_+ \cap U_-.$$

Clearly, if $a \leqslant -m$, then every $x \in \mathbb{C}^r \setminus 0$ has an a-section, whereas if $a \geqslant n+1$, then no $x \in \mathbb{C}^r \setminus 0$ has an a-section. We denote by $a(x)$, $x \in \mathbb{C}^r \setminus 0$, the largest integer a such that x has an a-section. Now we choose a basis $e_1, \ldots, e_r$ in $\mathbb{C}^r$ such that

$$a(e_j) = \max\{a(x) : x \in \mathbb{C}^r \setminus E_{j-1}\}. \tag{12.2}$$

where $E_0 := 0$ and, for $1 \leqslant j \leqslant r-1$, E_j is the subspace of $\mathbb{C}^r$ spanned by $e_1, \ldots, e_j$. Finally we choose $a(e_j)$-sections $(\varphi_j^+, \varphi_j^-)$ of e_j and denote by $M_\pm$ the matrices whose columns are $\varphi_j^\pm$. Then

$$DM_+ = MM_-, \tag{12.3}$$

where D is the diagonal matrix with entries $z^{a(e_j)}$. Now it remains to prove that for $z \in U_\pm$ the vectors $\varphi_1^\pm(z), \ldots, \varphi_r^\pm(z)$ are linearly independent.

To do this we first assume that for some $z_0 \in U_+$

$$\sum_{j=1}^{k} \lambda_j \varphi_j^+(z_0) = 0, \qquad \text{where } \lambda_k \neq 0. \tag{12.4}$$

Set

$$\psi_+(z) = \frac{1}{z - z_0} \sum_{j=1}^{k} \lambda_j \varphi_j^+(z), \qquad z \in U_+,$$

and

$$\psi_-(z) = \frac{z}{z - z_0} \sum_{j=1}^{k} \lambda_j z^{a(e_k) - a(e_j)} \varphi_j^-(z), \qquad z \in U_-.$$

Then, by (12.4), ψ_+ is holomorphic in U_+. Since $a(e_k) \leqslant a(e_j)$ for $j \leqslant k$, ψ_- is holomorphic in $U_- \setminus z_0$. Since by (12.3)

$$M(z)\psi_-(z) = z^{a(e_k)+1}\psi_+(z) \qquad \text{for } z \in U_+ \cap U_-, \tag{12.5}$$

it follows that ψ_- is also holomorphic at z_0 if $z_0 \in U_-$. Hence, by (12.5), (ψ_+, ψ_-) is an $(a(e_k)+1)$-section of the vector

$$\psi_-(\infty) = \lambda_k e_k + \sum_{1 \leqslant j < k,\, a(e_j) = a(e_k)} \lambda_j e_j.$$

Since $\lambda_k \neq 0$ and therefore $\psi_-(\infty) \in \mathbb{C}^r \setminus E_{k-1}$, this is a contradiction with (12.2).

Now we assume that for some $z_0 \in U_-$

$$\sum_{j=1}^{k} \lambda_j \varphi_j^-(z_0) = 0, \qquad \text{where } \lambda_k \neq 0. \tag{12.6}$$

Since $\varphi_j^-(\infty) = e_j$, then $z_0 \neq \infty$. Therefore we can define

$$\psi_-(z) = \frac{z}{z - z_0} \sum_{j=1}^{k} \lambda_j \varphi_j^-(z), \qquad z \in U_-,$$

and

$$\psi_+(z) = \frac{1}{z - z_0} \sum_{j=1}^{k} \lambda_j z^{a(e_j)-a(e_k)} \varphi_j^+(z), \qquad z \in U_+.$$

Then, by (12.6), ψ_- is holomorphic in U_-. Since $a(e_j) \geqslant a(e_k)$ for $j \leqslant k$, ψ_+ is holomorphic in $U_+ \setminus z_0$. Since by (12.3)

$$M(z)\psi_-(z) = z^{a(e_k)+1}\psi_+(z) \qquad \text{for } z \in U_+ \cap U_-, \tag{12.7}$$

it follows that ψ_+ is also holomorphic at z_0 if $z_0 \in U_+$. Hence, by (12.7), the pair (ψ_+, ψ_-) is an $(a(e_k) + 1)$-section of the vector

$$\psi_-(\infty) = \sum_{j=1}^{k} \lambda_j e_j \in \mathbb{C}^r \setminus E_{k-1},$$

which is a contradiction with (12.2).

II. *Uniqueness.* Assume the contrary, i.e. assume that there are collections $a_1 \geqslant \cdots \geqslant a_r$ and $b_1 \geqslant \cdots \geqslant b_r$ of integers such that, for some $1 \leqslant s \leqslant r$, $a_s < b_s$ and $D(a)M_- = M_+D(b)$, where $M_\pm \in \mathscr{A}^{\mathrm{GL}(r)}(U_\pm)$ and $D(a)$, $D(b)$ are the diagonal matrices with entries z^{a_j} and b^{a_j}. If $m_{ij}^\pm$ are the entries of the matrices $M_\pm$, then the equality $D(a)M_- = M_+D(b)$ can be written

$$m_{ij}^-(z) z^{a_i} = z^{b_j} m_{ij}^+(z), \qquad z \in U_+ \cap U_-.$$

Since $a_s < b_s$ and therefore $a_i < b_j$ for $1 \leqslant j \leqslant s \leqslant i \leqslant r$, it follows from Liouville's theorem that $m_{ij}^\pm \equiv 0$ if $1 \leqslant j \leqslant s \leqslant i \leqslant r$. This is impossible, because the matrices $M_\pm$ are invertible.

12.7. Remark. The numbers $a_1 \geqslant \cdots \geqslant a_r$ in Theorem 12.2 are called the *splitting type* of E.

If $a_1 - a_r \leqslant 1$, then for the endomorphism bundle $\operatorname{Ad} E \cong E \otimes E^*$ we have holomorphically

$$\operatorname{Ad} E \cong \mathcal{O}(b_1) \oplus \cdots \oplus \mathcal{O}(b_{r^2}),$$

where $-1 \leqslant b_i \leqslant 1$ for all $1 \leqslant i \leqslant r^2$. Therefore $H^1(\mathbb{P}^1, \mathcal{O}^{\operatorname{Ad} E}) = 0$ (the Laurent decomposition) and we conclude from Corollary 2.10 that all holomorphic vector bundles which are sufficiently "close" (in the sense of Corollary 2.10) to E have the same splitting type as E.

It is easy to see that this is not the case if $a_1 - a_r \geqslant 2$. For example, for each number $\lambda \in \mathbb{C}^1 \setminus 0$ one has

$$\begin{pmatrix} z & 0 \\ \lambda & z^{-1} \end{pmatrix} = \begin{pmatrix} z & -\lambda^{-1} \\ \lambda & 0 \end{pmatrix} \begin{pmatrix} 1 & \lambda^{-1} z^{-1} \\ 0 & 1 \end{pmatrix}, \qquad z \in \mathbb{C}^1 \setminus 0. \tag{12.8}$$

This means, the holomorphic vector bundle E_λ on $\mathbb{P}^1$ associated with the cocycle f^λ in $Z^1(\{\mathbb{P}^1_+, \mathbb{P}^1_-\}, \mathcal{O}^{\mathrm{GL}(2)})$ (see 1.7) defined by

$$f^\lambda_{+-}(z) = \begin{pmatrix} z & 0 \\ \lambda & z^{-1} \end{pmatrix}, \qquad z \in \mathbb{C}^1 \setminus 0,$$

is trivial for all $\lambda \neq 0$ whereas E_0 has the splitting type $(1, -1)$.

This shows also that Corollary 2.10 is not valid for arbitrary non-compact complex manifolds:

Counter-Example. Let $X = \mathbb{C}^1 \times \mathbb{P}^1$, $X_\pm = \mathbb{C}^1 \times \mathbb{P}^1_\pm$. For $\varepsilon \in \mathbb{C}^1$, let E_ε be a holomorphic vector bundle over X associated with the cocycle $f^\varepsilon \in Z^1(\{X_+, X_-\}, \mathcal{O}^{\mathrm{GL}(2)})$ defined by

$$f^\varepsilon_{+-}(\lambda, z) = \begin{pmatrix} z & 0 \\ e^\lambda - \varepsilon & z^{-1} \end{pmatrix}, \qquad \lambda \in \mathbb{C}^1, z \in \mathbb{P}^1_+ \cap \mathbb{P}^1_-.$$

Since $e^\lambda \neq 0$ for all $\lambda \in \mathbb{C}^1$, then, by (12.8), E_0 is holomorphically trivial. Hence $H^1(X, \mathcal{O}^{\mathrm{Ad}(E_0)}) = 0$ (the Laurent decomposition). However, there is no $\varepsilon \neq 0$ such that E_ε is holomorphically trivial, for if $\varepsilon \neq 0$ and $e^\lambda = \varepsilon$, then $E_\varepsilon|(\lambda \times \mathbb{P}^1)$ is holomorphically isomorphic to $\mathcal{O}(1) \oplus \mathcal{O}(-1)$.

§ 13. D-Uniform Bundles over Domains in $\mathbb{P}^n$ and their Radon-Penrose Transform

Denote by $\mathbb{P}^n$ the n-dimensional complex projective space, $n \geqslant 2$. The homogeneous coordinates of a point $z \in \mathbb{P}^n$ will be denoted by $[z^0 : \ldots : z^n]$. Set $\mathbb{P}^n_j = \{z \in \mathbb{P}^n : z^j \neq 0\}$. For each integer k, we denote by $\mathcal{O}(k)$ a holomorphic vector bundle over $\mathbb{P}^n$ associated with the cocycle $\{f_{ij}\} \in Z^1(\{\mathbb{P}^n_0, \ldots, \mathbb{P}^n_n\}, \mathcal{O}^{\mathbb{C}^1 \setminus 0})$ (see 1.7) defined by

$$f_{ij}(z) = \left(\frac{z^j}{z^i}\right)^k \qquad \text{for } z \in \mathbb{P}^n_i \cap \mathbb{P}^n_j.$$

A collection of integers $\tau = (a_1, r_1; \ldots; a_k, r_k)$ will be called a *splitting type* iff $a_1 > \cdots > a_r$ and $r_1, \ldots, r_k \geqslant 1$. For each splitting type $\tau = (a_1, r_1; \ldots; a_k, r_k)$ we set

$$\mathcal{O}(\tau) = \mathcal{O}(a_1)^{\oplus r_1} \oplus \cdots \oplus \mathcal{O}(a_k)^{\oplus r_k}.$$

Denote by $\mathrm{Gr}(1,n)$ the Grassmann manifold of complex projective lines $\mathbb{P}^1 \subseteq \mathbb{P}^n$. For each $U \subseteq \mathbb{P}^n$, we set $\mathrm{Gr}(1, U) = \{L \in \mathrm{Gr}(1,n) : L \subseteq U\}$, and, for each $D \subseteq \mathrm{Gr}(1,n)$, we set $D' = \bigcup_{L \in D} L$. It is clear that $D \subseteq \mathrm{Gr}(1, D')$, but, in general, $\mathrm{Gr}(1, D') \setminus D \neq \varnothing$. For each point $z \in \mathbb{P}^n$, we set

$$\alpha(z) = \{L \in \mathrm{Gr}(1,n) : z \in L\}. \tag{13.1}$$

Let E be a holomorphic vector bundle over an open set $U \subseteq \mathbb{P}^n$, and let $\mathrm{Gr}(1, U) \neq \varnothing$. Then, by Theorem 12.2, for each line $L \in \mathrm{Gr}(1, U)$, there is a unique splitting type $\tau(E, L)$ such that $E|L = \mathcal{O}(\tau(E, L))|L$. In general, this splitting type

$\tau(E,L)$ depends on $L \in \mathrm{Gr}(1,U)$. However, if V is a connected component of $\mathrm{Gr}(1,U)$, then there exist a splitting type τ_E and an analytic set $S_E \subseteq V$ such that $\tau(E,L) = \tau_E$ for $L \in V \backslash S_E$ and $\tau(E,L) \neq \tau_E$ for $L \in S_E$ (see, for instance, [50]; cf. also Remark 12.7 above). The lines in $V \backslash S_E$ are called *generic* or *general lines* of E, and the lines in S_E are called *jump lines* of E. If $\mathrm{Gr}(1,U)$ is connected, then the splitting type of $E|L$, where $L \in \mathrm{Gr}(1,U)$ is a generic line, is called the *generic splitting type* of E and will be denoted by τ_E.

13.1. Definition. Let $D = \mathrm{Gr}(1,n)$ be a connected open set, and let E be a holomorphic vector bundle over D'. E is called *D-uniform* iff $\tau_E = \tau(E,L)$ for all $L \in D$. For each splitting type τ, we denote by $\mathrm{Vect}_\tau(D)$ the set of holomorphic isomorphism classes of all D-uniform holomorphic vector bundles with this generic splitting type over D'.

Let $L \in \mathrm{Gr}(1,n)$ be a fixed line, and let τ be a splitting type. If $V, W \subseteq \mathrm{Gr}(1,n)$ are neighborhoods of L and $E \in \mathrm{Vect}_\tau(V)$, $F \in \mathrm{Vect}_\tau(W)$, then we shall say that E and F are *equivalent at* L iff there exists a neighborhood $Z \subseteq V \cap W$ of L such that $E|Z' = F|Z'$. The corresponding set of equivalence classes will be denoted by $\mathrm{Vect}_\tau(L)$. If $E \in \mathrm{Vect}_\tau(U)$, where $U \subseteq \mathrm{Gr}(1,n)$ is a neighborhood of L, then we denote by $\hat{E}(L)$ the class in $\mathrm{Vect}_\tau(L)$ defined by E.

Notice that the complex tangent bundle of $\mathbb{P}^n$ is an example for a bundle in $\mathrm{Vect}_{(2,1;1,n-1)}(\mathrm{Gr}(1,n))$ which is different from $\mathcal{O}((2,1;1,n-1))$. On the other hand, there do not exist non-trivial holomorphic vector bundles on $\mathbb{P}^n$ whose restrictions to all lines $L \in \mathrm{Gr}(1,n)$ are trivial, and, if $\tau = (a_1, r_1; \ldots; a_k, r_k)$ is a splitting type with $r_1 + \cdots + r_k < n$, then $\mathrm{Vect}_\tau(\mathrm{Gr}(1,n))$ contains only the bundle $\mathcal{O}(\tau)$. For these and further results on holomorphic vector bundles on $\mathbb{P}^n$, we refer to the book [50]. Here we restrict ourselves to a special point of view. By means of the Radon-Penrose transform, we study D-uniform holomorphic vector bundles, where, in general, $D \neq \mathrm{Gr}(1,n)$ and $D' \neq \mathbb{P}^n$. A motivation to do this is the fact that, for $n = 3$, domains of the form D' (where $D \subseteq \mathrm{Gr}(1,3) =$ compactified complexified Minkowski space) appear via the Radon-Penrose transform as twistor spaces for several classes of differential equations in mathematical physics (see, for instance, [1, 20, 10, 61]). In particular, via the Radon-Penrose transform, self-dual Yang-Mills fields on $D \subseteq \mathrm{Gr}(1,3)$ can be interpreted as D-uniform vector bundles whose restrictions to all lines in D are trivial [61] (see also [1]). Another motivation is the circumstance that each holomorphic vector bundle E over $\mathbb{P}^n$ becomes D-uniform if we restrict E to D', where D is the set of generic lines of E. In some cases (cf. Theorems 13.14, 13.17, 13.18, 13.19 below) this makes it possible to study holomorphic vector bundles on $\mathbb{P}^n$ by means of the Radon-Penrose transform in a neighborhood of a generic line.

13.2. The Global Radon-Penrose Transform. Let $D \subseteq \mathrm{Gr}(1,n)$ be an open set. Denote by $\mathbb{F}(D)$ the flag manifold of all pairs $(z,L) \in D' \times D$ with $z \in L$. Then we have the so-called standard diagram (see, for instance, [10, 50]):

$$\begin{array}{ccc} & \mathbb{F}(D) & \\ {}^{\mu}\swarrow & & \searrow^{\nu} \\ D' & & D \end{array}, \tag{13.2}$$

where $\mu(z, L) := z$ and $\nu(z, L) := L$. Denote by Ω_D^1, $\Omega_{D'}^1$, and $\Omega_{\mathbb{F}(D)}^1$ the sheaves of holomorphic 1-forms on D, D' and $\mathbb{F}(D)$. Let $\Omega_\mu^1 := \Omega_{\mathbb{F}(D)}^1/\mu^*\Omega_{D'}^1$ be the sheaf of μ-relative holomorphic 1-forms on $\mathbb{F}(D)$. Define $d_\mu := \pi_\mu \circ d$, where d is the exterior differential operator and $\pi_\mu : \Omega_{\mathbb{F}(D)}^1 \to \Omega_\mu^1$ is the quotient map.

Let $E \in \mathrm{Vect}_\tau(D)$, where $\tau = (a_1, r_1; \ldots; a_k, r_k)$. Set $E(-a_k) = E \otimes \mathcal{O}(-a_k)$. Since E is D-uniform, $\nu_*\mu^*E(-a_k)$ (the 0-th direct ν-image sheaf of μ^*E) is locally free and thus a holomorphic vector bundle over D.

Remark. We consider $\nu_*\mu^*E(-a_k)$ instead of $\nu_*\mu^*E$, for a_k is the largest integer m such that $\nu_*\mu^*E(-m)$ contains all information on E. (For instance, if $a_1 < 0$, then $\nu_*\mu^*E = 0$ contains no information.)

The operator d_μ induces a differential operator

$$\nabla_\mu : \mu^*E(-a_k) \to \mu^*E(-a_k) \otimes \Omega_\mu^1$$

as follows: Let $\mathcal{U} = \{U_i\}$ be an open covering of D', and let $\{f_{ij}\} \in Z^1(\mathcal{U}, \mathcal{O}^{\mathrm{GL}(r)})$ (see 1.5 and 1.7) be a cocycle which is associated with $E(-a_k)$. Then each holomorphic section s of $\mu^*E(-a_k)$ over an open set $W \subseteq \mathbb{F}(D)$ can be identified with a family of holomorphic maps $s_i : W \cap \mu^{-1}(U_i) \to \mathbb{C}^r$ such that $s_i = (f_{ij} \circ \mu)s_j$. Then $d_\mu s_i = (f_{ij} \circ \mu)\, d_\mu s_j$ and thus the family $d_\mu s_i \in \Omega_\mu^1(W \cap \mu^{-1}(U_i))$ defines a holomorphic section of $\mu^*E(-a_k) \otimes \Omega_\mu^1$ over W. This section will be denoted by $\nabla_\mu s$.

The operator ∇_μ induces an operator

$$\nabla_E : \nu_*\mu^*E(-a_k) \to \nu_*(\mu^*E(-a_k) \otimes \Omega_\mu^1).$$

Definition. The pair $\mathscr{P}(E) := (\nu_*\mu^*E(-a_k), \nabla_E)$ is called the *Radon-Penrose transform* of E.

Consider the case $k = 1$, i.e. $E(-a_k)|L$ is trivial for all $L \in D$. Then

$$\nu_*(\mu^*E(-a_k) \otimes \Omega_\mu^1) \cong \nu_*\mu^*E(-a_k) \otimes \Omega_D^1 \tag{13.3}$$

and ∇_E is a holomorphic connection in $\nu_*\mu^*E(-a_k)$, whose restriction to each $\alpha(z) \cap D$, $z \in D'$ (see (13.1)), is flat. (For $n = 3$ these are the self-dual Yang-Mills fields.)

13.2.1. Theorem [61]. *Let $D \subseteq \mathrm{Gr}(1, n)$ be an open set such that, for each $z \in D'$, the intersection $\alpha(z) \cap D$ is connected and simply connected. Then the Radon-Penrose transform sets up an 1-1 correspondence between* $\mathrm{Vect}_{(0,r)}(D)$ *and the set of isomorphism classes of holomorphic connections in rank r vector bundles over D which are flat over all $\alpha(z) \cap D$, $z \in D'$.*

For bundles $E \in \mathrm{Vect}_\tau(D)$ with $\tau = (a_1, r_1; \ldots; a_k, r_k)$, $k > 1$, in distinction to (13.3), we have only an epimorphism

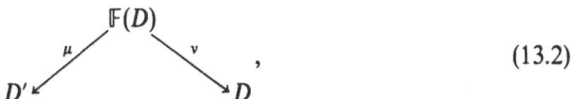

$$\mu^* E(-a_k) \otimes \Omega_D^1 \to \nu_*(\mu^* E(-a_k) \otimes \Omega_\mu^1)$$

with non-trivial kernel, and ∇_E is no longer a connection. But also for general splitting type it is possible to describe $\mathscr{P}(\mathrm{Vect}_\tau(D))$ more explicitly and to prove a generalization of Theorem 13.2.1 (see [44, 47] for a description in affine coordinates, and [62] for an invariant description). Here we restrict ourselves to a local version of the Radon-Penrose transform, which is simpler. First we introduce the notion of a *local τ-connection*, which will be the result of the local Penrose transform.

13.3. Local τ-Connections. Throughout this subsection, let $\tau = (a_1, r_1; \dots; a_k, r_k)$ be a fixed splitting type.

Denote by $P(\tau)$ the algebra of block matrices $M = (M_{ij})_{i,j=1}^k$, where M_{ij} are complex $r_i \times r_j$-matrices such that

$$M_{ij} = 0 \qquad \text{if } a_j - a_i \geqslant 2. \tag{13.4}$$

Denote by $G_\tau^1[\xi]$ the group of block matrices $M = (M_{ij})_{i,j=1}^k$, where M_{ij} are $r_i \times r_j$-matrices of complex polynomials in $\xi \in \mathbb{C}^1$ of degree $\leqslant a_i - a_j$ (in particular, $M_{ij} = 0$ for $i > j$ and $M_{ii} \in L(r_i)$) such that $\det M_{ii} \neq 0$ for all $i = 1, \dots, k$. Notice that $G_\tau^1[\xi]$ can be identified with the automorphism group of $\mathcal{O}(\tau)$ on $\mathbb{P}^1$.

The vectors $u \in \mathbb{C}^{2n-2}$ will be written as matrices

$$u = \begin{pmatrix} u_{01} \dots u_{0,n-1} \\ u_{11} \dots u_{1,n-1} \end{pmatrix}.$$

Let $0 \leqslant i < j < n$. For $t = [t^0, t^1] \in \mathbb{P}^1$ and $u \in \mathbb{C}^{2n-2}$, the point

$$[t^0 u_{01} + t^1 u_{11} : \dots : t^0 u_{0,i-1} + t^1 u_{1,i-1} : t^0 : t^0 u_{0,i+1} + t^1 u_{1,i+1} : \dots : t^0 u_{0,j-1} + t^1 u_{1,j-1} : t^1 : t^0 u_{0,j+1} + t^1 u_{1,j+1} : \dots : t^0 u_{0,n-1} + t^1 u_{1,n-1}]$$

in $\mathbb{P}^n$ will be denoted by $\mu_{ij}(u,t)$. For fixed $u \in \mathbb{C}^{2n-2}$, the line $\{\mu_{ij}(u,t) : t \in \mathbb{P}^1\} \subseteq \mathbb{P}^n$ will be denoted by $\Phi_{ij}(u)$. It is easy to see that $\mu_{ij}(\mathbb{C}^{2n-2} \times \mathbb{P}^1) = \mathbb{P}_i^n \cup \mathbb{P}_j^n$, and Φ_{ij} is biholomorphic from $\mathbb{C}^{2n-2}$ onto $\mathrm{Gr}(1, \mathbb{P}_i^n \cup \mathbb{P}_j^n)$.

A $(n-1)$-dimensional plane $X \subseteq \mathbb{C}^{2n-2}$ will be called an *α-plane* iff there exist a complex vector $f = (f_0, f_1) \neq 0$ and a vector $y \in \mathbb{C}^{n-1}$ such that $X = \{u \in \mathbb{C}^{2n-2} : fu = y\}$. A 2-dimensional plane $X \subseteq \mathbb{C}^{2n-2}$ will be called a *β-plane* iff there exist a complex $(n-1) \times (n-2)$-matrix f of maximal rank and $y \in \mathbb{C}^{2n-4}$ such that $X = \{u \in \mathbb{C}^{2n-2} : uf = y\}$.

A holomorphic 2-form F over an open set $U \subseteq \mathbb{C}^{2n-2}$ will be called an *α-form* (*β-form*) iff, for each α-plane (β-plane) $X \subseteq \mathbb{C}^{2n-2}$, $F | X \cap U = 0$. It is easy to see that each holomorphic 2-form F admits a unique decomposition $F = F_\alpha + F_\beta$, where F_α (F_β) is an α- (β-) form. (For $n = 3$, this is the decomposition into the self-dual and the anti-self-dual component with respect to the Minkowski metric in $\mathbb{C}^4$.)

A *τ-connection form* over an open set $U \subseteq \mathbb{C}^{2n-2}$ is by definition a differential form Θ on U such that:

(i) $\Theta = \Theta_1\, du_{01} + \cdots + \Theta_{n-1}\, du_{0,n-1}$, where $\Theta_1, \ldots, \Theta_{n-1}$ are holomorphic $P(\tau)$-valued maps on U;

(ii) $d\Theta + \Theta \wedge \Theta$ is an α-form on U.

Two τ-connection forms Θ and $\tilde{\Theta}$ over some neighborhoods U and $\tilde{U}$ of a point $u \in \mathbb{C}^{2n-2}$ will be called *τ-gauge equivalent* at u iff there exist a neighborhood $V \subseteq U \cap \tilde{U}$ of u and a holomorphic map $C: V \to G_\tau^1[\xi]$ such that

$$\tilde{\Theta} = C^{-1}\Theta C + C^{-1}(d_0^0 C - \xi d_1^0 C) \qquad \text{on } V,$$

where

$$d_j^0 C := \frac{\partial C}{\partial u_{j1}} du_{01} + \cdots + \frac{\partial C}{\partial u_{j,n-1}} du_{0,n-1} \qquad (j = 0, 1).$$

The corresponding set of equivalence classes will be denoted by $\mathrm{Conn}_\tau(u)$. The elements in $\mathrm{Conn}_\tau(u)$ will be called *local τ-connections* at u. If Θ is a τ-connection form over an open set $U \subseteq \mathbb{C}^{2n-2}$, then we denote by $\hat{\Theta}(u)$, $u \in U$, the local τ-connection at u defined by Θ.

13.4. The Local Penrose Transform. Fix a splitting type $\tau = (a_1, r_1; \ldots; a_k, r_k)$ and a pair $0 \leqslant i < j \leqslant n$. In [44, 47], for each line $L \in \mathrm{Gr}(1, \mathbb{P}_i^n \cup \mathbb{P}_j^n)$ a map

$$\mathscr{P}_{ij}: \mathrm{Vect}_\tau(L) \to \mathrm{Conn}_\tau(\Phi_{ij}^{-1}(L)) \tag{13.5}$$

is constructed such that the following theorem holds true:

13.4.1. Theorem. (i) (13.5) *is a bijection*;

(ii) $\mathscr{P}_{ij}(\mathcal{O}(\tau)(L)) = 0$;

(iii) *if $D \subseteq \mathrm{Gr}(1, \mathbb{P}_i^n \cup \mathbb{P}_j^n)$ is an open set and $E \in \mathrm{Vect}_\tau(D')$, then for each $L_0 \in D$ there is a neighborhood $U \subseteq D$ of L_0 and a τ-connection form Θ on U such that $\mathscr{P}_{ij}(\hat{E}(L)) = \hat{\Theta}(\Phi_{ij}^{-1}(L))$ for all $L \in U$.*

For a more complete explanation of this construction, we refer to [44, 47]. Here we give only the idea. Let $L \in \mathrm{Gr}(1, \mathbb{P}_i^n \cup \mathbb{P}_j^n)$, let $D \subseteq \mathrm{Gr}(1, \mathbb{P}_i^n \cup \mathbb{P}_j^n)$ be a neighborhood of L, and let $E \in \mathrm{Vect}_\tau(D')$. Assume for simplicity that E is trivial over $D' \cap \mathbb{P}_i^n$ and $D' \cap \mathbb{P}_j^n$, i.e. E is associated with some cocycle

$$\{f_{\nu\mu}\}_{\nu,\mu \in \{i,j\}} \in Z^1(\{\mathbb{P}_i^n \cap D', \mathbb{P}_j^n \cap D'\}, \mathcal{O}^{\mathrm{GL}(r_1 + \cdots + r_k)})$$

(see 1.5 and 1.7). Then for each $x \in D$ there exist holomorphic maps $f_0^x: \mathbb{P}^1 \setminus [0:1] \to \mathrm{GL}(r_1 + \cdots + r_k)$ and $f_1^x: \mathbb{P}^1 \setminus [1:0] \to \mathrm{GL}(r_1 + \cdots + r_k)$ such that

$$f_{ij}(\mu_{ij}(\Phi_{ij}^{-1}(x), t)) = f_1^x(t) D_\tau(t) f_0^x(t) \tag{13.6}$$

for $t \in \mathbb{P}^1 \setminus \{[0:1], [1:0]\}$, where $D_\tau(t)$ is the diagonal matrix with entries

$$\overbrace{\left(\frac{t^0}{t^1}\right)^{a_1}, \ldots, \left(\frac{t^0}{t^1}\right)^{a_1}}^{r_1 \text{ times}}, \ldots, \overbrace{\left(\frac{t^0}{t^1}\right)^{a_k}, \ldots, \left(\frac{t^0}{t^1}\right)^{a_k}}^{r_k \text{ times}}.$$

One can prove that after shrinking D and after a permutation of the columns of f_{ij}, this factorization can be chosen so that, for all $x \in D$,

$$D_\tau(t) f_0^x(t) D_\tau^{-1}(t) = 1 + 0\left(\left|\frac{t^1}{t^0}\right|\right) \qquad \text{for } t \to [1:0]. \tag{13.7}$$

If this permutation of the columns of f_{ij} is fixed, then the factorization (13.6) is uniquely determined by (13.7) and depends holomorphically on $x \in D$. Define

$$\Theta(u) = [f_1^{\Phi_{ij}(u)}([0:1])]^{-1} d_0^0 [f_1^{\Phi_{ij}(u)}([0:1])], \qquad u \in \Phi_{ij}^{-1}(D).$$

One can prove that Θ is a τ-connection form over $\Phi_{ij}^{-1}(D)$ and the local τ-connection $\hat{\Theta}(\Phi_{ij}^{-1}(L))$ depends only on $\hat{E}(L)$. Hence we can define $\mathscr{P}_{ij}(\hat{E}(L)) = \hat{\Theta}(\Phi_{ij}^{-1}(L))$.

Now we want to give some results on D-uniform bundles, which can be obtained by means of Theorem 13.6.1. It is not difficult to prove

13.5. Lemma. *Let Θ be a τ-connection form over a connected open set $U \subseteq \mathbb{C}^{2n-2}$. If, for some $u_0 \in U$, $\hat{\Theta}(u_0) = 0$, then $\hat{\Theta}(u) = 0$ for all $u \in U$.*

This lemma and Theorem 13.6.1 imply that if $D \subseteq \mathrm{Gr}(1, n)$ is a connected open set, $E \in \mathrm{Vect}_\tau(D)$ and

$$\hat{E}(L) = \widehat{\mathcal{O}(\tau)}(L) \tag{13.8}$$

at least for one $L \in D$, then (13.8) holds true for all $L \in D$. In other words:

13.6. Theorem. *Let $D \subseteq \mathrm{Gr}(1, n)$ be a connected open set, and let $E \in \mathrm{Vect}_\tau(D)$. If $E|U' \cong \mathcal{O}(\tau)|U'$ for some open set $U \subseteq D$, then there exists an open covering $\{U_i\}_{i \in I}$ of D such that $E|U_i' \cong \mathcal{O}(\tau)|U_i'$ for all $i \in I$.*

13.7. Definition. *Let $D \subseteq \mathrm{Gr}(1, n)$ be a connected open set. Then we denote by $\mathrm{Vect}_\tau^0(D)$ the set of all bundles of the type considered in Theorem 13.6.*

Notice the following well-known consequence of Hartog's extension theorem:

13.8. Proposition. *Let E and F be holomorphic vector bundles over $\mathbb{P}^n$. If $E|U' \cong F|U'$ for some open set $U \subseteq \mathrm{Gr}(1, n)$, then E and F are isomorphic over $\mathbb{P}^n$.*

Denote by G_τ^n the group of automorphisms of $\mathcal{O}(\tau)$ over $\mathbb{P}^n$. By means of Proposition 13.8 it is easy to prove

13.9. Lemma. *Let $D \subseteq \mathrm{Gr}(1, n)$ be a connected open set such that for each α-plane X $X \cap D$ is connected, let $E \in \mathrm{Vect}_\tau^0(D)$, and let $\{U_i\}_{i \in I}$ be an open covering of D such that there are isomorphisms $H_i : E|U_i' \to \mathcal{O}(\tau)|U_i'$. Then for all $i, j \in I$ with $U_i' \cap U_j' \neq \varnothing$ the automorphism $H_i H_j^{-1}$ of $\mathcal{O}(\tau)|(U_i' \cap U_j')$ admits a uniquely determined extension to an automorphism of $\mathcal{O}(\tau)$ over $\mathbb{P}^n$, i.e. $H_i H_j^{-1} \in G_\tau^n$.*

This lemma implies

13.10. Theorem. *For each connected open set $D \subseteq \mathrm{Gr}(1, n)$ such that, for all α-planes X, $D \cap X$ is connected, there is a natural bijection*

$$\mathrm{Vect}_\tau^0(D) \to H^1(D', G_\tau^n).$$

Together with Theorem 13.6 this gives the following

13.11. Corollary. *Let $D \subseteq \mathrm{Gr}(1,n)$ be a connected open set such that D' is simply connected and, for each α-plane X, $D \cap X$ is connected and, and let $E \in \mathrm{Vect}_\tau(D)$. If $E|U' \cong \mathcal{O}(\tau)|U'$ for some open set $U \subseteq D$, then $E \cong \mathcal{O}(\tau)|D'$.*

A holomorphic vector bundle over $\mathbb{P}^n$, whose restriction to some plane $\mathbb{P}^2 \subseteq \mathbb{P}^n$ is isomorphic to $\mathcal{O}(\tau)$, is isomorphic to $\mathcal{O}(\tau)$ over $\mathbb{P}^n$ (see, for instance, [50]). This fact admits the following generalization:

13.12. Theorem. *Let $D \subseteq \mathrm{Gr}(1,n)$ be a connected open set, and let E be a holomorphic vector bundle over D'. Suppose that D' contains a plane $\mathbb{P}^2$ and E restricted to this plane is isomorphic to $\mathcal{O}(\tau)$. Then $E \in \mathrm{Vect}_\tau^0(D)$, and if D' is simply connected, then $E \cong \mathcal{O}(\tau)|D'$.*

Sketch of proof. Denote by $\mathrm{Gr}(2,D')$ the Grassmann manifold of planes $\mathbb{P}^2 \subseteq D'$. Since $H^1(\mathbb{P}^2, \mathcal{O}(k)) = 0$ for all integers k, it follows from Corollary 2.10 and the hypotheses of the theorem that there is an open set $W \subseteq \mathrm{Gr}(2,D')$ such that $E|Z \cong \mathcal{O}(\tau)$ for all $Z \in W$. Fix $L_0 \in D$, $Z_0 \in W$ and $0 \leqslant k < j \leqslant n$ such that $L_0 \subseteq Z_0$ and $L_0 \subseteq \mathbb{P}_i^n \cup \mathbb{P}_j^n$. By Theorem 13.4.1 (iii) we can find a ball $U \subseteq \Phi_{ij}^{-1}(D)$ which is a neighborhood of $\Phi_{ij}^{-1}(L_0)$ and a τ-connection form Θ on U such that $\hat{\Theta}(u) = \mathscr{P}_{ij}(\hat{E}|\Phi_{ij}(u))$ for all $u \in U$. By Theorem 13.6, Corollary 13.11 and Theorem 13.4.1, it suffices to solve the equation

$$\Theta = C^{-1}(d_0^0 C - \xi d_1^0 C) \tag{13.9}$$

with a holomorphic map $C: U \to G_\tau^1[\xi]$. Theorem 13.4.1 applied to all $E|Z$, $Z \in W$, shows that this can be done over each β-plane $\Phi_{ij}^{-1}(\mathrm{Gr}(1,Z)) \cap U$, $Z \in W$. One can prove that this is sufficient for solvability of (13.9) on U.

By definition, τ-connection forms Θ have a block structure $\Theta = (\Theta_{\nu\mu})_{\nu,\mu=1}^k$. The blocks $\Theta_{\nu+1,\nu}$ have an invaraint meaning with respect to τ-gauge equivalence. In particular, for each $\vartheta \in \mathrm{Conn}_\tau(u)$ the equation $\vartheta_{\nu+1,\nu} = 0$ is well-defined.

13.13. Theorem [45]. *Let $D \subseteq \mathrm{Gr}(1,n)$ be a connected open set, and let $E \in \mathrm{Vect}_\tau(D)$, $\tau = (a_1, r_1; \ldots; a_k, r_k)$. Fix $1 \leqslant \nu \leqslant k-1$ and $L \in D$, and choose $0 \leqslant i < j \leqslant n$ so that $L \subseteq \mathbb{P}_i^n \cup \mathbb{P}_j^n$. Then the following two conditions are equivalent:*

(i) $(\mathscr{P}_{ij}(E(L)))_{\nu+1,\nu} = 0$.

(ii) *E contains a D-uniform subbundle of generic splitting type $(a_1, r_1; \ldots; a_\nu, r_\nu)$.*

By the same arguments as in [50], pages 203–204 (where the case $a_\nu - a_{\nu+1} \geqslant 2$ is treated), this implies

13.14. Theorem. *Let E be a holomorphic vector bundle over $\mathbb{P}^n$ of generic splitting type $(a_1, r_1; \ldots; a_k, r_k)$, let $1 \leqslant \nu \leqslant k-1$, and let $L \in \mathrm{Gr}(1,n)$ be a generic line of E. Then the following two conditions are equivalent:*

(i) $(\mathscr{P}_{ij}(E(L)))_{\nu+1,\nu} = 0$.

(ii) *E contains a coherent subsheaf of generic splitting type $(a_1, r_1; \ldots; a_\nu, r_\nu)$.*

By (13.4) this implies

13.15. Corollary (Grauert-Mülich-Spindler Theorem [25, 59, 50]). *If under the hypotheses of Theorem 13.14, $a_\nu - a_{\nu+1} \geqslant 2$, then E contains a coherent subsheaf of generic splitting type $(a_1, r_1; \ldots; a_\nu, r_\nu)$.*

Denote by $\mathrm{Gr}(2,n)$ the Grassmann manifold of planes $\mathbb{P}^2 \in \mathbb{P}^n$. It is easy to prove (see [45])

13.16. Lemma. *Let Θ be a τ-connection form over a connected open set $U \subseteq \mathbb{C}^{2n-2}$. Suppose there is an open set $W \subseteq \mathrm{Gr}(2,n)$ such that, for some $0 \leqslant i < j \leqslant n$ and all $Z \in W$, $U \cap \Phi_{ij}^{-1}(\mathrm{Gr}(1,Z)) \neq \emptyset$ and*

$$\Theta_{\nu+1,\nu} | U \cap \Phi_{ij}^{-1}(\mathrm{Gr}(1,Z)) = 0.$$

Then $\Theta_{\nu+1,\nu} = 0$.

From this lemma and Theorem 13.14 one can obtain

13.17. Theorem [45]. *Let E be a holomorphic vector bundle over $\mathbb{P}^n$ of generic splitting type $(a_1, r_1; \ldots; a_k, r_k)$, let $1 \leqslant \nu \leqslant k-1$, and let $W \subseteq \mathrm{Gr}(2,n)$ be an open set such that, for each $Z \in W$, $E|Z$ contains a coherent subsheaf of generic splitting type $(a_1, r_1; \ldots; a_\nu, r_\nu)$. Then E contains a coherent subsheaf of over $\mathbb{P}^n$ of this splitting type.*

Notice also the following result, which can be proved by means of Theorems 13.17 and 13.4.1 (see [45]):

13.18. Theorem. *Let E be a holomorphic vector bundle on $\mathbb{P}^n$ of generic splitting type $(a_1, r_1; \ldots; a_k, r_k)$ with $r_1 = \cdots = r_k = 1$. If E is stable (in the sense of Mumford and Takemoto; see, for instance, [50]), then the restriction of E to the general hyperplane $\mathbb{P}^{n-1} \subseteq \mathbb{P}^n$ is stable.*

For $k = 2$, this theorem is contained in a theorem of W. Barth [3] and M. Maruyama [48] (see also [50]). For $k = 3$ it is due to M. Schneider [57].

For $L \in \mathrm{Gr}(1,n)$ we denote by $L^{(m)}$ the infinitesimal neighborhood of order m of L in $\mathbb{P}^n$. By means of the Radon-Penrose transform one can prove also the following

13.19. Theorem ([47]). *Let E be a holomorphic vector bundle over $\mathbb{P}^n$, and let $\tau = (a_1, r_1; \ldots; a_k, r_k)$ be the generic splitting type of E. Then the following two statements hold true:*

(i) *E is holomorphically isomorphic to $\mathcal{O}(\tau)$ over $\mathbb{P}^n$ if and only if this is the case over the infinitesimal neighborhoods of order $a_1 - a_r + 2$ of the generic lines of E.*

(ii) *Let $1 \leqslant i \leqslant k-1$ such that $a_i - a_{i-1} = 1$. Suppose, for each generic line L of E, the restriction $E|L^{(1)}$ of E to the first-order infinitesimal neighborhood of L contains a locally free subsheaf $\mathscr{F}$ such that the factor $(E|L^{(1)})/\mathscr{F}$ is also locally free and the restriction of $\mathscr{F}$ to L is isomorphic to*

$$\mathcal{O}(a_1)^{\oplus r_1} \oplus \cdots \oplus \mathcal{O}(a_i)^{\oplus r_i}. \tag{13.10}$$

Then E contains a normal coherent subsheaf over $\mathbb{P}^n$ which is, over the generic lines, isomorphic to (13.10).

Remarks. If the generic splitting type of E is trivial (i.e. $k = 1$ and $a_1 = 0$) this theorem is well-known. In this case it follows from the fact that the curvature of the corresponding (via Radon–Penrose–Ward transform) Yang-Mills-connection can be identified with the family of the restrictions of E to the second-order infinitesimal neighborhoods.

The results collected in 13.14–13.19 admit generalizations to vector bundles over domains $X \subseteq \mathbb{P}^n$ satisfying the following two conditions: (1) For each $z \in X$ one can find a projective line $L \subseteq X$ with $z \in L$. (2) The set of all projective lines which are contained in X is connected (see [47]).

Bibliography*

1. Atiyah, M.F.: Geometry of Yang-Mills fields. Lezioni Fermiane, Pisa (1979). Zbl. 435.58001
2. Bănică, C., Forster, O.: Complete intersections in Stein manifolds. Manuscr. Math. 37, 343–356 (1982). Zbl. 491.32011
3. Barth, W.: Some properties of stable rank-2 vector bundles on $\mathbb{P}_n$. Math. Ann. 266, 125–150 (1977). Zbl. 332.32021
4. Bungart, L.: On analytic fiber bundles. I: Holomorphic fiber bundles with infinite dimensional fibers. Topology 7, 55–68 (1968). Zbl. 153.102
5. Cartan, H.: Sur les matrices holomorphes de n variables complexes. J. Math. Pures Appl. 19, 1–26 (1940). Zbl. 24.223
6. Cartan, H.: Espaces fibrés analytiques. Symp. Int. Topol. Algebraica. Mexico 97–121 (1958). Zbl. 121.305
7. Cornalba, M., Griffiths, P.: Analytic cycles and vector bundles on non-compact algebraic varieties. Invent. Math. 28, 1–106 (1975). Zbl. 293.32026
8. Douady, A.: Un espace de Banach dont le groupe lineare n'est pas connexe. Indagationes Math. 27, 787–789 (1965). Zbl. 178.264
9. Douady, A.: Le problème des modules pour sous-espaces analytiques compacts d'un espace analytique donné. Ann. Inst. Fourier 16, 1–95 (1966). Zbl. 146.311
10. Eastwood, M.G., Penrose, R., Wells, R.O., Jr.: Cohomology and massless fields. Commun. Math. Phys. 78, 305–351 (1981). Zbl. 465.58031
11. Forster, O.: Zur Theorie der Steinschen Algebren und Moduln. Math. Z. 97, 376–405 (1967). Zbl. 148.322
12. Forster, O.: Topologische Methoden in der Theorie Steinscher Räume. Actes Congrès Intern. Math. 2, 613–618 (1970). Zbl. 245.32006
13. Forster, O.: Riemannsche Flächen. Springer, Berlin-Heidelberg-New York (1977). Zbl. 381.30021
14. Forster, O., Knorr, K.: Über die Deformationen von Vektorraumbündeln auf kompakten komplexen Räumen. Math. Ann. 209, 291–346 (1974). Zbl. 272.32004
15. Forster, O., Ramspott, K.J.: Okasche Paare von Garben nichtabelscher Gruppen. Invent. Math. 1, 260–286 (1966). Zbl. 154.333
16. Forster, O., Ramspott, K.J.: Analytische Modulgarben und Endromisbündel. Invent. Math. 2, 145–170 (1966). Zbl. 154.334

*For the convenience of the reader, references to reviews in Zentralblatt für Mathematik (Zbl.), compiled using the MATH database, have, as far as possible, been included in this bibliography.

17. Forster, O., Ramspott, K.J.: Homotopieklassen von Idealbasen in Steinschen Algebren. Invent. Math. 5, 255–276 (1968). Zbl. 157.133
18. Forster, O., Ramspott, K.J.: Über die Anzahl der Erzeugenden von projektiven Steinschen Moduln. Arch. Math. 19, 417–422 (1968). Zbl. 162.385
19. Frenkel, J.: Cohomologie non abélienne et espaces fibrés. Bull. Soc. Math. Fr. 85, 135–220 (1957). Zbl. 82.377
20. Gindikin, S.G., Khenkin, M.G.: The Penrose transformation and complex integral geometry. Itogi Nauki Tekh., Ser. Sovrem. Probl. Mat., 17, 57–111 (1981). Zbl. 482.53053 English transl.: J. Soc. Math. 21, 508–551 (1983)
21. Grauert, H.: Approximationssätze für holomorphe Funktionen mit Werten in komplexen Räumen. Math. Ann. 133, 139–159 (1957). Zbl. 80.292
22. Grauert, H.: Holomorphe Funktionen mit Werten in komplexen Lieschen Gruppen. Math. Ann. 133, 450–472 (1957). Zbl. 80.292
23. Grauert, H.: Analytische Faserungen über holomorph-vollständigen Räumen. Math. Ann. 135, 263–273 (1958). Zbl. 81.74
24. Grauert, H., Kerner, H.: Approximation von Schnittflächen in Faserbündeln mit homogener Faser. Arch. Math. 14, 328–333 (1963). Zbl. 113.291
25. Grauert, H.: Mülich, G.: Vektorbündel vom Rang 2 über dem n-dimensionalen komplex-projektiven Raum. Manuscr. Math. 16, 75–100 (1975). Zbl. 318.32027
26. Grauert, H., Remmert, R.: Theory of Stein Spaces. Springer, Berlin–Heidelberg–New York (1979). Zbl. 433.32007 Grundlehren der Mathematischen Wissenschaften 236
27. Grauert, H., Remmert, R.: Coherent analytic sheaves. Springer, Berlin–Heidelberg–New York (1984). Zbl. 537.32001 Grundlehren der Mathematischen Wissenschaften 265
28. Grothendieck, A.: Sur la classification des fibrès holomorphes sur la sphère de Riemann. Am. J. Math. 79, 121–138 (1957). Zbl. 79.170
29. Gunning, R.C., Rossi, H.: Analytic functions of several complex variables. Prentice Hall, Englewood Cliffs, N.J. (1965). Zbl. 141.86
30. Hamm, H.A.: Zum Homotopietyp Steinscher Räume. J. Reine Angew. Math. 338, 121–135 (1983). Zbl. 491.32010
31. Hayes, S.: Homotopie für Okasche Paare von Garben homogener Räume. Math. Ann. 207 181–212 (1974). Zbl. 264.32007
32. Hayes, S.: Approximation für Okasche Paare von Garben homogener Räume. Manuscr. Math. 13, 153–173 (1974). Zbl. 299.32010
33. Hayes, S.: Über einige lokale Homotopieeigenschaften für reell- und komplex-analytische Okasche Paare. Manuscr. Math. 15, 193–209 (1975). Zbl. 439.32012
34. Khenkin, G.M., Leiterer, J.: Theory of functions on complex manifolds. Akademie-Verlag, Berlin (1983), Birkhäuser-Verlag, Basel-Boston-Stuttgart (1984). Zbl. 573.32001
34a. Khenkin, G.M., Leiterer, J.: Andreotti–Grauert theory by integral formulas, Akademie-Verlag Berlin 1988.
34b. Khenkin, G.M., Leiterer, J.: Proof of Grauert's Oka principle without induction over the basis dimension, Preprint P-Math-05/86 of the Karl–Weierstrass–Institut
34c. Heunemann, D.: An approximation theorem and Oka's principle for holomorphic vector bundles which are continuous on the boundary of strictly pseudoconvex domains. Math. Nachr. *127*, 275–280 (1986). Zbl. 607.32018
35. Hirzebruch, F.: Topological methods in algebraic geometry. Springer, Berlin–Heidelberg–New York (1966). Zbl. 138.420
36. Hörmander, L.: An introduction to complex analysis in several variables. D. van Nostrand, Princeton, N.J. (1966). Zbl. 138.62
37. Kajiwara, J.: Equivalence of Steinness and validity of Oka's principle for subdomains with continuous boundaries of a Stein manifold. Mem. Fac. Sci. Kyushu Univ. Ser. A. 33, 83–93 (1979). Zbl. 404.32007
38. Kajiwara, J, Kazama, H.: Oka's principle for relative cohomology sets. Mem. Fac. Sci., Kyushu Univ., Ser. A 23, 33–70 (1969). Zbl. 181.90

39. Kajiwara, J., Nishihara, M.: Charakterisierung der Steinschen Teilgebiete durch (das) Okasches Prinzip in zweidimensionaler Steinscher Mannigfaltigkeit. Mem. Fac. Sci., Kyushu Univ., Ser. A. 33 71–76 (1979). Zbl. 404.32006
40. Kuiper, N.H.: The homotopy type of the unitary group of Hilbert space. Topology 3, 19–30 (1965). Zbl. 129.389
41. Leiterer, J.: Analytische Faserbündel mit stetigem Rand über streng pseudokonvexen Gebieten. I. Garbentheoretische (Vorbereitungen). Math. Nachr. 71, 329–344 (1976); Zbl. 331.32023 II. Topologische Klassifizierung. Math. Nachr. 72, 201–217 (1976). Zbl. 341.32018
42. Leiterer, J.: The Oka principle for sheaves of infinite dimensional nonabelian groups. (Russian) In: Complex Analysis and its applications, Collect. Artic., Steklov Math. Inst. (Nauka, Moscow) 295–317 (1978). Zbl. 429.32037
43. Leiterer, J.: Equivalence of Steinness and validity of Oka's principle for subdomains of Stein manifolds. Math. Nachr. 89, 181–183 (1979). Zbl. 445.32012
44. Leiterer, J.: The Penrose transform for bundles non-trivial on the general line. Math. Nachr. 112, 35–67 (1983). Zbl. 548.32019
45. Leiterer, J.: Subsheaves in bundles on $\mathbb{P}_n$ and the Penrose transform. Springer Lect. Notes Math. 1039, 332–345 (1983). Zbl. 524.14019
46. Leiterer, J.: An implicit function theorem in Banach spaces. Ann. Pol. Math. 46, 171–175 (1985). Zbl. 625.46051
47. Leiterer, J.: On holomorphic vector bundles over linearly concave manifolds. Math. Ann. 274 391–417 (1986). Zbl. 572.58002
48. Maruyama, M.: Boundedness of semistable sheaves of small ranks. Nagoya Math. J. 78, 65–94 (1980). Zbl. 456.14011
49. Oka, K.: Sur les fonctions analytiques de plusieurs variables. III. Deuxième problème de Cousin. J. Sci. Hiroshima Univ., Ser. A 9, 7–19 (1939). Zbl. 20,240
50. Okonek, C., Schneider, M. Spindler, H.: Vector bundles on complex projective spaces. Birkhäuser, Boston–Basel–Stuttgart (1980). Zbl. 438.32016
51. Pankov, A.A.: On analytic bundles which are continuous on the boundary of a domain (Russian) Soobshch. Akad. Nauk Gruz. SSR 81, 277–280 (1976). Zbl. 319.32029
52. Ramspott, K.J.: Stetige und holomorphe Schnitte in Bündeln mit homogener Faser. Math. Z. 89, 234–246 (1965). Zbl. 163.321
53. Röhrl, H.: Das Riemann-Hilbertsche Problem der Theorie der linearen Differentialgleichungen. Math. Ann. 133, 1–25 (1957). Zbl. 88.60
54. Schneider, M.: Vollständige Durchschnitte in Steinschen Mannigfaltigkeiten. Math. Ann. 186, 191–200 (1970). Zbl. 188.392
55. Schneider, M.: Vollständige, fast-vollständige und mengen-theoretisch-vollständige Durchschnitte in Steinschen Mannigfaltigkeiten, Math. Ann. 260, 151–174 (1982). Zbl. 532.32005
56. Schneider, M.: On the number of equations needed to describe a variety. Complex analysis of several variables. Proc. Symp., Madison 1982. Proc. Symp. Pure Math. 41, 163–180 (1984). Zbl. 539.32009
57. Schneider, M.: Chernklassen semi-stabiler Vektorraumbündel vom Rang 3 auf dem komplex-projektiven Raum. J. Reine Angew. Math. 315, 211–220 (1980). Zbl. 432.14012
58. Serre, J.-P.: Application de la théorie générale à divers problèmes globaux. Séminaire Henri Cartan (1951–52), Exposé 20, Benjamin, New York (1967).
59. Spindler, H.: Der Satz von Grauert-Mülich für beliebige semistabile holomorphe Vektorraumbündel über dem n-dimensionalen komplex(en)-projektiven Raum. Math. Ann. 243, 131–141 (1979). Zbl. 435.32018
60. Vekua, N.P.: Systems of singular integral equations. (Russian) Nauka, Moscow (1970). Zbl. 202,118
61. Ward, R.S.: On self-dual gauge fields. Phys. Lett. A 61, 81–82 (1977).
62. Zaddach, B.: On the Penrose transform for vector bundles on $\mathbb{P}^n$. (to appear).

III. Deformations of Complex Spaces

V.P. Palamodov

Translated from the Russian
by J. Nunemacher

Contents

Introduction

The origin of deformation theory lies in the problem of moduli, first considered by Riemann. The problem in the theory of moduli can be described thus: to bring together all objects of a single type in analytic geometry, for example, all Riemann surfaces of given genus; to organize them by joining them into a fiber space; to describe the base of this space—the moduli space; to introduce on it an analytic structure; and to study the natural parameters, for example, the periods of holomorphic forms defined on Riemann surfaces. For nonsingular Riemann surfaces the moduli space was constructed in the theory of Teichmüller-Ahlfors-Bers, and the periods of holomorphic forms were described by the billinear relations of Riemann and the theorem of Rauch. But the problem of moduli for multidimensional complex manifolds turned out to be significantly more complicated. Its systematic study began with the work of Kodaira and Spencer in which the basic object of study was the analytic families of complex structures on a given smooth manifold. The concept of a complete effective family, due to Kodaira-Spencer, served as a substitute for the moduli space, but it had various peculiar properties: the structure of the fibers of this family can vary "continuously" along one curve in the base and have "jumps" along another, and the base itself can have singular points (Kuranishi).

The modern concept of deformation is due to Grothendieck. It includes complex spaces with arbitrary singularities as objects for deformation and as bases. In Grothendieck's approach an important role is played by infinitesimal deformations, i.e., deformations whose base consists of a single point but with a complicated structure algebra. The relation between the concept of families of structures in the sense of Kodaira-Spencer and that of infinitesimal deformation, is analogous to that between an analytic function and a finite segment of its Taylor series at some point. There is a close connection between infinitesimal deformations of complex spaces and the theory of extensions of commutative algebras. One of the basic concepts in deformation theory is that of the tangent cohomology of a complex space, which is a globalization of the cohomology of analytic algebras. Within the bounds of general complex spaces the theory of deformations has attained a closed form. In particular, in this theory (unlike the situation in algebraic geometry) any compact space has a versal deformation (which in the category of complex spaces is the substitute for the moduli space).

In connection with the development of deformation theory for spaces with singular points, a new set of problems arises which can be described as follows. Suppose that we are given a space with singularities, for example, a curve or critical level surface of a holomorphic function, which belongs to a family of nonsingular spaces (for example, neighboring level surfaces). We wish to find the relation between algebraic invariants of the singular points of the given space and topological or geometric characteristics of the neighboring family of nonsingular spaces. Related to these problems, in addition to the classical work of Picard-Lefshetz, is the work of Milnor on the singular points of complex hypersurfaces as well as a large part of the subsequent work in complex singularity theory. Also relevant is the theory of Griffiths-Schmid on limiting Hodge structures and its subsequent developments.

The problems and ideas of the deformation theory of complex spaces have many parallels in other mathematical areas: the stability and restructuring of smooth mappings, the theory of normal forms for vector fields, the moduli for solutions of gauge-invariant differential equations, extension theory for associative algebras, the general theory of quantization, etc. To some degree in each of these areas the language of deformation theory and its formal apparatus has proven to be relevant and in some cases has already been employed.

The purpose of the present survey is to give an account of the fundamentals of the modern theory of complex spaces and other structures in complex analytic geometry and from this position to describe and comment on various specific facts in this theory. For a more complete acquaintance with each portion of the theory the reader may consult the brief bibliography below and the references mentioned in these sources.

On The Subject of Deformation Theory. The theory of deformations of complex spaces is the result of the interaction and development of various ideas of geometric, algebraic, and analytic character. Before we begin to give an account of them, we pause to consider some simple examples and arguments which will clarify these ideas.

1. Complex Manifolds. Fundamental is the concept of Riemann surface, which appeared in the dissertation of Riemann as a surface "multi-sheetedly spread" over the extended complex plane. In modern language this geometric object is a holomorphic mapping from a one-dimensional complex manifold into $\mathbb{CP}_1$. The definition of such a manifold via a conformal gluing of domains in the complex plane was given by H. Weyl. Its generalization to the multi-dimensional situation led to the concept of a general complex manifold.

2. Complex Spaces (see Chapter 1, Section 1). In complete generality the definition of a complex space was given a quarter of a century ago by Grothendieck. An essential feature of this concept consists in allowing singular points and nilpotent elements in the fibers of the structure sheaf $\mathcal{O}_X$. The reason for the latter of these is already evident in the following simple example. We consider the mapping of the complex plane into itself defined by the formula $w = z^2$. The fiber of this mapping over a point $w \neq 0$ consists of two points; consequently, the algebra of holomorphic functions on this fiber is two-dimensional. The fiber over zero degenerates to the single point $z = 0$. If we view it as a simple point, i.e., as a zero-dimensional manifold, then the algebra of functions on this fiber reduces to the constants alone, i.e., it is one-dimensional. In the theory of complex spaces this fiber can be viewed as a multiple point—as the result of the "fusion" of two simple points. This multiplicity is expressed according to the general definition by endowing the single point $z = 0$ with the algebra $A = H_1/(z^2)$, where H_1 is the algebra of convergent power series in z. The algebra A is the algebra of double numbers; it can be represented as the ring of elements of the form $a + b\varepsilon$, where a and b are any complex numbers with multiplication subject to the relation $\varepsilon^2 = 0$. Thus all such elements with $a = 0$ have squares equal to zero, i.e., they are nilpotent. The algebra A is two-dimensional, and therefore with such a definition the dimension of the algebras of the fibers remains constant for all w.

The next example generalizes the one just given. We consider the mapping from $\mathbb{C}^n$ into itself defined by the formulas $w_i = \sigma_i(z_1, \ldots, z_n)$, where $\sigma_i, \ldots, \sigma_n$ are the elementary symmetric polynomials. The number of pre-images of a generic point w is equal to $n!$; more precisely, this is true under the condition $D(w_1, \ldots, w_n) \neq 0$, where $D(a_1, \ldots, a_n)$ is the discriminant of the polynomial of nth degree with coefficients $a_1, \ldots, a_n$. The number of geometric pre-images of the point w falls when w is a root of D, and when $w = 0$ there is only the single pre-image $z = 0$. Nonetheless, according to the general construction in the theory of complex spaces each fiber is endowed with its structure sheaf in such a way that the dimension of the space of its sections is the same for all w, i.e., it is equal to $n!$. Therefore the algebra corresponding to a "multiple" point of the fiber, i.e., to a point at which there are identical coordinates, must necessarily be nilpotent. In particular, the point $z = 0$ as the fiber over the origin is endowed with the algebra $A = H_n/(\sigma_1, \ldots, \sigma_n)$, where H_n is the algebra of convergent power series in $z_1, \ldots, z_n$ and $(\sigma_1, \ldots, \sigma_n)$ denotes the ideal in this algebra generated by the indicated polynomials. The reader may verify independently that $\dim A = n!$.

3. Flat Mappings. The property of preserving the dimension of the algebra of functions on the fibers in the example above has a far reaching generalization which pertains to proper mappings of complex spaces $f: X \to Y$. The role of the dimension of the algebra of functions is played by the Euler characteristic

$$\chi(X_y) = \sum_0^\infty (-1)^i \dim H^i(X_y, \mathcal{O}(X_y)),$$

where X_y is the fiber of this mapping over the point y and $\mathcal{O}(X_y)$ is the corresponding sheaf of algebras. According to a theorm of Grauert, the quantity $\chi(X_y)$ is locally constant as a function on Y under the condition that the mapping f is flat. The general definition of flatness is given in algebraic terms (Chapter 1, Section 3). It has numerous important properties which make it irreplaceable in deformation theory (see below).

4. Moduli. Riemann also investigated the question of how many nonequivalent compact surfaces there are of given genus g. His definition of equivalence differed from, but is equivalent to, the concept of conformal equivalence of Riemann surfaces. By parametrizing such surfaces using meromorphic functions on them Riemann deduced that the number of complex parameters, which he called the moduli, is equal to g if $g = 0$ or 1 and to $3g - 3$ if $g > 1$. His method of parametrization, however, did not provide a means of showing that Riemann surfaces in any sense depend holomorphically on these parameters. A rigorous meaning for these words was arrived at only one hundred years later in the theory of Ahlfors-Bers. The conclusion of this theory consists in the fact that for all classes of conformally equivalent Riemann surfaces of a given genus $g > 1$ one can construct a holomorphic family, whose base is called the moduli space (it is a quotient of the Teichmüller space relative to a discrete group). The complex structure on this space is uniquely determined from the property of analyticity of the period matrix (Chapter 3, Section 2; Chapter 5, Section 6). The theory of Riemann surfaces of genus 1 is in essence contained in the classical theory of elliptic functions. Each such surface is obtained by taking a quotient of the complex plane relative to a doubly periodic lattice. This lattice depends on one complex parameter which is locally the quotient of its periods. An absolute invariant for a surface of genus 1 is the modular function which injectively maps the set of such surfaces onto the complex plane (Chapter 3, Section 2.1).

In a similar way the moduli space for compact Riemann surfaces with a finite number of punctures can be constructed. For more general Riemann surfaces it has a more complicated structure. According to a theorem of Riemann-Hilbert any multiply-connected domain in the complex plane is conformally equivalent to the sphere $\mathbb{CP}_1$ with punctures and cuts along horizontal segments. One of these segments can be taken to be the positive x-axis. In this case the coordinates of the ends of the other segments and also the punctures are the moduli for the conformal classes of such domains. Here some of the moduli—the coordinates of the punctures and, let us say, the left ends of the segments are complex numbers, and the remaining ones, i.e., the lengths of the segments, are real. Thus the classes of non-simply-connected domains in general cannot be joined into holomorphic families. The reason for this difference lies in the fact that such domains are noncompact and even non-quasiprojective (i.e., not the completion of a submanifold in a projective manifold).

5. Families. The problem of describing the "moduli" for multi-dimensional complex manifold, differs in an essential manner from that of the one-dimensional case described above. One reason for this is that not every conformal class of such manifolds can be organized into a holomorphic family without repetitions, and therefore the moduli space does not exist even locally. This situation is demonstrated in the following simple example.

We consider the space M of all complex matrices of order 2. We shall call matrices A and B equivalent if there exists an invertible matrix Q so that $Q^{-1}AQ = B$. We investigate the question of whether there exists a moduli space for the equivalence classes of matrices; whether these classes can be parametrized by complex or real parameters in a manner consistent with the structure of the complex manifold M. We note that for the domain in M formed by those matrices whose spectrum lies in the annulus $\{0<|\lambda|<1\}$ this question is contained in the problem of moduli for Hopf surfaces (Chapter 3, Section 5). At the same time, it can be formulated purely algebraically: does the orbit space for the adjoint action of the group $\mathrm{Gl}(2, \mathbb{C})$ on M have the structure of a complex (or at least a real) manifold? The orbits of the group are characterized by the Jordan forms of their elements. There are three types of such forms:

$$\lambda E = \begin{pmatrix} \lambda & 0 \\ 0 & \lambda \end{pmatrix}, \quad \begin{pmatrix} \lambda & 1 \\ 0 & \lambda \end{pmatrix}, \quad \begin{pmatrix} \lambda & 0 \\ 0 & \mu \end{pmatrix} \lambda \neq \mu,$$

consequently there are three types of orbits. An orbit of the first type is zero-dimensional and reduces to the single matrix λE depending on the single complex number λ. Orbits of the other two types are two-dimensional, and the manifolds of these orbits of the second type are one-dimensional and of the third type are two-dimensional. The closure of an orbit of the second type contains an orbit λE. The moduli space in this example is the orbit space. How can it be described constructively? This can be done by selecting some subset $S \subset M$ which intersects each orbit in exactly one point. Can such a subset be a manifold? If it is, then at each point λE it can have real dimension no lower than 5, since otherwise it cannot intersect all nearby orbits of the second and third types. But on the other hand these orbits, as we have remarked, intersect S in manifolds of positive dimension. Therefore S cannot be a manifold and, moreover, it cannot have a Hausdorff topology.

Related to this phenomenon is another one, which is not characteristic of families of one-dimensional manifolds, namely, that of "jumps" in structure. In this example there is, closely connected with a given orbit, the space of matrices Γ which commute with a matrix A lying in the orbit. It is clear that the dimension of this space does not depend on the choice of the representative A of the orbit. In geometric language this space is the collection of all holomorphic tangent fields to the corresponding Hopf surface. For an orbit of the first type this dimension is equal to 4, which for an orbit of the second type which is "not separated" from it the dimension is equal to 2.

The representation of a continuous or holomorphic family of Riemann surfaces generalizes without difficulty to compact manifolds of any dimension: it forms the collection of fibers $X_s = f^{-1}(s)$ of a proper (the pre-image of a compact set is compact) regular mapping $f: X \to S$ of compact manifolds. The regularity condition means that for any point $x_0 \in X$ and coordinate system $y_1, \ldots, y_n$ in a neighborhood of the point $f(x_0)$ the functions $y_i(f(x))$ for $i = 1, \ldots, n$ can be completed to a coordinate system in a neighborhood of x_0. We note that the regularity condition is equivalent to requiring that all the fibers X_s be manifolds. The definition of a smooth family of compact manifolds can be given in a similar fashion. Such families of manifolds were studied systematically by Kodaira and Spencer. In their work the smooth structure of the fibers remains unchanged; consequently, the complex-analytic structure is being deformed, i.e., changed, on a fixed smooth manifold. The rate of this change can be characterized via an appropriate operator introduced by Frölicher and Nijenhuis, which has become known as the Kodaira-Spencer mapping (Chapter 2, Section 4). If we make the analogy of a family of manifolds to a function (smooth or holomorphic), then the Kodaira-Spencer mapping becomes the differential of this function. In particular, to the zero mapping there corresponds the trivial family, i.e., the family of the form $X = X_0 \times S$. The converse is, however, not true, but the analogy stands if we consider "higher" mappings, similar to higher differentials (Chapter 2, Section 4).

6. Deformations of Complex Spaces. What should we mean by a holomorphic family of complex spaces having singular points? In the definition of a family of complex manifolds the regularity condition must be replaced by some other local condition which must possess at a minimum the following properties:

1) The complexity of the singular points of the deformed space must not be restricted.

2) The condition must be preserved under changes of base for the family, for example, under the restriction of the family to some subspace of the base.

3) At points where the fiber X_s is a manifold, the condition must agree with the old regularity condition.

The natural condition which satisfies these requirements is that of flatness of the mapping, which we discussed above in 3. It is the natural replacement for the regularity condition also from other points of view. In particular, the dimension of the fiber of a flat mapping is always locally constant. The flatness condition has a particularly simple form when X is a connected manifold and S is a complex line. In this case a mapping $f: X \to S$ is flat if and only if $f \not\equiv$ const.

And so a deformation of compact complex spaces is any proper flat mapping of complex spaces $f: X \to S$; S is called the base of the deformation (Grothendieck).

When we speak of a deformation of a fixed space X_0, then the base is understood to be the germ of a complex space; thus we fix attention only on the fibers of the mapping f near to a distinguished fiber which is isomorphic to X_0. Related to the theory of deformations is, first of all, its "differential calculus" which includes the concepts of the differential (the generalization of the Kodaira-Spencer mapping), infinitesimal deformations (analogous to finite segments of the Taylor series of a holomorphic function), obstructions (which do not have an analog in function theory), etc. Fundamental here is the concept of tangent cohomology (Chapter 2).

How a compact space can vary under a deformation is shown by an example which is considered in Chapter 3, Section 2. In this example X_0 is the Riemann sphere with a point of transversal self-intersection. Under deformation this point is smoothed out and turns into a "vanishing" cycle, and the fiber becomes a Riemann surface of genus 1. The base of this deformation is the germ of a complex line, such that to each of its points except the distinguished one there correspond exactly these Riemann surfaces, By "gluing" this germ to the moduli space for surfaces of genus 1 (which, as we have observed, is isomorphic to $\mathbb{C}^1$), we obtain the compact moduli space $\mathbb{CP}_1$.

7. Versal Deformations. This kind of deformation is a local substitute for the moduli space in those cases when the moduli space does not exist, for example, as a Hausdorff topological space (see the example in 5 above). Heuristically a versal deformation of a compact space X_0 is a deformation of it such that among its fibers there occur all possible deformations of X_0 sufficiently near to the distinguished fiber. By minimal we mean a versal deformation which has a base spaces of minimum dimension. The germ S is defined uniquely up to isomorphism of germs, and so is the minimal versal deformation. A minimal versal deformation exists for any compact complex space, and it allows the introduction of a topology into the moduli space, i.e., the set of isomorphism classes of spaces of such type. The base of a minimal versal deformation of a space X_0 is the best approximation to the germ of the moduli space at the point corresponding to X_0. However, the base of the deformation is not identical to the germ of the moduli space in those cases when there is an infinite sequence of mutually isomorphic fibers of the deformation converging to the distinguished fiber. In fact in such cases there is an analytic subset in the base having positive dimension over which all the fibers are pairwise isomorphic. In these situations the topology of the germs of the moduli space is non-Hausdorff.

The base S of a minimal versal deformation in many cases can have singular points (Chapter 3, Section 7). The appearance of such points is related to the obstructions to the extension of infinitesimal deformations (Chapter 2, Section 5).

8. The Problem of Choosing Parameters in the Moduli Space. We assume that the base of the given deformation can serve nevertheless as the moduli space, i.e., the fibers of this deformation are pairwise nonisomorphic. There arises the question of how to choose natural coordinates on this base. In various cases this question is answered differently. For example, if the fibers of the deformation are

Riemann spheres with four punctures or four lines in the plane $\mathbb{C}^2$ which pass through the origin, then the natural parameter in the base of the deformation is the cross ratio of the corresponding four points in $\mathbb{CP}_1$. In the case of a family of Riemann surfaces of genus 1 the natural modulus is given by the classical absolute invariant J (see Chapter 3, Section 2.1). In the case of a family of compact manifolds a universal approach consists in selecting as parameters the integrals of holomorphic forms relative to fixed cycles, i.e., cycles which define constant homology classes in the fibers (taking into account the topological triviality of the family). Such integrals were studied already by Riemann, who laid the foundations for the theory. The development of Riemann's approach with regard to manifolds of arbitrary dimension led to the creation of the deep theory of variation of Hodge structure and mixed Hodge structures for algebraic varieties with singularities (Chapter 5, Sections 6–8).

9. Deformations of Other Objects. There are numerous parallel theories: for vector bundles, coherent analytic sheaves, subspaces, mappings of spaces, etc. They are either special cases of the theory of deformations of complex spaces or are completely analogous to this theory. They have the same characteristic features: the absence in the general case of the moduli space and instead of it the concept of versal deformation, jumps in structure, obstructions, etc.

Chapter 1. Basic Definitions

1. Complex Spaces

1.1. A very general and very simple concept is that of a $\mathbb{C}$-ringed space. We recall that an *algebra over the field* $\mathbb{C}$ of complex numbers, or for short a $\mathbb{C}$-*algebra*, is an associative ring A together with a ring imbedding $\mathbb{C} \to A$. Thus a $\mathbb{C}$-algebra contains a unity and is a vector space over $\mathbb{C}$. A homomorphism of $\mathbb{C}$-algebras is a ring homomorphism $\alpha : A \to B$ which when composed with the structure homomorphism $\mathbb{C} \to A$ yields the structure homomorphism $\mathbb{C} \to B$; in other words, α must be $\mathbb{C}$-linear. By a $\mathbb{C}$-*ringed space* we mean a topological space X (the underlying space) which has been given a sheaf $\mathcal{O}_X$ of non-zero $\mathbb{C}$-algebras (the structure sheaf).

A mapping of $\mathbb{C}$-ringed spaces $(X, \mathcal{O}_X) \to (Y, \mathcal{O}_Y)$ is a pair (f, φ) consisting of a continuous mapping $f: X \to Y$ and a homomorphism of sheaves of $\mathbb{C}$-algebras $\varphi : f^*(\mathcal{O}_Y) \to \mathcal{O}_X$. In what follows we shall often denote a mapping of $\mathbb{C}$-ringed spaces simply by the symbol $f: X \to Y$ with the existence of the structure sheaves and homomorphism φ being implicit. The concepts of the identity mapping $1(X)$, the composition of mappings, and isomorphism of $\mathbb{C}$-ringed spaces are defined in the natural manner.

An open subspace of a space $(X, \mathcal{O}_X)$ is a $\mathbb{C}$-ringed space consisting of an open subset $U \subset X$ together with the restriction of the sheaf $\mathcal{O}_X$ to U.

1.2. The linear space $\mathbb{C}^n$ endowed with the sheaf $\mathcal{H}_n$ of germs of holomorphic functions is a $\mathbb{C}$-ringed space. Any open subspace $(U, \mathcal{H}_n|U)$ is called a *model complex manifold*. An arbitrary *complex manifold* is by definition a $\mathbb{C}$-ringed space $(X, \mathcal{O}_X)$ which is locally isomorphic to a model complex manifold. In other words, X has an open covering $\{X_\alpha\}$ so that for any α there is defined an isomorphism of $\mathbb{C}$-ringed spaces $(X_\alpha, \mathcal{O}_{X_\alpha}) \cong (U_\alpha, H_{n_\alpha})$. The number n_α is called the *dimension of the manifold* X at any point $x \in X_\alpha$. This definition is equivalent to the usual construction of a complex manifold by gluing domains U_α using holomorphic mappings.

Let $(U, \mathcal{H})$ be a model manifold, $f_1, \dots, f_m$ certain holomorphic functions on U, and Z their set of common zeros. The germs of these functions at any point $z \in Z$ generate a proper ideal $\mathcal{I}_z$ in the algebra H_z. The collection of these ideals forms a subsheaf $\mathcal{I} = (f_1, \dots, f_m)$ of $\mathcal{H}$. We consider the quotient sheaf $\mathcal{A} = \mathcal{H}/\mathcal{I}$; its fibers are the nonzero $\mathbb{C}$-algebras $\mathcal{A}_z = \mathcal{H}_z/\mathcal{I}_z$. The $\mathbb{C}$-ringed space $(Z, \mathcal{A})$ is called a *model complex space*. We are now ready for the basic definition: a *complex space* is any $\mathbb{C}$-ringed space which is locally isomorphic to a model complex space [61], [87] (see also Article *I*). This definition is equivalent to the following construction: let there be given

1) a set I, called the index set;

2) for each $\alpha \in I$ a model complex space $(Z_\alpha, \mathcal{A}_\alpha)$, where $Z_\alpha \subset \mathbb{C}^{n_\alpha}$ and $\mathcal{A}_\alpha = \mathcal{H}_{n_\alpha}/\mathcal{I}_\alpha$ (a chart);

3) for each pair of indices α, β an open subset $Z_{\alpha\beta} \subset Z_\alpha$, a neighborhood of it $U_{\alpha\beta} \subset \mathbb{C}^{n_\alpha}$; and also a homomorphism $f_{\beta\alpha} : Z_{\alpha\beta} \to Z_{\beta\alpha}$ which extends to a holomorphic mapping $F_{\beta\alpha} : U_{\alpha\beta} \to \mathbb{C}^{n_\beta}$ (the gluing map). We require that the following conditions be satisfied:

I) For any α, β and element $a \in \mathcal{I}_\beta$ the function $F^*_{\alpha\beta}(a)(z) \equiv a(F_{\alpha\beta}(z))$ belongs to $\mathcal{I}_\alpha$.

II) If $\alpha = \beta$ then the mappings $f_{\beta\alpha}$ and $F_{\alpha\beta}$ are identity mappings.

III) For each triple α, β, γ and element $a \in \mathcal{H}_{n_\gamma}$ the relation $F^*_{\alpha\gamma}(a) - F^*_{\alpha\beta}F^*_{\beta\gamma}(a) \in \mathcal{I}_\alpha$ holds. In order to construct a complex space from this data, it is necessary to glue the sets Z_α using the homeomorphisms $f_{\alpha\beta}$ to form a single topological space X. It follows from II and III that these gluings are consistent. By I the mappings $F_{\alpha\beta}$ generate mappings of sheaves of $\mathbb{C}$-algebras $f^*_{\beta\alpha}(\mathcal{A}_\beta) \to \mathcal{A}_\alpha|Z_{\alpha\beta}$. With these mappings the sheaves $\mathcal{A}_\alpha$ are glued to form a single sheaf of $\mathbb{C}$-algebras; the consistency of the gluings again follows from II and III.

1.3. Let $(X, \mathcal{O}_X)$ be a complex space and $Y \subset X$ a subset. It is called an *analytic set* if for each point $y \in Y$ there is a local chart $Z \subset U \subset \mathbb{C}^n$ for the set X containing this point so that Y coincides with the set of common zeros of certain functions $g_1, \dots, g_p$ which are holomorphic on U. Every analytic set Y, upon the removal of an analytic subset $\operatorname{Sing}(Y)$ which is nowhere dense in it, becomes a complex manifold. In particular, the underlying space X in each chart is represented by some analytic subset $Z \subset \mathbb{C}^n$. Those points $x \in X$ in a neighborhood of which X cannot be represented as a complex manifold are called *singular* points of the underlying space. They form a closed analytic set $\mathrm{Sing}(X)$. The *dimension* of X at a point x which does not belong to $\mathrm{Sing}(X)$ is the dimension of the corresponding complex manifold. The *dimension* of X at a point $x \in \mathrm{Sing}(X)$ is by definition the maximum of the dimensions of X at points $y \in X \backslash \mathrm{Sing}(X)$ sufficiently near to x.

A complex space $(X, \mathcal{O}_X)$ is called *reduced* if there are no nilpotent elements in the sheaf $\mathcal{O}_X$, i.e., no nonzero elements a so that $a^p = 0$ for some natural number p. To represent a reduced complex space via the construction in Section 1.2, it suffices to define analytic sets Z_α, $Z_{\alpha\beta}$, and homeomorphisms $f_{\alpha\beta}$ which admit holomorphic extension. The subsheaves $\mathcal{I}_\alpha$ and the mappings $F_{\alpha\beta}$ are then uniquely determined.

1.4. Let X be a single point. Any complex space $(X, \mathcal{O}_X)$ in this case is defined by a single $\mathbb{C}$-algebra, which has the form H_n/I, where H_n is the algebra of convergent power series in n variables for $n = 0, 1, 2, \ldots$ and $I \subset H_n$ is an ideal of finite codimension. This kind of algebra is called Artinian. An essential feature of deformation theory is the use of such complex spaces as bases for deformations. The field $\mathbb{C}$ is the simplest Artinian algebra. A complex space consisting of the single point $*$ and the algebra $\mathbb{C}$ is a complex manifold of dimension zero. In the other cases complex spaces consisting of a single point are not manifolds. The simplest of these spaces is the *double point*, i.e., the complex space $(*, D)$, where $D = \mathbb{C}\{Z\}/(Z^2)$ is the algebra of dual numbers. It can be represented as the algebra of expressions of the form $a + b\varepsilon$ for $a, b \in \mathbb{C}$ with multiplication obeying the relation $\varepsilon^2 = 0$.

1.5. A *mapping of complex spaces* $(f, \varphi): (X, \mathcal{O}_X) \to (Y, \mathcal{O}_Y)$ is any mapping of the corresponding $\mathbb{C}$-ringed spaces. A *model mapping of complex spaces* is a mapping of model complex spaces $(f, \varphi): (Z, A) \to (Z', A')$ for which there exists a holomorphic mapping $F: U \to U'$ from some open neighborhood U of the analytic set $Z \subset \mathbb{C}^n$ into an open neighborhood U' of the set $Z' \subset \mathbb{C}^{n'}$ so that $F|Z = f$ and so that the diagram

$$\begin{array}{ccc} f^*(\mathcal{A}') & \xrightarrow{\varphi} & \mathcal{A} \\ \uparrow & & \uparrow \\ f^*(\mathcal{H}'|Z') & \xrightarrow{F^*} & \mathcal{H}|Z, \end{array}$$

commutes, where F^* is the mapping given by the substitution $a(Z') \mapsto a(F(Z))$. Any mapping of complex spaces is locally a model mapping.

1.6 Let $(X, \mathcal{O}_X)$ be any $\mathbb{C}$-ringed space. A closed subset of it is any pair $(X', \mathcal{O}_{X'})$ consisting of a closed topological subspace $X' \subset X$ and a sheaf $\mathcal{O}_{X'} = \mathcal{O}_X/\mathcal{I}$ with support X', where $\mathscr{I}$ is a sheaf of ideals in $\mathcal{O}_{X'}$. If $(X, \mathcal{O}_X)$ is a complex space, then a *closed complex subspace* of it is any closed subspace $(X', \mathcal{O}_{X'})$ of the $\mathbb{C}$-ringed space $(X, \mathcal{O}_X)$ satisfying the requirement that the sheaf of ideals $\mathscr{I}$ be locally finitely generated. This condition means that for any point $x \in X$ there exists a finite number of sections of $\mathscr{I}$ over a neighborhood of x which generate $\mathscr{I}$ over this neighborhood. The mapping of complex spaces $(X', \mathcal{O}_{X'}) \to (X, \mathcal{O}_X)$ which is generated by the natural mapping of sheaves $(\mathcal{O}_X | X') \to \mathcal{O}_X/\mathscr{I} | X'$ is canonically defined. This mapping of complex spaces is called a (closed) *imbedding*.

The *direct product* of the complex spaces $(X, \mathcal{O}_X)$ and $(Y, \mathcal{O}_X)$ is the complex space $(X \times Y, \mathcal{O}_{X \times Y})$ whose structure sheaf can be described as follows. Let $(Z_\alpha, \mathscr{H}_{m_\alpha}/\mathscr{I}_\alpha)$ and $(W_\beta, \mathscr{H}_{n_\beta}/\mathscr{I}_\beta)$ be model complex spaces giving charts for X and Y (see Section 2.1). The analytic sets $Z_\alpha \times W_\beta$ form an open covering of $X \times Y$. This set together with the sheaf

$$\mathscr{H}_{m_\alpha+n_\beta}/\mathscr{I}_\alpha \cdot \mathscr{H}_{m_\alpha+n_\beta} + \mathscr{I}_\beta \cdot \mathscr{H}_{m_\alpha+n_\beta}$$

is a model complex space. Using the gluing mappings for the charts on X and Y, we can define gluings of these model complex spaces. These gluings are consistent and give as a result the required sheaf $\mathcal{O}_{X \times Y}$. Mappings of complex spaces from $X \times Y$ to X and Y are well defined. They are defined locally by the model mappings generated by the coordinate projections from $\mathbb{C}^{m_\alpha+n_\beta}$ to $\mathbb{C}^{m_\alpha}$ and $\mathbb{C}^{n_\beta}$.

Suppose now that we are given mappings of complex spaces

$$(X, \mathcal{O}_X) \xrightarrow{(f,\varphi)} (S, \mathcal{O}_S) \xleftarrow{(g,\psi)} (Y, \mathcal{O}_Y).$$

Their *fiber product* is a closed complex subspace of the direct product $X \times Y$. The underlying space of the fiber product, denoted by $X \times_S Y$, is the set of pairs $(x, y) \in X \times Y$ such that $f(x) = g(y)$. Its structure sheaf is the restriction of the sheaf $\mathcal{O}_{X \times Y}/\mathscr{D}$ to $X \times_S Y$, where $\mathscr{D}$ is the sheaf of ideals defined as follows. Let σ be any section of the sheaf $\mathcal{O}_S$ over some open set $T \subset S$. It induces a section of the sheaf $f^*(\mathcal{O}_S)$ over $f^{-1}(T)$ and to this we can apply the mapping φ. We denote the result by $\varphi(\sigma)$; The section $\psi(\sigma)$ is defined similarly. Both of these can be viewed as sections of the sheaf $\mathcal{O}_{X \times Y}$ over $f^{-1}(T) \times q^{-1}(T)$, consequently, so can their difference $\varphi(\sigma) - \psi(\sigma)$. The sheaf $\mathscr{D}$ is generated over $\mathcal{O}_{X \times Y}$ by all such differences. It is locally finitely generated, since it suffices to restrict σ using coordinate functions to the model complex space describing S in a neighborhood of the given point s. Mappings from the fiber product onto the factors are defined naturally, and they form the commutative diagram

$$\begin{array}{ccccc} & & X & & \\ & \nearrow & & \searrow & \\ X \times_S Y & & & & S. \\ & \searrow & & \nearrow & \\ & & Y & & \end{array}$$

Let $f: X \to S$ be a mapping of complex spaces and s a point of S. This point can be viewed as a closed subspace of S with the algebra $\mathcal{O}_{S,s}/\mathfrak{m}_s \cong \mathbb{C}$, where $\mathfrak{m}_s$

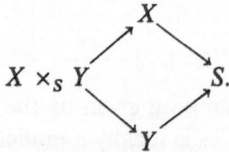

is the maximal ideal of the algebra $\mathcal{O}_{S,s}$. To this subspace there corresponds the imbedding $s \to S$. The fiber product of this imbedding and the mapping f is called the *fiber of the mapping* f over the point s; we shall denote it by X_s. The underlying space of the fiber is $f^{-1}(s)$ and the structure sheaf is isomorphic to $\mathcal{O}_X/m_s\mathcal{O}_X$. The fiber can have singular points and may not be reduced even when X and S are manifolds.

Thus to every mapping of complex spaces there is associated the family of complex spaces $\{X_s, s \in S\}$. This situation is however different from the concept of a deformation, since in the general case the relation among the fibers is too weak; in particular, neighboring fibers can have different dimensions.

2. Germs of Complex Spaces

2.1. An *analytic algebra* is a $\mathbb{C}$-algebra A which can be represented in the form $A \cong H_n/I$, where $H_n = \mathbb{C}\{z_1, \ldots, z_n\}$ is the algebra of convergent power series in n variables and I is some ideal in this algebra. In other words, A is an analytic algebra if there exists an epimorphism of $\mathbb{C}$-algebras $p: H_n \to A$. Every nonzero analytic algebra is local, i.e., it has a unique *maximal ideal* $m(A)$. In the algebra H_n this ideal consists of all series without constant terms, and $m(A)$ is the image of this ideal under the epimorphism mentioned above. The quotient algebra $A/m(A)$ is isomorphic to the field $\mathbb{C}$; thus there is defined an epimorphism of $\mathbb{C}$-algebras $r: A \to \mathbb{C}$ We shall call it the *numerical mapping*.

An analytic algebra A is called *regular* if it is isomorphic to H_n as a $\mathbb{C}$-algebra. Analytic algebras form a category. The morphisms of this category are homomorphisms of $\mathbb{C}$-algebras.

2.2. Let X be a complex space containing the point x. A *tangent vector* to X at the point x is any linear mapping $t: \mathcal{O}_{X,x} \to \mathbb{C}$ satisfying the requirement

$$t(ab) = t(a)r(b) + r(a)t(b)$$

for any $a, b \in \mathcal{O}_{X,x}$, where r is the numerical mapping. All tangent vectors at the point x form a vector space which is denoted $T_x(X)$ (the Zariski tangent space). Its dimension is the minimal integer n for which there exists an epimorphism of $\mathbb{C}$-algebras $H_n \to \mathcal{O}_{X,x}$.

Let $(f, \varphi): (X, \mathcal{O}_X) \to (Y, \mathcal{O}_Y)$ be a mapping of complex spaces. It defines for each point $x \in X$ a homomorphism of $\mathbb{C}$-algebras $\varphi_x: \mathcal{O}_{Y,f(x)} \to \mathcal{O}_{X,x}$. For any tangent vector $t \in T_x(X)$ the composition $t \cdot \varphi_x$ is a tangent vector to Y at the point $f(x)$. Consequently, a linear mapping $d\varphi_x: T_x(X) \to T_{f(x)}(Y)$ is defined, which is called the *differential* of (f, φ) at the point x. The rank of the mapping $d\varphi_x$ is called the *rank of the mapping* (f, φ) at x.

The tangent space $T(D)$ to the double point $D = (*, \mathbb{C}[\varepsilon])$ is one-dimensional. If a generator ε in the algebra of dual numbers is chosen, then a basis element $v \in T(D)$ can be singled out by the condition $v(\varepsilon) = 1$. Therefore to each mapping from D into a complex space X which sends the underlying point to x there corresponds a vector $t \in T_x(X)$ which is equal to the image of v. Conversely, to each tangent vector t to X at the point x there corresponds exactly one such mapping. Thus $T_x(X)$ is isomorphic to the set of all mappings from the double point into X which send * to x.

We note one more property of the differential which is necessary to have in view. Let $(f, \varphi): X \to Y$ be a mapping of complex spaces, where X is a complex manifold in a neighborhood of the point x and the mapping $d\varphi_x$ is epimorphic. Then the fiber X_s, where $s = f(x)$, and also all nearby fibers of f are manifolds in a neighborhood of x and $T_x(X_s) = \operatorname{Ker} d\varphi_x$.

2.3. A derivation in a $\mathbb{C}$-algebra A is any linear mapping $\delta: A \to A$ satisfying the Leibniz rule $\delta(ab) = \delta(a)b + a\delta(b)$. A *vector field* on a complex space $(X, \mathcal{O}_X)$ is any $\mathbb{C}$-linear mapping of sheaves $v: \mathcal{O}_X \to \mathcal{O}_X$ which is a derivation on each fiber $\mathcal{O}_{X,x}$ If a vector field v is defined on X, then its composition with any numerical mapping $r: \mathcal{O}_{X,x} \to \mathbb{C}$ is a tangent vector at the point x. (However, if x is a singular point, then not every tangent vector at x extends to a vector field in a neighborhood.)

2.4. Let $\mathscr{A}$ be the category of analytic algebras and their homomorphisms as $\mathbb{C}$-algebras. The opposite category $\mathscr{A}^0$ by definition consists of the same objects or, more precisely, of copies of them and the reversed arrows. The objects of the opposite category are called *germs of complex spaces.* Thus if $\varphi: A \to B$ is a homomorphism of analytic algebras, then there is defined a morphism $\varphi^0: B^0 \to A^0$ of the corresponding germs.

The geometric meaning of these definitions is as follows. A *representative of a germ* A^0 is any complex space X with distinguished point x together with an isomorphism of $\mathbb{C}$-algebras $\alpha: \mathcal{O}_{X,x} \xrightarrow{\sim} A$. Thus the germ A^0 can thought of as an equivalence class of triples of the form (X, x, α). Such a triple is equivalent to the triple (Y, y, β) if there exists an isomorphism from some neighborhood of x to some neighborhood of y which sends x to y and which gives rise to an isomorphism $\mathcal{O}_{X,x} \cong \mathcal{O}_{Y,y}$ compatible with α and β. Each class A^0 is nonempty. Indeed, if we choose a representative $A \cong H_n/I$ and a set of generators $f_1, \ldots, f_m$ for the ideal I, then the model complex space defined in a neighborhood of the point $x = 0$ in $\mathbb{C}^n$ by the functions $f_1, \ldots, f_m$ belongs to A^0.

If a morphism of germs $B^0 \to A^0$ (i.e., a homomorphism of $\mathbb{C}$-algebras $\varphi: A \to B$) is defined and representatives for them (Y, y, β) and (X, x, α) are chosen, then it is possible to find a neighborhood Y' of the point y and a mapping of complex spaces with distinguished points $(Y', y) \to (X, x)$ which generates a mapping of algebras $\psi: \mathcal{O}_{X,x} \to \mathcal{O}_{Y,y}$ so that $\beta\psi = \varphi\alpha$. This mapping of complex spaces is called a representative for the morphism of germs. Any morphism of germs $f: (X, x) \to (Y, y)$ has only one fiber, namely the fiber $f^{-1}(y)$ over the distinguished point (unlike a representative of this morphism which has a fiber over every point of Y). If $A \to B$ is the dual homomorphism of analytic algebras, then the algebra $B/m(A)\cdot B$ is dual to the fiber of the morphism of germs.

For every complex space X there corresponds to each point x the germ of a complex space which contains the representative $(X, x, 1)$, where 1 denotes the identity transformation of the algebra $\mathcal{O}_{X,x}$. For every mapping of complex spaces $X \to Y$ and point $x \in X$ there corresponds a morphism of germs for which a representative is given by the mapping $(X, x) \to (Y, f(x))$.

2.5. Let $(X, \mathcal{O}_X)$ be a complex space. An $\mathcal{O}_X$-sheaf is by definition a sheaf $\mathcal{F}$ of vector spaces on X in which each fiber has the structure of an $\mathcal{O}_{X,x}$-module such that this structure is consistent with the topologies of these sheaves. This condition means that if U is an open subset of X and a and f are sections of these sheaves over U, then the elements $a_x f_x$ define a section of the sheaf $\mathcal{F}$ over U (s_x denotes the germ of the section s at the point x). The support of an $\mathcal{O}_X$-sheaf $\mathcal{F}$ is the set supp $\mathcal{F}$ of all points $x \in X$ at which the fiber $\mathcal{F}_x$ of this sheaf is a nonzero space.

A mapping of $\mathcal{O}_X$-sheaves $\mathcal{F} \to \mathcal{G}$ is a mapping of sheaves of vector spaces which over each point $x \in X$ defines a mapping of $\mathcal{O}_{X,x}$-modules $\mathcal{F}_x \to \mathcal{G}_x$. An $\mathcal{O}_X$-sheaf $\mathcal{F}$ is called *coherent* if for each point x there is a neighborhood U and a sequence of mappings of $\mathcal{O}_U$-sheaves

$$\mathcal{O}_U^m \to \mathcal{O}_U^n \to \mathcal{F}|U \to 0,$$

which is exact in the fibers over each point $x \in U$. An $\mathcal{O}_X$-sheaf $\mathcal{F}$ is called a locally free sheaf of finite type if for each x one can find a sequence of $\mathcal{O}_U$-sheaves of this type with $m = 0$.

3. Flat Mappings

3.1. Let A be a commutative ring and M an A-module. M is called *flat* if for any exact sequence of A-modules $E \to F \to G$ the sequence

$$M \otimes_A E \to M \otimes_A F \to M \otimes_A G.$$

is exact. We note that if $G = 0$ then this condition is satisfied for any A-module M, and for the flatness of M it suffices that it be satisfied for $E = 0$. Using homological algebra ([9], [36]), the condition for the flatness of M can be rephrased in the form $\mathrm{Tor}_i{}^A(M, N) = 0$ for $i \geqslant 1$ and for all A-modules N. If A is an analytic algebra and M is an A-module having a finite number of generators, then the following criterion of Serre-Grothendieck holds: M is a flat A-module if $\mathrm{Tor}_i{}^A(M, \mathbb{C}) = 0$ (in which $\mathbb{C}$ is given the structure of an A-module by the numerical mapping r). In other words, M is a flat A-module if for any exact sequence of A-modules of the form

$$A^m \xrightarrow{b} A^n \xrightarrow{a} A \xrightarrow{r} \mathbb{C}$$

the sequence

$$M^m \xrightarrow{\tilde{b}} M^n \xrightarrow{\tilde{a}} M,$$

is exact in which $\tilde{b}$ and $\tilde{a}$ are the mappings generated by the matrices b and a.

A mapping of complex spaces $(f, \varphi): (X, \mathcal{O}_X) \to (Y, \mathcal{O}_Y)$ is called *flat* at a point x if the homomorphism $\varphi_x: \mathcal{O}_{Y,f(a)} \to \mathcal{O}_{X,x}$ turns $\mathcal{O}_{X,x}$ into a flat module over the algebra $\mathcal{O}_{Y,f(x)}$. We enumerate some of the properties of flat mappings [9], [26]:

I) The set of points x at which a given mapping of complex spaces $f: X \to Y$ is flat is open. The complement of this set is an analytic set.

II) If a mapping f is flat at x, then there is a neighborhood U of this point so that the mapping $f: U \to Y$ is open and at all points of U the dimension of the fibers of the mapping f is the same.

III) If a mapping f is flat at x, then for any mapping of complex spaces $g: Z \to Y$ the mapping $g^*(f): X \times_Y Z \to Z$ (see Chapter 1, Section 1) is flat at each point of the form (x, z) ($g^*(f)$ is called the *inverse image of the mapping* f).

IV) Let f be flat at x and let the fiber $X_{f(x)}$ be a manifold at this point. There exists a neighborhood $U \ni x$ which is isomorphic to a direct product of the open set $V = f(U)$ and the neighborhood $W = U \cap X_{f(x)}$ of the point x in the fiber, and this isomorphism fits into the commutative diagram

$$\begin{array}{ccccc} W & \xrightarrow{i} & U & \overset{f}{\searrow} & \\ \Vert\wr & & \Vert\wr & & V, \\ W \times \{f(x)\} & \xrightarrow{j} & W \times V & \nearrow & \end{array}$$

in which i and j are the natural imbeddings.

3.2. A mapping of complex spacies (f, φ) is called *proper* if the mapping f is proper, i.e., if $f^{-1}(K)$ is compact for each compact set K. The following general fact is due to Remmert: the image of any proper mapping of complex spaces is a closed analytic set. Grauert obtained a far reaching generalization of this result. One of his theorems [37] states that for any proper mapping $f: X \to Y$ and any coherent sheaf $\mathscr{F}$ on X all its direct images $R^i f(\mathscr{F})$ are coherent sheaves on Y. This result also generalizes a theorem of Cartan-Serre [61] on the finite-dimensionality of the cohomology of any coherent sheaf on a compact complex space. A discussion of questions related to this and an extensive bibliography are contained in the book [9].

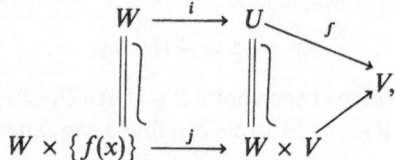

The theorem of Grauert is closely connected also with a theorem of Kodaira-Spencer [52] on semicontinuity in which an arbitrary family $f: X \to Y$ of compact manifolds is considered. By a family we mean a proper mapping of manifolds whose rank at each point equals the dimension of Y. The theorem states that for any q the function $h^q(y) = \dim H^q(X_y, \mathscr{F} | X_y)$ is upper semicontinuous if $\mathscr{F}$ is a locally free sheaf of finite type on X. Grauert established the semicontinuity of the functions $h^q(y)$ in a more general situation: f is any proper mapping of complex spaces and $\mathcal{F}$ is an arbitrary coherent sheaf on X which is flat over Y. This last condition means that for any point x the fiber $\mathcal{F}_x$ of this sheaf is a flat module over the algebra $\mathcal{O}_{Y,f(x)}$ (this algebra acts on $\mathcal{F}_x$ via the homomorphism $\varphi_x : \mathcal{O}_{Y,f(x)} \to \mathcal{O}_{X,x}$). The alternating sum $e(y) = \sum(-1)^q h^q(y)$ is called the Euler characteristic of the sheaf $\mathcal{F}|X_y$. Under the same hypotheses the function $e(y)$ is locally constant on Y.

The results of Grauert have been generalized in various directions. In particular, the theorem on the constancy of the Euler characteristic was generalized in [8]. It was proven, in particular, that for any two coherent sheaves $\mathcal{F}$ and $\mathcal{G}$ on X which are flat over Y their intersection index on the fiber X_y is a locally constant function on Y. The intersection index is defined to be the alternating sum of the dimensions of the vector spaces $\mathrm{Tor}_i^{\mathcal{O}_{X_y}}(\mathcal{F}|X_y, \mathcal{G}|X_y)$ (under the condition that this sum is finite). A similar result also holds for the dimensions of $\mathrm{Ext}^i(\mathcal{F}|X_y, \mathcal{G}|X_y)$.

4. Deformations

As defined by Grothendieck [87], a *deformation* is any proper flat mapping of complex spaces $f : X \to S$. The space S is called the *base* of the deformation and X the deformation space. The fibers X_s of this mapping for $s \in S$ are the spaces "undergoing" deformation. For any mapping of complex spaces $g : T \to S$ the inverse image $g^*(f)$ is again a deformation (Chapter 1, Section 3, Property III). In particular, if g is an imbedding of the space T into S, then $g^*(f)$ is called the *restriction* of f to this subspace and is denoted by $f|T$. The dimension of the fibers of a deformation over a connected base is constant (Chapter 1, Section 3, Property II). If the base S is a manifold and one of the fibers is also a manifold, then all neighboring fibers are manifolds and the rank of the mapping f on these fibers is equal to the dimension of S (Chapter 1, Section 3, Property IV). In this case f defines a family of compact manifolds in the sense of Kodaira-Spencer [51].

Let S be a complex space with distinguished point $*$. A deformation of an arbitrary complex space X_0 with base $(S, *)$ is any mapping of complex spaces $f : X \to S$ together with a fixed isomorphism $i : X_0 \xrightarrow{\sim} f^{-1}(x)$ which is flat at each point of the fiber $f^{-1}(*)$ (this fiber is called the distinguished fiber). Now let another deformation $\tilde{f} : \tilde{X} \to S$, $\tilde{i} : X_0 \to \tilde{f}^{-1}(*)$ of the same space X_0 with the same base. We say that they are equivalent if there is a neighborhood X' of the distinguished fiber in X and neighborhood $\tilde{X}'$ of the distinguished fiber in $\tilde{X}$ and an isomorphism a so that the diagram

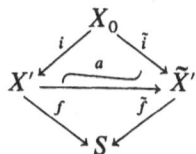

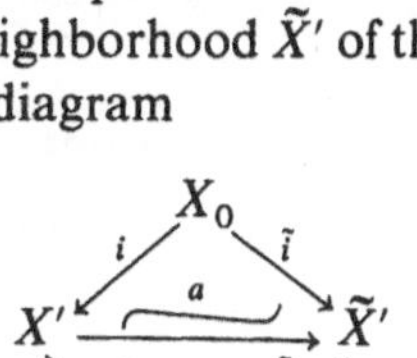

commutes. We can then pass to equivalence classes of such deformations. It is clear that each class is still defined if the whole space S is not given but only its germ $\tilde{S}$ at the distinguished point. Therefore we can make the following definition: a *deformation of a complex space X_0 over a germ $\tilde{S}$* is any equivalence class of deformations of X_0 defined for some representative S of this germ. Each such deformation we shall call a *representative of the given deformation* over the germ $\tilde{S}$ and this germ *the base of the deformation.*

This definition is consistent with the preceding one. Each deformation of a compact complex space has a representative $f: X \to S$ which is a proper flat mapping (Chapter 1, Section 3.1), i.e., it is a deformation in the sense of Grothendieck.

The *trivial deformation* is the canonical mapping $X_0 \times S \to S$ or the equivalence class containing this mapping. A deformation with a zero-dimensional base is called *infinitesimal.*

Let X_0 be a germ of a complex space. A *deformation* of this germ is any flat mapping of germs $f: X \to S$ together with an isomorphism of germs $i: X_0 \xrightarrow{\sim} f^{-1}(*)$. In dual terms a deformation of germs is defined as follows: given are an analytic algebra A (dual to X_0) and a flat mapping $\varphi: \Gamma \to B$ of analytic algebras (i.e., so that B becomes a flat Γ-module) together with an isomorphism of $\mathbb{C}$-algebras $A \cong B/m(\Gamma)\cdot B$. An isomorphism of deformations of germs and the inverse image of such a deformation are defined in the natural fashion.

A deformation of a germ of a complex space can be described using model complex spaces. We assume first that X_0 is the *germ of a complete intersection.* This means that it has a model representative $(Z, \mathscr{H}_n/(f_1, \ldots, f_m))$ possessing the following property: $\dim Z = n - m$ (in the general case $\dim Z \geqslant n - m$). For simplicity let the base S be the germ of $\mathbb{C}^k$ at the origin. A deformation of X_0 with this base has as its deformation space the germ of a model complex space X defined in $\mathbb{C}^n \times \mathbb{C}^k$ by functions $F_1(z, s), \ldots, F_m(z, s)$ holomorphic in a neighborhood of $Z \times \{0\}$ and subject to the single requirement that $F_i(z, 0) = f_i(z)$ for $i = 1, \ldots, m$. In the general case by the Serre-Grothendieck criterion the functions $F_i(z)$ define by the same rule a deformation of X_0 if and only if for any functions $g_1, \ldots, g_m \in H_n$ such that $g_1 f_1 + \cdots + g_m f_m = 0$ there exist functions $G_1, \ldots, G_m \in H_{n+k}$ so that $G_1 F_1 + \cdots + G_m F_m = 0$ and $G_i(z, 0) = g_i(z)$. In other words, any relation among the f_i must extend to a relation among the F_i.

Chapter 2. The General Theory of Deformations

1. The Resolution and Tangent Cohomology of an Analytic Algebra

1.1. One of the basic concepts in deformation theory is that of the Tyurina resolution (see [69]). It is the analog in the category of analytic algebras of the Tate resolution and the DGA-resolution of Quillen ([79]). In all cases the construction of a free metacommutative graded algebra[1] over a commutative ring R is basic. We begin the description of this construction with a more elementary concept—that of a graded tensor algebra over a ring R. We fix a set E, called a set of generators, in which there is associated to each element $e \in E$ an integer $\deg e$ called its degree. The tensor algebra $R[E]$ is by definition the free R-module with basis formed by the elements $e_1 \otimes \cdots \otimes e_m$ for $e_i \in E$. To each such element there is associated a degree equal to $\deg e_1 + \cdots + \deg e_m$. Any linear combination over R of basis elements of degree k is called a homogeneous element of this degree. The collection of all homogeneous elements of degree k is clearly a submodule of $R[E]$; it is denoted by $R_k[E]$. A multiplication is defined on the basis elements by the formula

$$(e_1 \otimes \cdots \otimes e_k) \cdot (e'_1 \otimes \cdots \otimes e'_l) = e_1 \otimes \cdots \otimes e_k \otimes e'_1 \otimes \cdots \otimes e'_l$$

and this extends by R-linearity to the whole tensor algebra. It is clear that the product of homogeneous elements is a homogeneous element, whose degree is the sum of the degrees of the factors. $R[E]$ is a graded associative R-algebra.

Let J be the two-sided ideal in this algebra generated by all elements of the form

$$a = e_2 \otimes e_1 - (-1)^{\deg e_1 \cdot \deg e_2} e_1 \otimes e_2, \quad e_1, e_2 \in E,$$

i.e., J consists of sums of elements of the form $r \cdot a \cdot s$ for $r, s \in R[E]$. The quotient algebra $R\{E\} = R[E]/J$ is called the free metacommutative graded R-algebra with generating set E. The image of the element $e_1 \otimes \cdots \otimes e_k \in R[E]$ in this quotient algebra is denoted by $e_1 \cdot \ldots \cdot e_k$. Such elements generate it as an R-module, but they do not form a basis, since they are dependent. Indeed, the element $e_1 \cdot \ldots \cdot e_k$ does not change under any permutation of the factors e_i of even degree and changes sign under a permutation of neighboring factors of odd degree. The grading is defined by the expansion $R\{E\} = \bigoplus_k R_k\{E\}$, where $R_k\{E\}$ is the image in $R\{E\}$ of the module $R_k[E]$. An element a of $R_k\{E\}$ is called homogeneous of degree $k = \deg a$. The commutativity property can be written as:

$$b \cdot a = (-1)^{\deg a \cdot \deg b} a \cdot b$$

The significance of the definition that we have just given consists in the following. Let S be any graded metacommutative R-algebra. In order to construct a homomorphism of graded R-algebras $\varphi : R\{E\} \to S$, it suffices to specify the values of φ on the generating set E, these values being subject only to the condition that $\deg \varphi(e) = \deg e$.

1.2. By a *Tyurina resolution* of an analytic algebra A we mean the following collection of objects:

[1] In [69] these algebras are called free commutative graded algebras.

1) an epimorphism of $\mathbb{C}$-algebras $p: R_0 \to A$, where R_0 is a regular analytic algebra, i.e., an algebra of convergent power series in r_0 variables (these variables are to be thought of as the generators of degree zero);

2) a free metacommutative R_0-algebra R generated by some generating set E with negative degrees, the number in each degree being finite;

3) a differential s in the R_0-module R of degree 1 (i.e., $J: R_k \to R_{k+1}$) which is a derivation, i.e., which satisfies the relation

$$s(a \cdot b) = s(a) \cdot b + (-1)^{\deg a} a \cdot s(b), \tag{2.1.1}$$

so that the sequence

$$\cdots \to R_{-2} \xrightarrow{s} R_{-1} \xrightarrow{s} R_0 \xrightarrow{p} A \to 0.$$

is exact. Each analytic algebra has at least one (in fact infinitely many) Tyurina resolution. The construction of this resolution is not complicated, and we shall give it now.

If the algebra A is itself regular, then we may set $R = R_0 = A$. Otherwise, we choose some epimorphism $p: R_0 \to A$ (the dual operation is to choose some imbedding of germs $X_0 \to (\mathbb{C}^{r_0}, 0)$, where X_0 is dual to the algebra A). The kernel of p is an ideal in the algebra R_0, consequently, it has a finite number of generators $f = (f_1, \ldots, f_{r_1})$ (equations defining X_0 in $\mathbb{C}^{r_0}$). We form the Koszul complex $K(R_0, f)$. This is the metacommutative graded R_0-algebra with generators e_1, $\ldots$, e_{r_1} of degree -1 and differential s_1 which satisfies the relation (2.1.1) and is defined on the generators by the formula $s_1(e_i) = f_i$. If the germ X_0 is a complete intersection and the number of equations r_1 is chosen to be the minimum possible, i.e., $\dim X_0 = r_0 - r_1$, then the Koszul complex $R^1 = K(R_0, f)$ is a resolution of A, since $H^n(R^1) = 0$ for $n < 0$ and $H^0(R^1) \cong A$. Conversely, if $H^{-1}(R^1) = 0$ then X_0 is a complete intersection of dimension $r_0 - r_1$. These results are contained in the classical work of Serre [88]. Since the Koszul complex by construction is a free metacommutative graded R_0-algebra and the differential satisfies (2.1.1), it can also serve as a Tyurina resolution for the analytic algebra A, which is dual to the complete intersection X_0 (speaking loosely, we sometimes say that A is a complete intersection).

We now assume that A is not a complete intersection. Then already we have $H^{-1}(R^1) \neq 0$; therefore to construct a Tyurina resolution we need to add generators $e'_1, \ldots, e'_{r_2}$ of degree -2 and define on them values of the differential $s(e'_i) \in R^1_{-1}$ so that the R_0-span of these values together with the image of the differential $s: R^1_{-2} \to R^1_{-1}$ is equal to the kernel of $s: R^1_{-1} \to R_0$. This is possible, since this kernel has a finite number of generators over R_0. The free metacommutative graded R_0-algebra R^2 which is spanned by the generators of degree -1 and -2 is the second step in the construction of the resolution. Its differential is already defined on the generators and by (2.1.1) it extends uniquely to the whole algebra. Here we have $H^{-1}(R^2) = 0$. If $H^{-2}(R^2) \neq 0$ then we add a set of generators of degree -3 and define on them the differential s so that in the extended R_0-algebra R^3 we have $H^{-2}(R^3) = 0$. As before the cohomology of R^3 in dimension -1 is equal to zero. Continuing this construction, we add necessary (or superfluous) generators of degrees -4, -5, etc.

1.3. The Tangent Cohomology. Suppose that for some analytic algebra A we have chosen a Tyurina resolution. A derivation of degree k in the $\mathbb{C}$-algebra R is any linear mapping $u: R \to R$ which raises the grading of all elements by k and satisfies the relation

$$u(a \cdot b) = u(a) \cdot b + (-1)^{k \cdot \deg a} a \cdot u(b) \tag{2.1.2}$$

for homogeneous elements a and b. In particular, the differential s in R is by (2.1.1) a derivation of degree one. The collection of all derivations on R forms a graded vector space $\operatorname{Der}(R) = \bigoplus_{-\infty}^{\infty} \operatorname{Der}_k(R)$. We introduce in it the bilinear operation

$$[u, v] = u \circ v - (-1)^{\deg u \cdot \deg v} v \circ u,$$

where $u \circ v$ denotes the composition of endomorphisms of the space R. This operation clearly satisfies the relations

$$[v, u] = -(-1)^{\deg u \cdot \deg v}[u, v], \tag{2.1.3}$$

$$[u, [v, w]] = [[u, v], w] + (-1)^{\deg u \cdot \deg v}[v, [u, w]], \tag{2.1.4}$$

i.e., it turns $\operatorname{Der}(R)$ into a graded Lie meta-algebra. In this algebra there is the linear operator $d: u \mapsto [u, s]$ of degree 1. It is a differential, i.e., $d^2 = 0$, since by (2.1.3) and (2.1.4)

$$2d^2 u = 2[[u, s], s] = [u, [s, s]] = 0,$$

($[s, s] = 2s^2 = 0$). The cohomology of the complex $(\operatorname{Der}(R), d)$ is concentrated in the nonnegative degrees and is denoted by $T^*(A) = \oplus T^k(A)$ or $T^*(X_0)$. It is called the *tangent cohomology of the analytic algebra A* and of the corresponding *germ of a complex space* X_0. The bracket described above lifts to the tangent cohomology. Indeed, it follows easily from (2.1.4) that the bracket of cycles in the complex $(\operatorname{Der}(R), d)$ is a cycle and the bracket of a cycle and a boundary is a boundary. Therefore there is defined the operation $[\operatorname{cl}(u), \operatorname{cl}(v)] = \operatorname{cl}[u, v]$ for any tangent cohomology classes. It also satisfies (2.1.3) and (2.1.4), consequently, $T^*(A)$ is a graded Lie meta-algebra. It is important to note that this structure does not depend on the choice of the Tyurina resolution. In $T^*(A)$ there is also defined invariantly the structure of an A-module. This structure is related to the bracket by the formula $[u, av] = u(a) \cdot v + a[u, v]$, where $u(a)$ is the action of the derivation $u \in T^0(A) \cong \operatorname{Der}(A)$ on the element $a \in A$ and $u(a) = 0$ if $\deg u > 0$.

1.4. Let A be a regular analytic algebra, i.e., $A \cong \mathbb{C}\{z_1, \dots z_n\}$. Then $T^0(A)$ is the A-module of operators in A of the form $\sum_1^n a_i(z)\partial/\partial z_i$ for $a_i \in A$, and $T^i(A) = 0$ if $i \geqslant 1$.

If A is an arbitrary analytic algebra, then the following device, which follows from Section 1.3, is useful in the computation of the tangent cohomology. We select a regular analytic algebra R_0 and an algebra epimorphism $p: R_0 \to A$. Let I denote its kernel. We have the exact sequence of R_0-modules

$$0 \to T^0(A) \to \mathrm{Der}(R_0) \xrightarrow{d} \mathrm{Hom}_{R_0}(I, A) \to T^1(A) \to 0, \qquad (2.1.5)$$

where d acts according to the rule $t \mapsto p \circ t|I$. We assume that A is a complete intersection. According to what was said in Section 1.2, R_0 can be completed to a Tyurina resolution (R, s) for the algebra A with generators $e_1, \ldots, e_m$ of degree -1, where the elements $s(e_i) = f_i$ for $i = 1, \ldots, m$ are a minimal generating set for the ideal I. It is clear that the complex $\mathrm{Der}(R)$ is concentrated in degrees 0 and 1, consequently, $T^0(A) = 0$ for $i > 1$. To calculate $T^1(A)$, we use (2.1.5) and describe $\mathrm{Hom}(I, A)$. We consider the imbedding of A-modules

$$\mathrm{Hom}(I, A) \to A^m,$$

which maps the homomorphism h to the vector $(h(f_1), \ldots, h(f_m))$. This mapping is an epimorphism, since for every vector $(b_1, \ldots, b_m)$ we can construct a homomorphism $h: I \to A$ by setting $h(\Sigma a_i f_i) = \Sigma a_i b_i$. This is a consistent definition, since it follows from $\Sigma a_i f_i = 0$ that the vector $(a_1, \ldots, a_m)$ belongs to the image of the differential $s: R_{-2} \to R_{-1}$, consequently, $a_i \in I$ for $i = 1, \ldots, m$. Thus there is an isomorphism of A-modules

$$T^1(A) \cong A^m / df \cdot A^n,$$

where df denotes the Jacobian matrix of the mapping $f = (f_1, \ldots, f_m): \mathbb{C}^n \to \mathbb{C}^m$.

2. The Resolution of a Complex Space

2.1. The Resolution of a Model Space. Let $(Z, \mathcal{O}_Z)$ be a model complex space defined as a closed subspace of an open set $V \subset \mathbb{C}^n$. By definition there is an epimorphism of sheaves of analytic algebras $\mathcal{H}_n | V \to \mathcal{O}_Z$. We set $R_0 = \mathcal{H}_n | V$, the complex of R_0-sheaves

$$\cdots \to \mathcal{R}_{-2} \xrightarrow{s} \mathcal{R}_{-1} \xrightarrow{s} \mathcal{R}_0$$

on the set V is called a *resolution of the model complex space* $(Z, \mathcal{O}_Z)$ if the fiber of this complex over each point $z \in Z$ is a Tyurina resolution for the algebra $\mathcal{O}_{Z,z}$ and some subset e of sections over V of this complex defines in each fiber a distinguished set of generators for this resolution. For each model complex space Z we can construct a resolution for the open subspace $Z \cap V'$ if the closure of V' is compact in V. For the proof it suffices to repeat the construction of the Tyurina resolution with the obvious changes. It follows that for any model complex space, and thus also for any complex space X, the union $\mathcal{T}^k(X) = \bigcup_{x \in X} \mathcal{T}^k(\mathcal{O}_{X,x})$ for any k is a coherent $\mathcal{O}_X$-sheaf; we call it the *kth tangent sheaf* on X. In particular, $\mathcal{T}^0(X)$ is the sheaf of holomorphic tangent fields.

The resolutions of model complex spaces form a category. If two resolutions $(V, \mathscr{R})$ and $(V', \mathscr{R}')$ are given, then a morphism from the first into the second is a pair consisting of a mapping of complex manifolds $f: V \to V'$, $\varphi: f^*(\mathscr{R}'_0) \to \mathscr{R}_0$ and a mapping of sheaves $\psi: f^*(\mathscr{R}') \to \mathscr{R}$, which

a) on the sheaf $f^*(\mathscr{R}'_0)$ agrees with φ,

b) at each point $z \in V$ is a homomorphism of graded algebras, and

c) is a mapping of complexes, i.e., commutes with the differentials. These properties can be expressed briefly as follows: ψ is a homomorphism of differential graded algebras over the homomorphism φ.

2.2. Polyhedra and Coverings. A *polyhedron in a complex space X* is any relatively compact subset P of X together with a proper holomorphic imbedding $\varphi: U \to V \subset \mathbb{C}^N$, which is defined in some neighborhood $U \supset \bar{P}$ so that V is a neighborhood of the closure of the unit coordinate polydisc D^N and $P = \varphi^{-1}(D^N)$ (φ is called the barrier mapping for P). With each polyhedron there is associated the complex space $(\varphi(U), \varphi_*(\mathcal{O}_X(U))$ which is a closed subspace of V. It is a model space in a neighborhood of $\bar{D}^N$.

Now let $P = \{P_\alpha, \alpha \in \mathscr{A}\}$ be a covering of the space X formed by polyhedra. A simplex of this covering is any finite subset $A \subset \mathscr{A}$ so that the intersection $P_A = \bigcap_{\alpha \in A} P_\alpha$ is nonempty. This intersection is a polyhedron with barrier mapping $\varphi_A(x) = (\varphi_{\alpha_0}(x), \ldots, \varphi_{\alpha_n}(x))$, where $A = (\alpha_0, \ldots, \alpha_n)$. The number n is called the dimension of the simplex A. The *nerve of the covering* is the collection $\mathscr{N}$ of all simplices. We can view the nerve as a category whose objects are simplices and whose morphisms are inclusions of simplices $v_A^B: A \hookrightarrow B$. If $\mathscr{K}$ is any category, then a covariant functor $F: \mathscr{N} \to \mathscr{K}$ is by definition a collection of objects $F(A)$ of this category indexed by the simplices of the covering and of morphisms $f_A^B: F(A) \to F(B)$ defined for each pair of simplices $A \subset B$. Two conditions only need to be satisfied: f_A^A must be the identity morphism, and for any simplices A, B, B', and Γ such that $A \subset B \cap B'$ and $B \cup B' \subset \Gamma$ we must have the relation $f_{B'}^\Gamma \circ f_A^{B'} = f_B^\Gamma \circ f_A^B$. Contravariant functors $\mathscr{N} \Rightarrow \mathscr{K}$ can be described in a similar fashion. With each polyhedral covering $\mathscr{P} = \{P_\alpha\}$ there is associated a contravariant functor from its nerve into the category of complex manifolds, which associates to a simplex $A \in \mathscr{N}$ the polydisc $D_A = D^{N_A}$ with $N_A = \sum_{\alpha \in A} N_\alpha$ endowed with the sheaf $\mathscr{H}_A$ of germs of holomorphic functions in N_A variables. An inclusion

$A \subset B$ there corresponds the coordinate projection $p_A^B: D_B \to D_A$ together with the natural imbedding $(p_A^B)^*(\mathscr{H}_A) \hookrightarrow \mathscr{H}_B$.

2.3. Formulation of the Basic Definition [69]. Suppose that we are given a covering of a complex space $(X, \mathcal{O}_X)$ by polyhedra P_α with $\alpha \in \mathscr{A}$, and let $\mathscr{N}$ be its nerve. A *resolution of the space X* with this covering is a contravariant functor $\mathscr{R}$ from $\mathscr{N}$ into the category of resolutions of model spaces which satisfies the following conditions:

I) for any simplex A, $\mathscr{R}(A)$ is a resolution of the model space $\varphi_A(P_A, \mathcal{O}_X|_{P_A})$ on the polydisc D_A with generating set $e(A)$.

II) for any simplices $A \subset B$ the morphism $\mathscr{R}(v_A^B)$ consists of the coordinate projection $p_A^B: D_B \to D_A$ and the morphism $r_A^B: (p_A^B)^*(\mathscr{R}(A)) \to \mathscr{R}(B)$ of sheaves of differential graded algebras, which on elements of degree zero coincides with the imbedding $(p_A^B)^*(\mathscr{H}_A) \hookrightarrow \mathscr{H}(B)$ with generators going to generators, i.e., $r_A^B(e(A)) \subset e(B)$.[2]

A more general definition can be given for mappings of complex spaces $X \to Y$. In this case all constructions become relative. In particular, a relative polyhedron in X/Y is defined to be a relatively closed subset $P \subset X$ with a barrier imbedding $\varphi: U \to V \times Y$, which is a mapping over Y; polydiscs are replaced by spaces of the form $D^N \times Y$, etc. Thus we obtain the definition of a resolution of the relative space X/Y, see [70].

Every complex space has a resolution on any polyhedral covering. For any mapping $X \to Y$ a resolution of X/Y exists.

3. The Tangent Complex and Cohomology

3.1. Let X be a complex space and let $\mathscr{R}$ be a resolution of X on a covering with nerve $\mathscr{N}$. A derivation of $\mathscr{R}$ of degree k is any mapping of functors $u: \mathscr{R} \to \mathscr{R}$ whose value on each object $\mathscr{R}(A)$ is a derivation of the given degree. In other words, u is a set of mappings of degree k of graded sheaves $u_A: \mathscr{R}(A) \to \mathscr{R}(A)$, which are defined for all $A \subset \mathscr{N}$, and connected by the relations $u_B \circ r_A^B = r_A^B \circ (p_A^B)^* u_A$ for every pair of simplices $A \subset B$, so that for every $A \in \mathscr{N}$ and point $z \in D_A$ the mapping u_A acts on the fiber $\mathscr{R}_z(A)$ as a derivation, i.e., it satisfies (2.1.2). The set of all derivations of $\mathscr{R}$ forms a graded vector space, which is denoted by $T^*(\mathscr{R})$. The collection of differentials $s = \{s_A\}$ of the complexes $\mathscr{R}(A)$ is by definition a derivation of degree 1, i.e., it belongs to $T^1(R)$. The composition of any two derivations is defined as a mapping of functors $\mathscr{R} \rightsquigarrow \mathscr{R}$. Therefore, we can define the graded metacommutator

$$[u, v] = u \circ v - (-1)^{\deg u \cdot \deg v} v \circ u.$$

It is not difficult to verify that it is a derivation of degree $\deg u + \deg v$ and that the identities of the form (2.1.3) and (2.1.4) are satisfied, i.e., $T^*(\mathscr{R})$ becomes a graded Lie meta-algebra. The operator $d = [\cdot, s]$ is a differential in this algebra. The cohomology of the complex $(T^*(\mathscr{R}), d)$ is called the *tangent cohomology of the complex space* X and is denoted by $T^*(X)$. It inherits the structure of a graded Lie meta-algebra. The tangent cohomology and the Lie bracket on it do not depend of the choice of covering of the complex space X or of the resolution.

---

[2] In [69] this construction is called a resolution of the sheaf $\mathcal{O}_X$.

3.2. Properties of the Tangent Cohomology. It is concentrated in the nonnegative degrees, i.e., $T^*(X) = \bigoplus_0^\infty T^k(X)$. The first three homogeneous terms have interpretations in terms of deformation theory.

The zero-dimensional cohomology $T^0(X)$ is the space of holomorphic vector fields on X. The Lie operation on the tangent cohomology described above acts on $T^0(X)$ as the ordinary bracket of tangent fields.

There is an isomorphism between $T^1(X)$ and the set $\mathrm{Def}_D(X)$ of isomorphism classes of germs of deformations of X with base equal to the double point D. The zero of the space $T^1(X)$ corresponds to the germ of the trivial deformation. This isomorphism is not difficult to describe explicitly. If we are given an element $t \in T^1(X)$, we select a representative for it $u \in T^1(\mathscr{R})$, where $\mathscr{R}$ is a resolution of X on a covering formed by polyhedra. It consists of sheaves $\mathscr{R}(A)|\bar{D}_A$. We consider the sheaves $\tilde{\mathscr{R}}(A) = \mathscr{R}(A) \otimes_{\mathbb{C}} \mathscr{O}_D$ on the complex space $X \times D$. We realize $\mathscr{O}_D$ as an algebra of quantities of the form $a + \varepsilon b$ for $a, b \in \mathbb{C}$, where $\varepsilon^2 = 0$. In each sheaf $\tilde{\mathscr{R}}(A)$ we define a differential $\tilde{s}_A = s_A + \varepsilon u_A$, where u_A is the value of u on the simplex A. The equation $s_A^2 = 0$ together with the cycle condition $[u_A, s_A] = 0$ implies that $\tilde{s}_A^2 = 0$. The complex $(\tilde{\mathscr{R}}(A), \tilde{s}_A)$ is acyclic in the negative degrees. The cohomology of degree zero is a sheaf of analytic algebras. We denote it by $\tilde{\mathscr{O}}_A$ and view it as a sheaf on $\bar{D}_A$. From the fact that $A \rightsquigarrow (\bar{D}_A, \tilde{\mathscr{R}}(A))$ is a contravariant functor, it is not difficult to conclude that the sheaves $\tilde{\mathscr{O}}_A|\bar{D}_A$ glue together into a single sheaf $\tilde{\mathscr{O}}$ over X, which defines on the topological space X the structure of a complex space $\tilde{X}$. The sheaf $\mathscr{O}_D$ maps naturally onto the subsheaf of $\tilde{\mathscr{O}}$ of sections which do not depend on x. Thus we have a mapping $\tilde{X} \to D$, and it is easy to see that this mapping is flat. If we take the quotient of $\tilde{\mathscr{O}}$ relative to the sheaf of ideals generated by ε, we obtain the sheaf $\mathscr{O}_X$, consequently, X is the fiber of this mapping over the point $* \in D$. Hence $\tilde{X}|D$ is a deformation of X. Different representatives of the class t lead to isomorphic deformations. The correspondence $t \mapsto \mathrm{cl}(\tilde{X}|D)$ defines the required isomorphism $T^1(X) \cong \mathrm{Def}_D(X)$. A more general construction will be described in Chapter 2, Section 4.

The space $T^2(X)$ contains the values of obstructions to the extension of infinitesimal deformations (see Chapter 2, Section 5). The role of the Lie bracket will be discussed in this chapter.

3.3 On any complex space there is a spectral sequence which converges to $T^*(X)$ with second term

$$E_2^{pq} = H^p(X, \mathscr{T}^q), \tag{2.3.1}$$

where $\mathscr{T}^q = \mathscr{T}^q(X)$ are the sheaves of tangent cohomology of the structure sheaf $\mathscr{O}_X$ (Chapter 2, Section 2). In particular, there is an exact sequence

$$0 \to H^1(X, \mathscr{T}^0) \to T^1(X) \to \Gamma(X, \mathscr{T}^1) \xrightarrow{d_2} H^2(X, \mathscr{T}^0) \to \dots \tag{2.3.2}$$

The image of $H^1(X, \mathscr{T}^0)$ in $T^1(X)$ corresponds to those deformations of X over D for which each germ (X, x) is deformed trivially. If X is a manifold, then each analytic algebra $\mathcal{O}_{X,x}$ is regular; consequently, each deformation of the germ (X, x) is trivial. Therefore $\mathrm{Def}_D(X) \cong H^1(X, \mathcal{T}^0)$, which agrees with the Kodaria-Spencer theory [51].

A deformation of a complex space X which deforms nontrivially a singular point x has nonzero image in $\Gamma(X, \mathcal{T}^1)$. However, not every section $\sigma \in \Gamma(X, \mathcal{T}^1)$ corresponds to a deformation of X. The obstruction is the class $d_2\sigma \in H^2(X, \mathcal{T}^0)$. An example of a nonzero mapping d_2 will be given in Chapter 3, Section 7. Grothendieck's formal theory of deformation of schemes is discussed in [3&] and [7&].

3.4. Semicontinuity Theorems. Let $f: X \to S$ be a proper flat mapping of complex spaces with X_s its fiber over the point s. Each of the functions $\tau^q(s) = \dim T^q(X_s)$ is upper semicontinuous. If for some $n < N$ the functions $\tau^n(s)$ and $\tau^N(s)$ are constant in a neighborhood of a point $s_0 \in S$, then the function $\tau_n^N(s) = \sum_n^N (-1)^q \tau^q(s)$ is also constant. These results are corollaries of a more general fact: if the numbers n and N have the same parity, then the function τ_n^N is upper semicontinuous for n even and lower semicontinuous for n odd (see [69]).

If S and all the fibers X_s are manifolds, then $T^q(X_s) = H^q(X_s, \mathcal{T}^0)$. In this case the semicontinuity was established in [52].

4. The Kodaira- Spencer Mapping and the Differential of a Deformation

The foundations of deformation theory for compact manifolds were laid in the work of Kodaira and Spencer [51]. In these articles they study families of such manifolds, i.e., families of fibers of a proper regular mapping of manifolds $f: X \to S$ (regularity means that the rank of the differential of f is equal at every point to $\dim S$). The linear mapping

$$T_s(S) \to H^1(X_s, \mathcal{T}^0(X_s)), \quad X_s = f^{-1}(s),$$

from the tangent space to S at an arbitrary point s, which is constructed in these articles, has fundamental significance for the theory. A construction of the *Kodaira-Spencer mapping* can be given using an appropriate covering $\{U_\alpha\}$ of the complex manifold X_s; for example, the U_α can be the holomorphic images of polydiscs. Let $\varphi_{\beta\alpha}(z)$ be the gluing mappings of coordinate systems on these polydiscs (see Chapter 1, Section 1). By the compactness of X_s and the regularity of f, they can be included in a holomorphic family of holomorphic gluings $\Phi_{\beta\alpha}(z, s')$ which is defined for some covering $\{U'_\alpha\}$ such that $U'_\alpha \Subset U_\alpha$, where the parameter s' varies in some neighborhood of the point s, and the gluing of the domain U'_α via $\Phi_{\beta\alpha}(z, s')$ is the fiber $X_{s'}$ of the same family. Let v be an arbitrary tangent vector to S at the point s. The derivative $v(\Phi_{\beta\alpha})$ is a holomorphic vector field on $U'_\alpha \cap U'_\beta$. The collection of such fields is a cocycle on the covering $\{U'_\alpha\}$ with values in the tangent sheaf $T^0(X_s)$. Let $h \in H^1(X_s, T_0)$ be the corresponding cohomology class. By definition the value of the Kodaira-Spencer mapping on the vector v is the class h.

Generalizations of this construction to the case of spaces with singularities were given in the articles [87], [85], and [69]. In [69] the concept of the differential of a deformation $f: X \to (S, *)$ was introduced. The *differential of a deformation* is a linear mapping

$$Df: T(S) \to T^1(X_0), \quad X_0 = f^{-1}(*),$$

where $T(S)$ is the tangent space to the germ S (Chapter 1, Section 2.2). As we noted in Chapter 1, Section 2.2, there is a one-to-one correspondence between tangent vectors $v \in T(S)$ and morphisms of germs $u: D \to S$. For each such morphism u, $u^*(f)$ is a deformation of X_0 over D; it corresponds to a class $c \in \mathrm{Def}_D(X_0)$. Let $t \in T^1(X_0)$ be the image of this class under the isomorphism described in Chapter 2, Section 3.2. The differential of the deformation f is defined by the equation $Df(v) = t$.

The differential has the following properties:

1) if $X = X_0 \times S$ and f is the canonical projection onto S, then $Df = 0$.

2) in the case of $S = D$ the converse is also true: the equation $Df = 0$ implies that the deformation f is trivial.

3) if $\varphi: S' \to S$ is any mapping of germs, then $D\varphi^*(f) = Df \circ d\varphi$.

This last property underlines the analogy between differentials of mappings and differentials of deformations.

A more general construction, which we now describe, is the analog of higher differentials or, more precisely, of finite segments of Taylor expansions of mappings [71]. A germ of a complex space Y is called zero-dimensional if the underlying space is a single point; in this case the corresponding analytic algebra $\mathcal{O}_Y$ is finite-dimensional (and, consequently, Artinian). A zero-dimensional germ Y is called an *extension of a germ Z with ideal I* if $\mathcal{O}_Z \cong \mathcal{O}_Y/I$. The extension is called *small* if $m(Y) \cdot I = 0$. Let Y be a small extension of a germ Z with ideal I, and let f and f' be deformations of some complex space X_0 with base Y so that there exists an isomorphism $i: f/Z \xrightarrow{\sim} f'/Z$ of their restrictions to Z. Then the element

$$D_Z(f', f, i) \in T^1(X_0) \otimes_{\mathbb{C}} I,$$

is well defined and has the following properties:

I) $D_Z(f', f, i) = 0$ if and only if i extends to an isomorphism $f \xrightarrow{\sim} f'$.

II) If f'' is another deformation of X_0 with base Y and an isomorphism $i': f'|Z \xrightarrow{\sim} f''|Z$ is given, then

$$D_Z(f'', f, i' \circ i) = D_Z(f'', f', i') + D_Z(f', f, i),$$

from which it follows that $D_Z(f, f', i^{-1}) = -D_Z(f', f, i)$.

III) The element $D_Z(f', f, i)$ is functorial relative to the pair (Z, Y).

IV) For any deformation f of the space X_0 with base Y and any element $t \in T^1(X_0) \otimes I$ we can construct a deformation f' with the same base and an isomorphism $i: f|Z \to f'|Z$ so that $D_Z(f', f, i) = t$.

V) If Y' is a small extension of a zero-dimensional germ Z' with ideal I' and $g_1, g_2 : Y' \to Y$ are morphisms which agree on the subspace Z' and map it to Z, then

$$D_{Z'}(g_2^*(f), g_1^*(f), 1) = (Df \otimes 1)h,$$

where Df is the differential of the deformation f and $h = g_2 - g_1 \in T(Y) \otimes I'$. The tensor $D_Z(f', f, i)$ is the called the distinguishing element.

In the particular case when Z is a simple point, the assignment of the isomorphism i becomes superfluous, $I = m(Y)$, and $m^2(Y) = 0$ (the condition for a small extension). Therefore I is the space dual to $T(Y)$ and $T^1(X_0) \otimes I$ is isomorphic to the space of linear mappings $T(Y) \to T^1(X_0)$. It follows from V that the mapping corresponding to the tensor $D_Z(f, e)$, where e is the trivial deformation, is the differential Df. By II $D_Z(f', f)$ is equal to the difference of the differentials Df' and Df.

In the general case the heuristic meaning of the tensor $D_Z(f', f, i)$ lies in the fact that it is a measure of the "divergence" of the deformations f' and f on Y which are "joined" on Z by the isomorphism i. We describe the dependence of this "divergence" on the choice of i. Let Z be a small extension of a subspace W with ideal J and let g be any deformation of X_0 with base Z. The collection of all automorphisms of g, i.e., of isomorphisms $g \rightrightarrows g$ is a group, which we denote by Aut(g). We consider the subgroup consisting of the automorphisms which when restricted to $g|W$ coincide with the identity automorphism of this deformation. This subgroup is isomorphic to $T^0(X_0) \otimes J$ (and belongs to the center), since $T^0(X_0)$ is the set of vector fields on X_0, i.e., of infinitesimal automorphisms of X_0. The following relation describes the change in the tensor D_Z under the variation of the isomorphism i not affecting $i|W$.

VI) For any element $v \in T^0(X_0) \otimes J \subset \mathrm{Aut}(f|z)$ we have

$$D_Z(f', f, i \circ v) - D_Z(f', f, i) = [v, Df].$$

Here the bracket denotes the combination of the Lie operation $T^0(X_0) \times T^1(X_0) \to T^1(X_0)$ and the bilinear mapping $J \otimes m(Y)/m^2(Y) \to I$, generated by multiplication in $\mathcal{O}_Y$. It is clear that the choice of $i|W$ has also an influence on the "divergence" of f' and f; the description of this dependence requires in addition to the Lie operation the higher operations of Massey in the tangent cohomology.

5. Obstructions to the Extension of Deformations

Let $f: X \to D$ be a deformation of a complex space X_0 over the double point. The triple point T is the small extension of D with algebra $\mathcal{O}_T = \mathbb{C}\{\zeta\}/(\zeta^3)$. We undertake to extend the deformation f to a deformation over the triple point.

Let $\mathscr{R} : A \mapsto (\mathscr{R}_A, s_A)$ be a resolution of X on some covering. According to the construction in Chapter 2, Section 3.2, the relative space X/D has a resolution which is generated by the complexes of sheaves $\mathscr{R}_A \otimes D$ on the same covering with differentials $\tilde{s}_A = s_A + \varepsilon u_A$, where $u : A \mapsto u_A$ is the cycle in $T^1(\mathscr{R})$ whose image in $T^1(X_0)$ corresponds to the class of the deformation f. To extend it to T, it is necessary and sufficient to define in the sheaves $\mathscr{R}_A \otimes T$ differentials of the form $\tilde{\tilde{s}}_A = s_A + \zeta u_A + \zeta^2 v_A$, where $v : A \mapsto v_A$ is an element of $T^1(\mathscr{R})$, in such a way that $\tilde{\tilde{s}}_A^2 = 0 \bmod (\zeta^3)$. We have

$$\tilde{\tilde{s}}_A^2 = \zeta^2(u_A \circ u_A + [v_A, s_A]).$$

The coefficient of ζ^2 is cohomologous to the cycle $u \circ u = \frac{1}{2}[u, u] \in T^2(R)$. The class of this cycle does not depend on v and is the obstruction to extending f to a deformation over the triple point. It depends only on the class of the deformation f, i.e., on the element $t = \mathrm{cl}(u) \in T^1(X_0)$ and is equal to $\frac{1}{2}[t, t] \in T^2(X_0)$, this quantity is called the *first obstruction to the extension of* f.

We now describe a more general construction. Let f be a deformation of a complex space X_0 with zero-dimensional base Z, and let Y be a small extension of Z with ideal J. In this case there is defined a tensor $\mathrm{Ob}(f, Y) \in T^2(X_0) \otimes J$, which is functorial relative to the pair (Z, Y) and which vanishes if and only if there exists a deformation g of the complex space X_0 with base Y so that $g|Z \cong f$. It is called *the obstruction to the extension* of f to Y. If $T^2(X_0) = 0$ then all obstructions are equal to zero.

The constructions in this and the preceding paragraph come together in an exact sequence which we now describe (see [71]). We recall that a pair of mappings of sets with distinguished points $F \xrightarrow{f} G \xrightarrow{g} H$ is called exact if $g^{-1}(h) = f(F)$, where h is the distinguished point of H. If one of these is a group, then the distinguished point is assumed to be the identity element. Let f be a deformation of a complex space X_0; its automorphisms are the isomorphisms $f \rightrightarrows f$ (Chapter 1, Section 4). The automorphisms of f form a group, which is denoted by $\mathrm{Aut}(f)$. If Y is the base for f and Z is a subgerm of Y, any automorphism of f generates an automorphism of the restriction $f|Z$ on this subgerm. This defines a group homomorphism $\mathrm{Aut}(f) \to \mathrm{Aut}(f|Z)$. Let Y be a small extension of a zero-dimensional germ Z with ideal J. For any deformation f of the complex space X_0 with base Y and any automorphism a of this deformation there is the exact sequence of mappings

$$0 \to T^0(X_0) \otimes J \xrightarrow{\varepsilon_a} \mathrm{Aut}(f) \xrightarrow{\rho_0} \mathrm{Aut}(f|Z) \xrightarrow{\pi_0} T^1(X_0) \otimes J \xrightarrow{\varepsilon_f} \mathrm{Def}_Y(X_0)$$
$$\xrightarrow{\rho_1} \mathrm{Def}_Z(X_0) \xrightarrow{\pi_1} T^2(X_0) \otimes J,$$

in which the first three mappings are group homomorphisms and the last three are set mappings with the distinguished points $\mathrm{cl}(f)$, $\mathrm{cl}(f|Z)$, and 0 respectively. Only the first mapping depends on a, and the last two do not depend on f. This exact sequence has, in particular, the following properties:

I) $\pi_0(h) = D_Z(f, f, h)$.

II) For any deformation f' of the space X_0 with base Y and any isomorphism $i: f'|Z \to f|Z$ we have the equation $\varepsilon_f \circ D_Z(f', f, i) = \mathrm{cl}(f')$.

III) ρ_1 sends the class of the deformation to the class of its restriction to Z.

IV) $\pi_1\,(\mathrm{cl}(g)) = \mathrm{Ob}(g, Y)$.

The construction described above relates to the so-called formal theory of deformations. The next step in the theory consists of general existence theorems.

6. Versal Deformations

A deformation $f: X \to S$ of a space X_0 is called *versal* if every other deformation $Y \to T$ of the same space is isomorphic to the inverse image of f under some mapping of the bases $\varphi: T \to S$. Thus a versal deformation, if it exists, "contains" all deformations of the given complex space. For example, every trivial deformation $X_0 \times T$ is contained in a versal deformation in the sense that it can be obtained from the versal deformation as the inverse image under the mapping of the entire germ T to the distinguished point $* \in S$. In particular, every fiber of an arbitrary deformation $Y \to T$ which is sufficiently near to the distinguished point occurs among the fibers of a versal deformation.

By definition any deformation of X_0 over the double point must be induced from a versal deformation. Therefore the differential Df of a versal deformation must be an epimorphism. If it acts isomorphically, then the versal deformation is called *minimal*. Any two minimal versal deformations are isomorphic (see property 3 in Chapter 2, Section 4). In the Kodaira-Spencer theory such deformations were called complete effective. In the article [53] a complete effective family was constructed for any compact manifold X_0 under the hypothesis that $H^2(X_0, \mathcal{T}^0) = 0$ (which guarantees the absence of obstructions to extension). This hypothesis was removed by Kuranishi [55]; in this case the base is singular if any of the tensors $\mathrm{Ob}(g, Y) \in H^2(X_0, \mathcal{T}^0) \otimes J$ is different from zero for some infinitesimal deformation g of the complex space X_0 and some small extension Y of its base with ideal J.

It was established in the articles [28], [39], [69] and [2&] that a minimal versal deformation exists for any compact complex space X_0. According to [69], the base for such a deformation can be realized as the pre-image of zero under some holomorphic mapping $\Phi: V \to T^2(X_0)$ defined in a neighborhood of zero $V \subset T^1(X_0)$. In other words, the base S is a model subspace of V defined by the equations $\Phi_1 = \cdots = \Phi_\sigma = 0$, where the Φ_i are the components of the mapping Φ. Here $\Phi(0) = 0$, $d\Phi(0) = 0$, and $d^2\Phi(u) = \frac{1}{2}[u, u]$ (the second differential contains all first obstructions), and the other differentials of Φ are related to higher obstructions and can be calculated in terms of Massey's operations [69], [6&]. The base S of a versal deformation is the germ of a manifold if and only if all the tensors $\mathrm{Ob}(g, Y)$ are equal to zero. A versal deformation of a complex space X_0 is versal for each of its fibers which are sufficiently near to X_0.

Versal deformations for germs of a complex space X_0 were constructed in [96], [23], [38] and [4&]. The hypothesis dim $T^1(X_0) < \infty$ is necessary and sufficient for the base of a minimal versal deformation to be the germ of a finite-dimensional complex space. It is satisfied if all points aside from the distinguished point are nonsingular. The base of such a deformation can be described similarly as in the case of a compact complex space.

A Sufficient Condition for Versality. Let X_0 be a compact complex space or a germ of a complex space for which dim $T^1(X_0) < \infty$. Let $g: Y \to Z$ be a deformation of it with nonsingular (i.e., Z is the germ of a manifold). If the differential $Dg: T(Z) \to T^1(X_0)$ is an epimorphic mapping or an isomorphism, then the deformation g is versal or, respectively, minimal versal. Indeed, let $f: X \to S$ be a minimal versal deformation, which exists by the theorem discussed above. We consider a morphism $\varphi: Z \to S$ such that $\varphi^*(f) \cong g$. By Chapter 2, Section 4 we have $Dg = Df \circ d\varphi$. Since Df is an isomorphism, and Dg is by hypothesis an epimorphism, then $d\varphi$ is also an epimorphism. It is easy to conclude from this that the germ S is also nonsingular and that there exists a morphism $\psi: S \to Z$ such that $\varphi \circ \psi = 1$. Consequently $f \cong \psi^*(g)$, which implies that the deformation g is versal.

7. Modular Deformations

7.1. Let $f: X \to S$ be a deformation of a complex space X_0. It is called *universal* if for every deformation $g: Y \to T$ of the same space there exists exactly one morphism of germs $\varphi: T \to S$ for which the induced deformation $\varphi^*(f)$ is isomorphic to g. In other words f is universal if it is versal and for any germ of a complex space T different morphisms $\varphi, \Psi: T \to S$ induce from f nonisomorphic deformations with base T. It is clear that universality implies that the versal deformation f is minimal. However, far from every minimal versal deformation is universal. For example, if X_0 is the germ of a complete intersection with a single singular point then a minimal versal deformation of X_0 cannot be universal.

In those cases when a minimal versal deformation exists, we can single out a subgerm in its base S for which the universality condition holds. Thus a germ $M \subset S$ is called *modular* if for any germ Z and morphisms $\varphi: Z \to M$ and $\Psi: Z \to S$ the deformation $\varphi^*(f)$ and $\Psi^*(f)$ are isomorphic if and only if $\varphi = \Psi$. If $M = S$ then the deformation is universal. It is shown in [71] that for any compact complex space X_0 there exists in the base of a minimal versal deformation a maximal modular germ M possessing the following properties:

1) the support of M consists of the set of points $s \in S$ for which f is a minimal versal deformation of X_s (at the remaining points f is only a versal deformation of the fiber);

2) the support of M coincides with the set of points $s \in S$ at which dim $T^0(X_s) =$ dim $T^0(X_0)$;

3) the tangent space $T(M)$ coincides with the space of vectors $t \in T(S)$ such that $[Df(t), v] = 0$ for all $v \in T^0(X)$.

The restriction of a deformation f to a modular germ M is called *modular deformation.* We denote it by $f_M : X_M \to M$. It follows from 1) that for every point $s \in M$ there is a neighborhood $V \subset M$ so that the fibers X_σ for $\sigma \in V$ of a modular deformation are not isomorphic to X_s. However, some fibers of a modular deformations can be isomorphic. These fibers can be included in one-dimensional families; i.e., there exist (distinct) morphisms of germs $\varphi, \Psi : (\mathbb{C}, 0) \to M$ so that the fibers of the deformations $\varphi^*(f_M)$ and $\Psi^*(f_M)$ are pairwise isomorphic, and these isomorphisms form a holomorphic family over $(\mathbb{C}, 0)$. Here the isomorphism of the distinguished fibers $a : X_0 \rightrightarrows X_0$, of course, is not the identity; otherwise it would contradict the modularity of the deformation. Moreover, a does not belong to the connected component of the identity $\mathrm{Aut}_e(X_0)$ in the complex Lie group $\mathrm{Aut}(X_0)$ of holomorphic automorphisms of X_0. Conversely, each automorphism $a \in \mathrm{Aut}(X_0)$ corresponds to exactly one morphism $a_M : M \to M$ such that the deformation $a_M^*(f_M)$ is isomorphic to (f, a). It is clear that a_M is a biholomorphic mapping from the germ M to itself. Thus the action of the group $\mathrm{Aut}(X_0)$ lifts to an action in the bundle f_M. It turns out that the subgroup $\mathrm{Aut}_e(X_0)$ acts trivially on the base M. Consequently, the discrete group $\Gamma = \mathrm{Aut}(X_0)/\mathrm{Aut}_e(X_0)$ acts on the base. It is not difficult to show that for any points $s, t \in M$ which are sufficiently near the distinguished point an isomorphism $X_s \cong X_t$ exists if and only if these points lie in the same orbit of the action of Γ on M. It follows that M/Γ can serve as a germ of the moduli space for compact spaces.

The definition and properties 1) and 3) of modular germs carry over without significant changes to the case when X_0 is a germ of a complex space. If X_0 is a complete intersection, then 2) is also valid if we replace the infinite-dimensional Lie algebra $T^0(X_0)$ by a finite-dimensional quotient algebra $\tilde{T}^0(X_0)$ relative to the ideal which is the image of $T^0(X/S)$. Here $X \to S$ is a minimal versal deformation of X_0 (it is assumed that $\dim T^1(X_0) < \infty$) and $T^0(X/S)$ is the space of vertical vector fields (Chapter 4, Section 3), i.e., derivations of $\mathcal{O}_X$ which equal zero on the image of the structure homomorphism $\mathcal{O}_S \to \mathcal{O}_X$. We note that the bracket of elements of this ideal with elements of $T^1(X_0)$ is always equal to zero. Modular deformations of germs of hypersurfaces are studied in [56].

7.2. Examples. 1. The family of complex tori of dimension n is complete effective in the sense of [51], i.e., it is a minimal versal deformation of each of its fibers. Therefore it is a modular deformation.

2. Compact Riemann surfaces X_g of genus $g > 1$ form a holomorphic family with base M_g of dimension $3g - 3$ (see Chapter 3, Section 2). It is a modular deformation since $T^0(X_g) = 0$.

3. Deformations of Hopf manifolds are described in Chapter 3, Section 5. It follows from this description that modular deformation of a linear Hopf manifold V_A is formed also by linear Hopf manifolds V_B for which the dimension of the Lie

algebra $\mathfrak{g}$ is the same as that for V_A. This means that for the corresponding matrices B all the same resonance relations $\lambda_k = \lambda_1^{d_1} \dots \lambda_m^{d_m}$ are satisfied as hold for A and the corresponding Jordan blocks have the same dimensions. Here $\sum d_i \geqslant 1$, i.e., it is necessary to consider both linear and nonlinear resonances.

4. Any family of parabolic germs of hypersurfaces (Chapter 5, Section 2) is modular.

7.3. We consider the relation between properties 2) and 3) for more general deformations. Let $f: X \to Y = (\mathbb{C}, 0)$ be any nontrivial deformation of a compact complex space X_0 with one-dimensional base. According to Chapter 2, Section 4, there is a maximal integer k so that the restriction $f|Y_{k-1}$ is a trivial deformation. We denote by Y_k the infinitesimal neighborhood of the point $y = 0$ defined by the ideal $m^{k+1}(Y)$, where $m(Y)$ is the maximal ideal in $\mathcal{O}_Y$. Thus there is defined a nonzero element $D_{Y_{k-1}}(f|Y_k, e, i)$, where e denotes the trivial deformation of X_0 over Y_k. This element belongs to $T^1(X_0) \otimes m^k(Y)/m^{k+1}(Y)$; consequently it can be written in the form $\delta_k \otimes \mathrm{cl}(y^k)$, where y is the generator of $m(Y)$. The class $\delta_k \in T^1(X_0)$ acting on $T^*(X_0)$ via bracket defines a linear operator of degree 1. The image of the action of this operator on the nth term of the grading we denote by $[\delta_k, T^n(X_0)]$. We can estimate the jump in dimension of the tangent cohomology when we pass from the fiber X_0 to any nearby fiber X_y for $y \neq 0$ by the dimension of these images: Namely, for any $n \geqslant 0$ we have the inequality ([71]):

$$\dim T^n(X_0) - \dim T^n(X_y) \geqslant \dim[\delta_k, T^{n-1}(X_0)] + \dim[\delta_k, T^n(X_0)].$$

Chapter 3. Some Examples and Special Questions

1. Versal Deformations of Germs

1.1. Let X_0 be the germ of a complete intersection with a unique singular point. We specify it as a model subspace of the ball $U \subset \mathbb{C}^n$ defined by equations $f_1 = \dots = f_m = 0$ in such a way that the distinguished point of X_0 coincides with the origin. The analytic algebra dual to this germ is $A = H/I$, where H is the algebra of convergent power series in the coordinates $z_1, \dots, z_n$ of $\mathbb{C}^n$ and I is the ideal generated by the functions $f_1, \dots, f_m$. According to Chapter 2, Section 1 we have

$$T^1(A) \cong H^m/(df \cdot H^n + I \cdot H^n). \tag{3.1.1}$$

This vector space is finite-dimensional, since the singular point $z = 0$ is unique in X_0. Let τ denote its dimension. We select elements $e_1, \dots, e_\tau \in H^m$ whose images $\bar{e}_1, \dots, \bar{e}_\tau$ in $T^1(A)$ constitute a basis for this space. We have: $e_i = (e_{i1}, \dots, e_{im})$, where the e_{ij} are holomorphic functions in some neighborhood of zero U'; it is always possible to choose the e_i so that the e_{ij} are monomials. We consider the functions

$$F_j(z, s) = f_j(z) + \sum_{i=1}^{\tau} s_i e_{ij}(z), \qquad j = 1, \ldots, m,$$

restricted to $U' \times \mathbb{C}^\tau$. They define a model space $X \subset U' \times \mathbb{C}^\tau$ whose projection onto $\mathbb{C}^\tau$ generates a mapping of germs $p: (X, 0) \to (\mathbb{C}^\tau, 0)$. It is clear that $p^{-1}(0) \cong X_0$. We verify that the mapping p is flat. By Chapter 1, Section 4 it suffices to show that any relation among the functions $f_1, \ldots, f_m$ extends to a relation among $F_1, \ldots, F_m$. The collection of all such relations is the kernel of the morphism $H^n \to H$ defined by the formula $(a_1, \ldots, a_m) \mapsto \sum a_i f_i$. Since the functions $f_1, \ldots, f_m$ define a complete intersection, the kernel of this morphism is generated as an H-module by elements of the form

$$(0, \ldots, 0, \overset{i}{f_j}, 0, \ldots, 0, \overset{j}{-f_i}, 0, \ldots, 0), \qquad i < j,$$

(see Chapter 2, Section 1). Each such element extends to a vector $(0, \ldots, 0, F_j, 0, \ldots, 0, -F_i, 0, \ldots, 0)$ defining a relation among $F_1, \ldots, F_m$; thus any elements of the kernel extends to a relation among these functions. Thus p is a deformation of the germ X_0.

We verify that this deformation is minimal versal. According to Chapter 2, Section 6, it suffices to show that the differential of this deformation

$$Dp: T_0(\mathbb{C}^\tau) \to T^1(A)$$

is an isomorphism. For any vector $v \in T_0(\mathbb{C}^\tau)$ its image $Dp(v)$ is equal to the image of the vector $(v(F_1), \ldots, v(F_m))$ in $T^1(A)$. The vectors $v = \frac{\partial}{\partial s_i}$ for $i = 1, \ldots, \tau$ form a basis for $T_0(\mathbb{C}^\tau)$, and $Dp\left(\frac{\partial}{\partial s_i}\right) = \bar{e}_i$, i.e., the basis maps to a basis. Consequently Dp is an isomorphism and p is a minimal versal deformation.

1.2. We consider the germ of the incomplete intersection which is defined in $\mathbb{C}^2$ by the two functions $f = zw$, $g = w^2$. This germ X_0 is represented by the axis $w = 0$ with the distinguished singular point $(0, 0)$. It corresponds to the analytic algebra $A = H/I$, where H is the algebra of convergent power series in z and w and $I = (f, g)$. To compute $T^1(A)$ we employ the exact sequence (2.1.5). All relations between f and g are generated by the relation $wf = zg$. Therefore $\operatorname{Hom}(I, A)$ is isomorphic to the submodule of A^2 generated by pairs (a, b) such that $wa = zb$. The image of $\operatorname{Der}(H)$ in this submodule is generated by the pairs $(w, 0)$ and $(z, 2w)$. Therefore the space $T^1(A)$ is one-dimensional and is generated by the vector $(0,w)$. The functions

$$F(z, w, s) = zw, \qquad G(z, w, s) = w^2 + sw$$

define a model complex space $X \subset \mathbb{C}^2 \times S$, where $S = \mathbb{C}$. Let p be the projection from X onto S. This mapping defines a deformation of the germ X_0. We have $p^{-1}(0) \cong X_0$, and the relation between f and g indicated above extends to the relation $(w + s)F = zG$. This deformation is minimal versal since Dp is an isomorphism.

The fiber of the deformation $p: X \to S$ over a point $s \neq 0$ is a complex space which consists of the one-dimensional manifold $w = 0$ and the isolated point $(0, -s)$ (endowed with the algebra $\mathbb{C}$). Thus the versal deformation of the germ X_0 causes the singular point $(0, 0)$ to "split off and jump away from" the line $w = 0$.

1.3. Let X_0 be the origin in $\mathbb{C}^2$ endowed with the algebra $A = H/m^2(H)$. We have $\dim A = 3$, and the ideal $m^2(H)$ is generated by the monomials z^2, zw, w^2. Calculations similar to those in Section 1.2 show that X_0 has a minimal versal deformation with base $S = (\mathbb{C}^4, 0)$ which is defined by the polynomials

$$F_1(z, w, s) = z^2 + az + bw + bd, \qquad F_2(z, w, s) = zw - bc,$$

$$F_3(z, w, s) = w^2 + cz + dw + ac.$$

The fiber of this deformation over a point $s = (a, b, c, d)$ is represented by three simple points if s is not a solution of the equation

$$4(3ac - d^2)(3bd - a^2) - (ad - 9bc)^2 = 0,$$

which defines the discriminant set (Chapter 5, Section 1). If $s \neq 0$ belongs to this set, then the fiber of the deformation is the union of a simple point and a double point (Chapter 1, Section 1) or a triple point.

2. Deformations of Compact Riemann Surfaces

2.1. A *Riemann surface* is a connected reduced one-dimensional complex space. We assume that X is a nonsingular Riemann surface. The underlying space of X is a sphere with g handles; the number g is called the genus of this Riemann surface. A Riemann surface of genus zero is $\mathbb{CP}_1$. It has no nontrivial deformations. A Riemann surface of genus 1 as a topological space is a two-dimensional torus. All the structures of a complex manifold on the torus are parametrized by the complex plane. In order to introduce this parameter, we realize these manifolds as surfaces in $\mathbb{CP}_2$ defined by the equations

$$X_s: \xi_0\xi_2^2 - \xi_1^3 - s_1\xi_1\xi_0^2 - s_0\xi_0^3 = 0. \tag{3.2.1}$$

In nonhomogeneous coordinates these equations turn out to be simpler:

$$w^2 = z^3 + s_1 z + s_0, \qquad z = \frac{\xi_1}{\xi_0}, \qquad w = \frac{\xi_2}{\xi_0}.$$

Under the scaling transformation $w = t^3 w'$, $z = t^2 z'$ the surface X_s maps isomorphically to another surface of the same family, but the quantity s_1^3/s_0^2 remains invariant. The values of this ratio are different for nonisomorphic surfaces; consequently, this function defines an imbedding of the set of isomorphism classes of surfaces X_s into the extended complex plane $\mathbb{CP}_1$. The value $s_1^3/s_0^2 = -27/4$ corresponds to singular surfaces X_s and the other values to nonsingular ones. Thus the set of isomorphism classes of nonsingular Riemann surfaces of genus 1 is isomorphic to the finite plane $\mathbb{C} = \mathbb{CP}_1 \setminus \{-27/4\}$. Moreover, there exists a family $X_1 \to \mathbb{C}$ of such surfaces which forms a universal deformation for each fiber in which representatives of all such classes occur.

In the case of $g > 1$ there is also a family $X_g \to M_g$ of Riemann surfaces of genus g having similar properties. Here $\mathcal{M}_g = T_g/\Gamma_g$, where T_g is the so-called Teichmüller space, which is a complex manifold of dimension $3g - 3$ which is homeomorphic to a ball, and Γ_g is a certain group which acts discretely on T_g (see also Chapter 5, Section 6 and [1], [10], [11]).

2.2. Now let X be any compact one-dimensional reduced complex space. It can possess only a finite number of singular points $x_1, \ldots, x_p$. We assume that at each of these points the germ of the complex space X is a complete intersection. In this situation the sheaf $\mathcal{T}^2(X)$ is everywhere equal to zero, and the sheaf $\mathcal{T}^1(X)$ is concentrated on these points and can be calculated by the formula (3.1.1). It follows that

$$H^0(X, \mathcal{T}^2) = H^1(X, \mathcal{T}^1) = H^2(X, \mathcal{T}^0) = 0,$$

since the space X is one-dimensional. Thus in the spectral sequence (2.3.1) all terms of second degree are equal to zero. It follows that $T^2(X) = 0$, and according to Chapter 2, Section 6, X has a minimal versal deformation f with base $S = (T^1(X), 0)$. The exactness of (2.3.2) implies the exactness of the sequence

$$0 \to H^1(X, \mathcal{T}^0) \to T^1(X) \to \Gamma(X, \mathcal{T}^1) \to 0, \tag{3.2.2}$$

where

$$\Gamma(X, \mathcal{T}^1) = \bigoplus_{i=1}^{p} \mathcal{T}^1(X, x_i),$$

consequently all the singular points x_i give independent contributions to the base of the versal deformation. More precisely, in S we can distinguish the germ of a submanifold S_1 so that the restriction of the deformation to S_1 defines a minimal versal deformation of the disjoint union of germs $(X, x_1), \ldots, (X, x_p)$. There is also in S the germ of a submanifold S_0, complementary to S_1, so that the restriction $f|S_0$ is a locally trivial deformation, i.e., a deformation in which all the germs (X, x_i) are undeformed (all remaining germs of X are nonsingular and thus cannot be deformed). The germ S_0 is tangent to the image of $H^1(X, \mathcal{T}^0)$ in $T^1(X)$. It belongs to the discriminant set D, i.e., the set of points $s \in S$ such that the fiber X_s has at least one singular point. This set is a hypersurface, i.e., it is defined by a single nontrivial holomorphic equation in S.

If $s \in S \backslash D$, then the fiber X_s is a nonsingular compact Riemann surface. Its genus g is defined by the equation $3g-3=\dim H^1(X_s, \mathcal{T}^0)-\dim H^0(X_s, \mathcal{T}^0)$, which follows from the Riemann-Roch Theorem. Since $T^2(X_s)=0$, we can employ the theorem in Chapter 2, Section 3, which asserts that the alternating sum of dimensions of the tangent cohomology of a deformation is constant. In view of (3.2.2) this gives the equation

$$3g-3=\dim H^1(X, \mathcal{T}^0)+\dim H^0(X, \mathcal{T}^1)-\dim H^0(X, \mathcal{T}^0),$$

where g is the genus of any Riemann surface which arises under the deformation of the singular surface X.

As an example, we consider the family of surfaces (3.2.1). For $s=(0,0)$ this surface as a topological space is a sphere with the single singular point $(1,0,0)$. For $s \neq (0,0)$ and $s_1^3/s_0^2=-27/4$ this is a sphere with one singular point of transversal self-intersection. The other surfaces in this family, as we have noted, are nonsingular tori. This family is a minimal versal (but not universal) deformation of the fiber $X_{(0,0)}$; for the other fibers this deformation is versal but not minimal.

3. Rigid Germs and Spaces

3.1. A complex space or germ X is called (infinitesimally) *rigid* if every (infinitesimal) deformation of X is trivial, i.e., is isomorphic to a deformation of the form $X \times S \to S$. A criterion for infinitesimal rigidity is the equation $T^1(X)=0$, which follows from properties of the distinguishing element (Chapter 2, Section 4). If X is a compact complex space or germ, then the condition $T^1(X)=0$ is sufficient (and clearly necessary) for its rigidity. Indeed, it follows from the description (2.6.1) of the base of a minimal versal deformation of X that under this condition the base is a simple point. The inverse image of such a deformation is always trivial.

We consider some examples.

I) The nonsingular germ $X=(\mathbb{C}^n, 0)$ is rigid. In general the germ of a complete intersection is rigid if and only if it is nonsingular. This follows easily from formula (3.1.1) and the criterion for rigidity.

II) The next example is taken from the article [40]. The germ $X \subset \mathbb{C}^{2n}$ for $n \geqslant 3$ is defined by the condition rank $M \leqslant 1$, where

$$M=\begin{pmatrix} z_1 \ldots z_n \\ w_1 \ldots w_n \end{pmatrix}$$

is the matrix formed from the coordinate functions. A calculation using (2.1.5) shows that $T^1(X)=0$. Thus the germ is rigid although it has a singular point.

III) $X=\mathbb{CP}_n$. In this case $T^1(X)=H^1(\mathbb{CP}_n, \mathcal{T}^0)=0$; consequently, projective space is rigid.

3.2. Let $f: X \to S$ be any proper flat mapping of a complex space. If one of its fibers X_{s_0} is rigid, then by definition the restriction of f to some neighborhood of the point $s_0 \in S$ is trivial. One can show that there exists a proper closed complex subspace $T \subset S$ such that all the fibers X_s for $s \in S \setminus T$ are isomorphic to X_{s_0}. However, some fibers of f can be nonisomorphic to X_{s_0}; consequently, not every deformation in the sense of Grothendieck of a rigid space has fibers isomorphic to each other. We consider the following example.

IV) Let $p: V \to \mathbb{CP}_1$ be a vector bundle corresponding to the sheaf $\mathcal{O}(1)$. The space V is obtained from two copies of the space $\mathbb{C}^2$ with coordinates (z, λ) and (z', λ') glued via the mapping

$$z' = \frac{1}{z}, \qquad \lambda' = \frac{\lambda}{z}, \qquad z \neq 0.$$

In the structure sheaf $\mathcal{O}_V$ we consider the sheaf of ideals $\mathcal{I}_0$ generated by the function λ^2, resp. λ'^2. Let X_0 be the closed complex subspace of V with underlying space $\mathbb{CP}_1$ and structure sheaf $\mathcal{O}_{X_0} = \mathcal{O}_V/\mathcal{I}_0$. We consider the deformation of it $\pi: X \to \mathbb{C}$, where X is the subspace of $V \times \mathbb{C}$ with sheaf $\mathcal{O}_X = \mathcal{O}_{V \times \mathbb{C}}/I$ and $\mathcal{I}$ is the sheaf of ideals in $\mathcal{O}_{V \times \mathbb{C}}$ generated by the function $\lambda^2 + sz$, resp. $\lambda'^2 + sz'$, where s is the projection of $V \times \mathbb{C}$ onto $\mathbb{C}$. This is well defined, since for $z \neq 0$ we have

$$\lambda'^2 + sz' = \frac{\lambda^2}{z^2} + \frac{s}{z} = \frac{\lambda^2 + sz}{z^2},$$

i.e., the generators differ by a nonzero factor. Since $\mathcal{I}$ locally has only one generator, the mapping π is flat. It is clearly proper, and the fiber $\pi^{-1}(0)$ is isomorphic to X_0. Thus π is a deformation of X_0 with base $\mathbb{C}$. For $s \neq 0$ the fiber X_s is a manifold, since $\frac{\partial}{\partial z}(\lambda^2 + sz) \neq 0$. Consequently, $X_s \hookrightarrow V \xrightarrow{p} \mathbb{CP}_1$ is a mapping of compact Riemann surfaces. Its degree is equal to 2, and there are two critical points $z = 0$ and $z' = 0$ of order 1. Therefore by the Riemann–Hurwitz Theorem the genus of X_s is equal to zero, i.e., $X_s \cong \mathbb{CP}_1$. Thus all fibers of the deformation π are isomorphic to $\mathbb{CP}_1$ with the exception of $X_0 \ncong \mathbb{CP}_1$.

4. Deformations of Projective Complete Intersections

A projective complete intersection X is a subspace of $\mathbb{CP}_n$ defined by equations

$$p_1(z) = \cdots = p_k(z) = 0,$$

where $p_1, \ldots, p_k$ are any homogeneous polynomials in the homogeneous coordinates $z_0, \ldots, z_n$ under the requirement that $\dim X = n - k$. It is endowed with the sheaf $\mathcal{O}_X = \mathcal{O}/(p_1, \ldots, p_k)$, where $\mathcal{O} = \mathcal{O}_{\mathbb{CP}_n}$. In the case of $k = 1$ such a subspace is called a projective hypersurface. We describe the versal deformation of any projective complete intersection, following [71] which considered hypersurfaces. Another method was used in [105]. Deformations of nonsingular projective complete intersections were studied by Kodaira and Spencer [51] (hypersurfaces) and Sernesi. More general results are due to Borcea [13].

Let $m_1, \ldots, m_k$ denote the degrees of the polynomials $p_1, \ldots, p_k$. We consider the space $\mathbb{P}_m$ of all homogeneous polynomials in $z_0, \ldots, z_n$ of degree m and set $\mathbb{P} = \mathbb{P}_{m_1} \times \cdots \times \mathbb{P}_{m_k}$. We define the mapping of manifolds $\pi : \mathscr{X} \to \mathbb{P}$, where

$$\mathscr{X} = \{(z, q_1, \ldots, q_k), z \in \mathbb{CP}_n, q_i \in \mathbb{P}_{m_i}, q_1(z) = \cdots = q_k(z) = 0\},$$

so that X is a fiber of the mapping. In the base $\mathbb{P}$ there is a proper closed subspace N so that all the fibers of π over $\mathbb{P}' = \mathbb{P} \backslash N$ are complete intersections. If $\mathscr{X}' = \mathscr{X} \backslash \pi^{-1}(N)$, then the mapping $\pi : \mathscr{X}' \to \mathbb{P}'$ is a deformation of X. We investigate in which cases this deformation is versal. Without loss of generality we may assume that $m_i > 1$ for $i = 1, \ldots, k$.

Theorem 1. *This deformation is versal for any complete intersection X of dimension greater than* 1 *with the exception of the following three cases*:

I) $n = 3, k = 1, m_1 = 4$ (*one equation of fourth degree*);

II) $n = 4, k = 2, m_1 = 3, m_2 = 2$ (*the intersection of a cubic and a quadric*),

III) $n = 5, k = 3, m_1 = m_2 = m_3 = 2$ (*the intersection of three quadrics*).

In these cases all nonsingular fibers of the mapping π are surfaces of type $K3$.

We recall that a surface of type $K3$ is a connected simply connected two-dimensional complex manifold with trivial canonical bundle. From [74] it is known that under a small deformation of any projective surface of type $K3$ nonalgebraic surfaces arise which do not have even a single divisor. According to Theorem 1 in all other situations small deformations of projective complete intersections are again projective manifolds. For manifolds of dimension 1 this follows from the theory of Riemann surfaces. Below we shall consider particular versal deformations for the exceptional types of surfaces.

We move now to the proof of the theorem. We recall that $\mathcal{O}(m)$ is an invertible sheaf on $\mathbb{CP}_n$, whose sections on an open subset U can be identified with holomorphic functions $f(z_0, \ldots, z_n)$ defined in the domain $\mathbb{C}^{n+1}$, where $(z_0, \ldots, z_n) \in U$, having degree of homogeneity m. In particular, $\mathcal{O}(0) = \mathcal{O}$. For any $\mathcal{O}$-sheaf $\mathscr{F}$ on $\mathbb{CP}_n$ we set

$$\mathscr{F}(m) = \mathscr{F} \otimes_{\mathcal{O}} \mathcal{O}(m), \qquad \mathscr{F}_X = \mathscr{F} \otimes_{\mathcal{O}} \mathcal{O}_X, \qquad H^i(\mathscr{F}) = H^i(\mathbb{CP}_n, \mathscr{F}).$$

We denote by Θ the tangent sheaf $\mathscr{T}^0(\mathbb{CP}_n)$ and define the mapping of $\mathcal{O}$-sheaves

$$dp : \Theta_X \to \bigoplus_1^k \mathcal{O}_X(m_i),$$

in which the germ of a field v is mapped to the vector $(v(p_1), \ldots, v(p_k))$. It generates the mapping of cohomology

$$h^i : H^i(\Theta_X) \to H^i\left(\bigoplus_1^k \mathcal{O}_X(m_j)\right), \qquad i = 0, 1, \ldots$$

Lemma 1. *For any projective complete intersection X the tangent cohomology is described by the formulas*

$$T^i(X) \cong \operatorname{Ker} h^i \oplus \operatorname{Coker} h^{i-1} \quad \text{for } i = 0, 1, \ldots. \tag{3.4.1}$$

Proof. We denote by $\mathscr{T}^i$ for $i \geqslant 0$ the tangent sheaves of the space X. According to (2.1.5) there is an exact sequence

$$0 \to \mathscr{T}^0 \to \Theta_X \to \operatorname{Hom}(\mathscr{I}, \mathcal{O}_X) \to \mathscr{T}^1 \to 0, \tag{3.4.2}$$

where $\mathscr{I} = (p_1, \ldots, p_k)$ is the sheaf of ideals defining X. Since X is a complete intersection, $\mathscr{I}$ has the resolution

$$\cdots \to \bigoplus_{i<j} \mathcal{O}(-m_i - m_j) \to \bigoplus_i \mathcal{O}(-m_i) \to \mathscr{I} \to 0, \tag{3.4.3}$$

which is the Koszul complex constructed on the mappings $p_i : \mathcal{O}(-m_i) \to \mathcal{O}$ for $i = 1, \ldots, k$. Applying to this resolution the functor $\operatorname{Hom}(\cdot, \mathcal{O}_X)$, we obtain that $\operatorname{Hom}(\mathscr{I}, \mathcal{O}_X) \cong \bigoplus \mathcal{O}_X(m_i)$. Putting this into (3.4.2) and passing to cohomology, we obtain a spectral sequence in which the term $E_2^{*,q}$ is the cohomology of the complex

$$0 \to H^q(\mathscr{T}^0) \to H^q(\Theta_X) \to \bigoplus H^q(\mathcal{O}_X(m_i)) \to H^q(\mathscr{T}^1) \to 0,$$

and the limit is equal to zero. In it only the differentials d_2 and d_3 are different from zero, and the last is the correspondence between $H^{q+2}(\mathscr{T}^0)$ and $H^q(\mathscr{T}^1)$. It is not difficult to show that the inverse correspondence is a single-valued linear mapping $e^q : H^q(\mathscr{T}^1) \to H^{q+2}(\mathscr{T}^0)$. It follows that there is an isomorphism $H^0(\mathscr{T}^0) \cong \operatorname{Ker} h^0$ and an exact sequence

$$\begin{aligned} 0 &\to H^1(\mathscr{T}^0) \to \operatorname{Coker} h^0 \oplus \operatorname{Ker} h^1 \to H^0(\mathscr{T}^1) \xrightarrow{e_0} H^2(\mathscr{T}^0) \\ &\to \operatorname{Coker} h^1 \oplus \operatorname{Ker} h^2 \to H^1(\mathscr{T}^1) \xrightarrow{e_1} H^3(\mathscr{I}^0) \to \cdots. \end{aligned} \tag{3.4.4}$$

On the other hand, since X is a complete intersection at each point, we have $\mathscr{T}^i = 0$ for $i \geqslant 2$. Therefore the spectral sequence (2.3.1) gives the isomorphism $T^0(X) \cong H^0(\mathscr{T}^0)$ and the exact sequence

$$0 \to H^1(\mathscr{T}^0) \to T^1(X) \to H^0(\mathscr{T}^1) \xrightarrow{d_2} H^2(\mathscr{T}^0) \to T^2(X) \to H^1(\mathscr{T}^1) \xrightarrow{d_2} H^3(\mathscr{T}^0) \to \cdots$$

A routine verification shows that the mapping d_2 coincides with e_q. Therefore a comparison with (3.4.4) leads to (3.4.1).

Lemma 2. *For any m the natural mapping $H^0(\mathcal{O}(m)) \to H^0(\mathcal{O}_X(m))$ is an epimorphism and $H^1(\mathcal{O}_X(m)) = 0$.*

Proof. Replacing the sheaf $\mathscr{I}$ by the sheaf $\mathcal{O}$ in (3.4.3), we obtain a resolution $\mathscr{R} = \bigoplus_{-\infty}^{0} \mathscr{R}_p$ for the sheaf $\mathcal{O}_X$. The complex $\mathscr{R}(m) = \bigoplus_p \mathscr{R}_p(m)$ is a resolution for $\mathcal{O}_X(m)$. The terms of this complex are direct sums of sheaves of the form $\mathcal{O}(\ell)$; consequently their cohomology vanishes in dimension q for $0 < q < n$. With this complex there is associated a spectral sequence which converges to $H^*(\mathcal{O}_X(m))$, whose first term is equal to $E_1^{pq} = H^q(\mathscr{R}_p(m))$. We consider the groups E_1^{pq} such that $p + q \leqslant 1$. Since $p \geqslant -k \geqslant -n + 2$, we have $q < n$. Therefore these terms differ from zero only when $q = 0$. Consequently, the limit of this spectral sequence in dimension zero is isomorphic to the cokernel of the mapping $H^0(\mathscr{R}_{-1}(m)) \to H^0(\mathscr{O}(m))$ and is equal to zero in dimension 1. From this both assertions of the lemma follow.

The tangent space to $\mathbb{P}$ can be identified with $\bigoplus H^0(\mathscr{O}(m_i))$. By what has been said, there are the mappings

$$\bigoplus_{i=1}^{k} H^0(\mathscr{O}(m_i)) \to \operatorname{Coker} h^0 \to T^1(X). \tag{3.4.5}$$

By Lemma 2 the first mapping is an epimorphism. The cokernel of the second is $\operatorname{Ker} h^1$ by Lemma 1. It is not hard to see that their composition coincides with the differential $D\pi$ of the deformation π at the point p which corresponds to the space X. Thus our problem reduces to describing all cases in which $\operatorname{Ker} h^1 = 0$.

Lemma 3. *For any $n > 2$ and $m \geqslant 0$ we have*

$$\dim H^0(\Theta(-m)) = \begin{cases} n^2 + 2n, & m = 0, \\ n + 1, & m = 1, \\ 0, & m > 1, \end{cases}$$

$$H^i(\Theta(-m)) = 0, \qquad 0 < i < n - 1,$$

$$\dim H^{n-1}(\Theta(-m)) = \begin{cases} 0, & m \neq n + 1, \\ 1, & m = n + 1. \end{cases}$$

Proof. There is the exact sequence

$$0 \to \mathscr{O}(-m) \xrightarrow{\varepsilon} [\mathscr{O}(1 - m)]^{n+1} \xrightarrow{\sigma} \Theta(-m) \to 0, \tag{3.4.6}$$

where the mappings are defined by the formulas

$$\varepsilon(a) = (z_0 a, \dots, z_n a), \qquad \sigma(a_0, \dots, a_n) = \sum a_i \frac{\partial}{\partial z_i}.$$

Since the cohomology of the sheaves $\mathscr{O}(l)$ vanishes in dimension q for $0 < q < n$, we obtain from (3.4.6) the exact sequences

$$0 \to H^0(\mathscr{O}(-m)) \xrightarrow{\varepsilon} [H^0(\mathscr{O}(1 - m))]^{n+1} \xrightarrow{\sigma} H^0(\Theta(-m)) \to 0, \tag{3.4.7}$$

$$0 \to H^{n-1}(\Theta(-m)) \to H^n(\mathscr{O}(-m)) \xrightarrow{\varepsilon} [H^n(\mathscr{O}(1 - m))]^{n+1} \xrightarrow{\sigma} H^n(\Theta(-m)) \to 0, \tag{3.4.8}$$

and also deduce that the cohomology of the sheaf $\Theta(-m)$ vanishes in dimension p for $0 < p < n - 1$. The space $H^0(\mathscr{O}(l))$ is equal to zero for $l < 0$, and for $l \geqslant 0$ it is isomorphic to the space of homogeneous polynomials of degree l in $n + 1$ variables. Therefore a formula for the dimension of $H^0(\Theta(-m))$ follows from (3.4.7). Further, we can employ Serre duality between $H^n(\mathscr{O}(l))$ and $H^0(\mathscr{O}(-n - l - 1))$.

It is functorial; consequently, (3.4.8) is dual to the exact sequence

$$0 \leftarrow \operatorname{Coker} \varepsilon' \leftarrow H^0(\mathcal{O}(m-n-1)) \overset{\varepsilon'}{\leftarrow} [H^0(\mathcal{O}(m-n-2))]^{n+1} \leftarrow \operatorname{Ker} \varepsilon' \leftarrow 0,$$

in which ε' is the operator dual to the mapping ε in (3.4.8), i.e., it acts via the formula $\varepsilon'(a_0, \dots, a_n) = \sum z_i a_i$. Therefore the space $H^{n-1}(\Theta(-m))$ is dual to $\operatorname{Coker} \varepsilon'$. It is easy to show that the latter space is different from zero only in the case when $H^0(\mathcal{O}(m-n-1))$ is the space of polynomials of zero degree, i.e., when $m = n + 1$. This yields the required formula for the dimension of $H^{n-1}(\Theta(-m))$.

Lemma 4. $H^1(\Theta_X) = 0$ *for all* X *except for those cases mentioned in the statement of the theorem. In these cases* $\dim H^1(\Theta_X) = 1$ *and* $H^2(\Theta_X) = 0$.

Proof. We again consider the resolution $\mathcal{R}$ of the sheaf $\mathcal{O}_X$. Since the sheaf Θ is locally free and therefore flat, $\mathcal{R} \otimes \Theta$ is a resolution for Θ_X. Therefore there is a spectral sequence with first term $E_1^{pq} = H^q(\mathcal{R}_q \otimes \Theta)$ which converges to $H^*(\Theta_X)$. Since for any $p > 0$ $\mathcal{R}_p \otimes \Theta$ is a direct sum of sheaves of the form $\Theta(-m)$ with $m > 1$, the group $H^q(\mathcal{R}_p \otimes \Theta)$ can be nonzero by Lemma 3 only when $p = q = 0$ or when $p > 0$ and $q = n - 1$ or n. We assume that $p + q = 1$; in this case $q = n$ is not possible, since $p \geqslant -k \geqslant -n + 2$ by the condition $\dim X \geqslant 2$. If $q = n - 1$ then $p = -k = -n + 2$, thus

$$\bigoplus_{p+q=1} E_1^{pq} \cong H^{n-1}\left(\Theta\left(-\sum_1^k m_i\right)\right).$$

By Lemma 3 this space is nonzero only under the condition $\sum m_i = n + 1$, and in this case it is one-dimensional. Since $k = n - 2$ we obtain the equivalent relation $\sum (m_i - 1) = 3$. Solving this and recalling that $m_i > 1$, we arrive at cases I–III. This implies the first assertion of the lemma. The second can be verified in a similar fashion.

We observe that Theorem 1 follows from the first assertion of the lemma.

Theorem 2. *If a complex space* X *belongs to one of the types* I–III, *then it has a versal deformation with base* $(\mathbb{P} \times \mathbb{C}, p \times \{0\})$, *whose restriction to the germ* $(\mathbb{P} \times \{0\}, p \times \{0\})$ *is isomorphic to* π.

Proof. It follows from Lemma 2 that $\operatorname{Coker} h^1 = 0$ and from Lemma 4 that $\operatorname{Ker} h^2 = 0$. Therefore by Lemma 1 we have $T^2(X) = 0$; consequently, by the general theorem in Chapter 2, Section 6, the complex space X has a minimal versal deformation f with base $(T^1(X), 0)$. Let $\varphi : (\mathbb{P}, p) \to (T^1(X), 0)$ be a mapping of the base under which f generates π, i.e., $\pi \cong \varphi^*(f)$. According to Chapter 2, Section 4 $Df \circ d\varphi = D\pi$ and Df is an isomorphism, and the mapping $D\pi$ is the composition of the mappings in (3.4.5). Therefore the image of $D\varphi$ is the direct summand $\operatorname{Coker} h^0$ in $T^1(X)$. The complementary term $\operatorname{Ker} h^1$ is one-dimensional by Lemmas 2 and 4. We assign the isomorphism of germs $\Psi : \mathbb{P} \times \mathbb{C} \to T^1(X)$ in such a way that its restriction to $\mathbb{P} \times \{0\}$ coincides with φ. The deformation $\Psi^*(f)$ has the required properties.

It is clear from the construction of this deformation that its fibers over points not belonging to $\mathbb{P} \times \{0\}$ are not subspaces of $\mathbb{CP}_n$. This does not mean that they cannot be algebraic manifolds, since they can be imbedded into a projective space of dimension higher than n. However, the points of the base $\mathbb{P} \times \mathbb{C}$ over which the fibers are algebraic manifolds form only a countable union of analytic subsets of codimensional 1. In fact these fibers are surfaces of type $K3$. The structure of the minimal versal deformations of such surfaces was investigated in detail by G.N. Tyurina [74]. The base S of such a deformation has dimension 20 and its points corresponding to fibers which are algebraic manifolds and form a countable union of analytic subsets of dimension 19. The original deformation is versal for the fibers over the points y which are sufficiently near to the distinguished point. Therefore it is generated by a minimal versal deformation of the fiber X_y for some mapping of the bases $\varphi : (\mathbb{P} \times \mathbb{C}, y) \to S$, whose rank is equal to 20. Under this mapping the inverse image of an analytic set in S of dimension 19 is an analytic subset of $\mathbb{P} \times \mathbb{C}$ of codimension 1.

5. Deformations of Hopf Manifolds

5.1. We consider a group G_A which is generated by a linear transformation A on $\mathbb{C}^n$ whose eigenvalues satisfy the inequalities

$$0 < |\lambda_i| < 1, \quad i = 1, \ldots, m. \tag{3.5.1}$$

The quotient space $V_A = \mathbb{C}^n \backslash \{0\}/G_A$ is called a (linear) *Hopf manifold*. It is a compact complex manifold which is homeomorphic to $S^1 \times S^{2n-1}$. Hopf manifolds are the fibers of a mapping of complex manifolds $V \to M$, where M is a domain in $\mathbb{C}^{n \times n}$ formed by the matrices A which satisfy (3.5.1). For an open dense subset of M this mapping is a versal deformation of the corresponding fibers. But for some fibers it is not. Borcea [12], Dabrowski [21], and Wehler [104] discovered that in the case of $n = 2$ the deformation V is not versal for the Hopf surface V_A if the corresponding eigenvalues satisfy the relation $\lambda_1 = \lambda_2^d$, where $d \geqslant 2$ is some integer. Such a surface can be deformed using a family of nonlinear Hopf surfaces $V(s) = \mathbb{C}^2 \backslash \{0\}/G(s)$, where $G(s)$ is the group generated by the nonlinear transformations

$$A_s(z_1, z_2) = (\lambda_1 z_1 + s z_2^d, \lambda_2 z_2),$$

and $s \in \mathbb{C}$ is the parameter of the deformation. In the general case nonlinear deformations of Hopf manifolds can arise if a relation of the form

$$\lambda_k = \lambda_1^{d_1} \cdot \ldots \cdot \lambda_n^{d_n}, \qquad \sum d_i \geqslant 2, \qquad d_k = 0. \tag{3.5.2}$$

is satisfied for the eigenvalues of the operator A. Such relations are called *resonance relations*.

In order to describe versal deformations for resonance Hopf manifolds, we consider first a parallel problem. We describe the connected component of the identity in the Lie group of holomorphic automorphisms of V_A. Its Lie algebra is $\mathfrak{g} = H^0(V_A, \Theta)$, where Θ is the sheaf of tangent fields on V_A. For all Hopf manifolds we have the equation

$$\dim H^0(V_A, \Theta) = \dim H^1(V_A, \Theta), \qquad H^2(V_A, \Theta) = 0. \tag{3.5.3}$$

Consequently "jumps" in the dimension of the automorphism group of V_A correspond to jumps in the dimension of the base of a minimal versal deformation of this Hopf manifold. The dimension of the subgroup generated by the linear fields on the covering space $\mathbb{C}^n$ is equal to the dimension of the submanifold in the base of this deformation over which the fibers are linear Hopf manifolds. Nonlinear vector fields on Hopf surfaces were discovered by Namba [66] and have the form $v = sz_2^d\partial/\partial z_1$.

In the general case $\mathfrak{g} = \mathfrak{g}_0 \oplus \mathfrak{g}_+$, where $\mathfrak{g}_0$ is the Lie subalgebra of linear fields on $\mathbb{C}^n$ and $\mathfrak{g}_+$ is the nilpotent ideal. In other words, $\mathfrak{g}_0$ can be realized as the algebra of matrices of order n which commute with A and $\mathfrak{g}_+$ as the collection of vector fields V on $\mathbb{C}^n$ with polynomial coefficients having no constant or linear terms, which satisfy the relation $AV(z) = V(Az)$. As a vector space $\mathfrak{g}_+$ decomposes into a direct sum of subspaces $\Gamma_k^{d_1,\dots,d_m}$, each corresponding to a single resonance relation (3.5.2), where $\lambda_1, \dots, \lambda_m$ denote the eigenvalues of the operator A on the irreducible eigenspaces $\mathbb{C}^n = E_1 \oplus \cdots \oplus E_m$ (some of these numbers may be equal). Let $e_k^0, \dots, e_k^{r_k}$ be a basis in E_k such that $(A - \lambda_k)e_k^i = e_k^{i-1}$ for $i = 0, \dots, r_k$, and $e_k^{-1} = 0$. According to [72], the subspace $\Gamma_k^{d_1,\dots,d_m}$ is formed by fields of the form

$$v = \sum_0^{r_k} f_i \frac{\partial}{\partial z_k^i}, \tag{3.5.4}$$

whose coefficients satisfy the relations

$$\Lambda_k f_0 = f_1, \qquad \Lambda_k f_1 = f_2, \dots, \Lambda_k f_{r_k} = 0, \tag{3.5.5}$$

where $\Lambda_k f(z) = \lambda_k f(z) - f(Az)$. Such a field is defined by the "highest" coefficient f_0, which must satisfy the equation

$$\Lambda_k^s f_0 = 0, \qquad s = r_k + 1. \tag{3.5.6}$$

Let z_k^i be coordinates in $\mathbb{C}^n$ such that $z = \sum_{i,k} z_k^i e_k^i$. The function f_0 satisfying (3.5.6) is a polynomial in these coordinates. In addition, it is required that this polynomial be homogeneous relative to the group of variables $z_j = (z_j^0, \dots, z_j^{r_j})$ of degree d_j for $j = 1, \dots, m$.

5.2. The infinitesimal analog of equation (3.5.6) is the differential equation

$$V_+^s g = 0, \tag{3.5.7}$$

where

$$V_+ = \sum_j \sum_i z_j^{i+1} \frac{\partial}{\partial z_j^i}.$$

Its solution is characterized by the following property: if $\varphi_1(t), \ldots, \varphi_m(t)$ are arbitrary polynomials in one variable of degrees $r_1, \ldots, r_m$, then under the substitution $z_j^i = \dfrac{d^i\varphi_j}{dt^i}$ for $i = 0, \ldots, r_j$ and $j = 1, \ldots, m$ we obtain a function $h(t) = g(D^{r_1}\varphi, \ldots, D^{r_m}\varphi)$ of degree no higher than $s - 1$. Conversely, if a polynomial g has this property, then it is a solution of (3.5.7). We consider the transformation

$$f(z_1, \ldots, z_m) = g(l_1(z_1), \ldots, l_m(z_m)),$$

where $w_j = l_j(z_j)$ for $j = 1, \ldots, m$ are polynomial mappings, which can be described symbolically

$$w_j^0 = z_j^0, \qquad w_j^{(i)} = \left(\lambda_j \ln\left(\frac{z_j}{\lambda_j} + 1\right)\right)^i = \left(z_j - \frac{z_j^2}{2\lambda_j} + \frac{z_j^3}{3\lambda_j^2} - \cdots\right)^i. \qquad (3.5.8)$$

Here on the right it is necessary to expand in powers of z_j and then to replace the power by an upper index, i.e., $(z_j)^i$ becomes z_j^i, and to discard degrees higher than r_j. If the polynomial g satisfies (3.5.7) and has degree of homogeneity d_j relative to z_j for $j = 1, \ldots, m$, then after this substitution we obtain a polynomial f satisfying (3.5.6), and all solutions of (3.5.6) can be obtained in this manner.

In the theory of invariants there is the Hilbert identity

$$\sum (-1)^i \varphi^{(i)}(t) \varphi^{(n-i)}(t) \equiv \text{const} \qquad (3.5.9)$$

for polynomials φ of even degree n. To it there corresponds the solution $g(w) = \sum (-1)^i w_m^{(i)} w_m^{(n-i)}$ of equation (3.5.7) having degree of homogeneity $(0, \ldots, 0, 2)$. After the substitution (3.5.8) we obtain the polynomial

$$f(z) = \sum (-1)^i \left(z_m - \frac{z_m^2}{2\lambda_m} + \cdots\right)^i \cdot \left(z_m - \frac{z_m^2}{2\lambda_m} + \cdots\right)^{n-i},$$

which satisfies (3.5.6) with $s = 1$. Consequently the field $v = f\dfrac{\partial}{\partial z_1^0}$ generates a section of the sheaf Θ on V_A if there is the resonance relation $\lambda_1 = \lambda_m^2$, and the Jordan block for the eigenvalue λ_m has order $n + 1$. In the case of odd n there is the analogous identity

$$\sum (-1)^i (n - 2i) \varphi^{(i)} \varphi^{(n-i)} = c_1 t + c_0,$$

which gives a solution of equation (3.5.7) with $s = 2$. Under the substitution (3.5.8) we obtain from it a solution of (3.5.6) and consequently a tangent field to the Hopf manifold in the case when the dimension of E_1 is not smaller than 2.

Proposition. *If* $s > 1$ *then any solution of equation* (3.5.7) *can be written in the form* $g = h' + V_- h''$, *where* h' *and* h'' *are solutions of the equation* $V_+^{s-1} h = 0$ *and*

$$V_- = \sum_j \sum_i (r_j - i + 1) i z_j^{i-1} \frac{\partial}{\partial z_j^i}.$$

Proof. The operators V_+, V_-, and $H = \frac{1}{2}[V_+, V_-]$ form a representation for the Lie algebra $\mathfrak{gl}(2, \mathbb{C})$ in the space of polynomials. Here H is the weight operator; every monomial is an eigenvector for it. Expanding this representation into a direct sum of irreducibles, we obtained the result.

Thus all solutions of (3.5.7) can be obtained from solutions of the first order equation $V_+g = 0$ by applying a power of the operator V_-. The solutions of this equation form a subalgebra $R^{r_1,\ldots,r_m}$ of the algebra of all polynomials in z. By a theorem of Weitenbek this algebra always has a finite number of generators.

5.3. Examples I. Let the matrix A be semisimple, i.e., all $r_k = 0$. Then $V_+ = 0$, and for any resonance the nonlinear tangent fields on V_A are generated by fields of the form $f\dfrac{\partial}{\partial z_k}$, where f is any polynomial homogeneous in all variables with degree d_j in z_j.

II. Let the matrix A satisfy the single resonance relation $\lambda_1 = \lambda_2^2$. The corresponding tangent fields have the form (3.5.4)–(3.5.5) with $k = 1$. The coefficient f_0 is obtained from a solution g of equation (3.5.7) under the substitution $w = l_2(z_2)$ which has the form $g = g_0 + V_-g_1 + \cdots + V_-^{r_1}g_{r_1}$, where g_i are solutions of equation (3.5.7) with $s = 1$. The space of solutions of the latter equation can be described explicitly: there is a basis for it formed by the polynomials

$$w^{(0)}w^{(n)} - w^{(1)}w^{(n-1)} + \cdots + w^{(n)}w^{(0)},$$

$$w^{(2)}w^{(n)} - w^{(3)}w^{(n-1)} + \cdots + w^{(n)}w^{(2)},$$

$$\cdots\cdot$$

$$(w^{(n)})^2,$$

where n is the maximal even number which does not exceed r_1. They all correspond to the Hilbert identity (3.5.9) applied to $\varphi, \varphi^{\mathrm{II}}, \varphi^{\mathrm{IV}}, \ldots$.

5.4. By (3.5.3) the base for a minimal versal deformation is the germ of a manifold whose dimension is equal to the dimension of the space of tangent fields on V_A, i.e., to the sum of the dimensions of the spaces $\Gamma_k^{d_1,\ldots,d_m}$ and $\mathfrak{g}_0$. We can construct this deformation as follows: let A^* be the Hermitian adjoint to A. For any field $h \in H^0(V_{A^*}\Theta)$ we consider the polynomial transformation $A_h(z) = Az + h(z)$ in $\mathbb{C}^n$, where $h(z)$ is the vector formed by the coefficients of the field h. If we restrict this transformation to the domain $P = \{0 < |z| < 1\}$, and the field is chosen from a sufficiently small neighborhood of zero $Q \subset H^0(V_{A^*}, \Theta)$, then the transformation A_k will act properly and discretely. Moreover, the quotient space of the domain $P \times Q$ relative to the group Γ of transformations generated by the family of mappings A_h will have the structure of a complex manifold which is proper over Q. The mapping $P \times Q/\Gamma \to Q$ is a minimal versal deformation of the Hopf manifold V_A. Any nonlinear Hopf manifold $V_{\mathscr{A}}$ is isomorphic to a fiber of this deformation if $\mathscr{A}$ is a transformation acting in a neighborhood of zero in $\mathbb{C}^n$ with linear part A. The family A_h contains the normal forms of all such transformations (for details see [72]). The normal forms are analogous to the normal forms of resonance vector fields (the Poincaré-Dulac Theorem).

6. Singular Klein Germs

6.1. Let us suppose that a finite group Γ of holomorphic transformations acts on $\mathbb{C}^n$. On the orbit space $X = \mathbb{C}^n/\Gamma$ we can introduce the structure of a complex space. The elements of a fiber $\mathcal{O}_{X,x}$ in the structure sheaf of this space are holomorphic functions defined in a neighborhood of the orbit of $x \subset \mathbb{C}^n$ and invariant relative to the action of the group Γ. If Γ consists of linear transformations, then each such function is a holomorphic function of invariant polynomials on $\mathbb{C}^n$. The algebra of invariant polynomials has a finite number of homogeneous generators $\varphi_1, \ldots, \varphi_N$, which define a holomorphic imbedding $X \to \mathbb{C}^N$. A relation among these generators is a polynomial $F(w_1, \ldots, w_N)$ such that $F(\varphi_1(z), \ldots, \varphi_N(z)) \equiv 0$. The set of all relations forms an ideal I in the algebra of polynomials in N variables, and $\mathcal{O}_X \cong \mathcal{H}_N/I \cdot \mathcal{H}_N$.

Klein [49] determined the structure of the quotient space $\mathbb{C}^2/\Gamma$, where Γ is any finite subgroup of $SL(2, \mathbb{C})$. For each such subgroup Γ the algebra of invariant polynomials has three generators which are related by a single equation. Thus $\mathbb{C}^2/\Gamma$ can be realized as a subspace of $\mathbb{C}^3$ defined by a single equation. Here the origin is the unique singular point. We indicate these equations below:

$$A_n : x^2 + y^2 + z^{n+1} = 0, \qquad n \geqslant 1;$$

$$D_n : x^2 + y^{n-1} + yz^2 = 0, \qquad n \geqslant 4;$$

$$E_6 : x^2 + y^3 + z^4 = 0;$$

$$E_7 : x^2 + y^3 + yz^3 = 0;$$

$$E_8 : x^2 + y^3 + z^5 = 0.$$

They correspond to Γ being a cyclic group, a dihedral group, and the groups of the tetrahedron, the octahedron, and the icoscahedron respectively, and they define a class of two-dimensional germs with a unique singular point consisting of the two series A_n and D_n and the three germs E_6, E_7, and E_8. This class coincides with the set of binary rational points on surfaces [6]. For any deformation of a germ in this class all singularities of nearby fibers again belong to the same class. There are various other descriptions of the singular Klein germs (see also Chapter 5, Section 2).

6.2. Let X be a reduced complex space and $\mathrm{Sing}(X)$ the set of its singular points. A *resolution of the singularities of* X is any complex manifold $\tilde{X}$ together with a proper mapping $\pi : \tilde{X} \to X$ which is an isomorphism on $\tilde{X} \backslash \pi^{-1}$ (Sing(X). It has been known for a long time that the singularities of a two-dimensional complex space (and also germs of complex spaces of any dimension) can be resolved. The question of the existence of resolutions of deformations of a complex space or germ is more complicated. Let $X \to S$ be a mapping of complex spaces in which S is a manifold with distinguished point. In Brieskorn's terminology a resolution of this mapping is a commutative diagram of mappings of complex spaces

$$\begin{array}{ccc} Y & \longrightarrow & X \\ \downarrow & & \downarrow \\ T & \xrightarrow{\varphi} & S, \end{array}$$

in which T is also a manifold with distinguished point and φ is a finite epimorphic mapping of the corresponding germs, i.e., a proper mapping in which the pre-image of any point is a finite nonempty set, the mapping $Y \to T$ has rank equal to $\dim T$ at each point, and for $t \in T$ the mapping of fibers $Y_t \to X_{\varphi(t)}$ is a resolution of the singularities of $X_{\varphi(t)}$. It is known that not every mapping of complex spaces has a resolution. Atiyah first observed that a deformation of a singular germ of type A_1 can be resolved if φ is taken to be the two-sheeted covering $\varphi(t) = t^2$. In other words, the deformation formed by the surfaces $X_t = \{x^2 + y^2 + z^2 = t^2\}$ can be resolved without change of base. Brieskorn [14] and G.N. Tyurina [97] established that a versal deformation for any Klein germ admits a resolution, and consequently so does any other deformation of such a germ. G.N. Tyurina observed that the Galois group of the epimorphism φ corresponding to a resolution of the versal deformation coincides with the Weyl group of the complex simple Lie algebra of the corresponding type (in the case of germs of types A and D).

Brieskorn's report [15] elucidated the mysterious parallel between binary rational points and one-dimensional simple Lie algebras. It contained the following result which confirmed a conjecture of Grothendieck. Let G be a complex simple Lie group of type A, D, or E, let $\tau : G \to T/W$ be the canonical mapping, where T is a maximal torus in G and W is the Weyl group. Any unipotent element $g \in G$ under this mapping is sent to the class of the identity element e. Then such an element g is subregular if and only if the germ of the mapping $(G, g) \to (T, e)/W$ is the composition of a mapping of germs of manifolds $(G, g) \to (X, x)$ of rank equal to $\dim X$ and a versal deformation $(X, x) \to (T, e)/W$ of a binary rational point of the same type as G. We recall that an element of a Lie group G is called *subregular* if the dimension of its centralizer is greater by two than the minimal dimension of the centralizers of elements of G.

This result reduces to the resolution of deformations of Klein germs via the general construction of Springer-Grothendieck

$$\begin{array}{ccc} Y & \longrightarrow & G \\ \downarrow & & \downarrow{\scriptstyle\tau} \\ T & \longrightarrow & T/W, \end{array}$$

which resolves the mapping τ in the large. Here Y is a submanifold in $G \times B$, where B is the projective space of all Borel subgroups of G. It consists of pairs (g, b) such that $g \in b$. The Weyl group occurs explicitly in this construction as the Galois group of the covering of the base, and this explains the remark of G.N. Tyurina mentioned above.

The work of Brieskorn was extended by Slodovy [89]. The resolution of deformations with more complicated singular germs was studied in work of Laufer [57], M. Artin [7], Wahl [101], [102]. The work of Elkik [31] is also relevant.

7. Examples of Obstructions

As we saw in Chapter 2, Section 5, the obstructions to the extension of infinitesimal deformations of a space X_0 take values in $T^2(X_0)$. According to Chapter 2, Section 6, the base for a minimal versal deformation of a compact complex space X_0 is defined in $T^1(X_0)$ by holomorphic equations, the number of these being equal to dim $T^2(X_0)$. However, for a particular X_0 the minimal number of such equations can be smaller. Such equations can in general be absent when $T^2(X_0) \neq 0$. This means, in particular, that there are no obstructions for any infinitesimal extensions of deformations.

If X_0 is a one-dimensional compact complex space with a finite number of singular points at which its germs are complete intersections, then $T^2(X_0) = 0$. Therefore in this case there are also no obstructions. In the cases next in complication, when X_0 is either a one-dimensional germ which is not a complete intersection or a compact two-dimensional manifold, obstructions appear in situations which are not too complicated. We now consider examples of such obstructions.

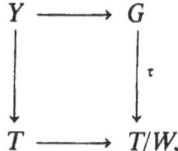

Example 1. Let X_0 be the germ at the origin of the four coordinate axes in $\mathbb{C}^4$. The corresponding analytic algebra is $A = H_4/I$, where I is the ideal in H_4 generated by the six functions $z_1z_2, z_3z_4, z_1z_3, z_2z_4, z_1z_4, z_2z_3$. We denote them by f_{ij}, where (i,j) runs over all pairs from $\{1, 2, 3, 4\}$. We compute $T^1(X_0)$ using (2.1.5). The module of relations R among these generators is generated by the 12 vectors r_k, where $r_1 = (z_3, 0, -z_2, 0, 0, 0)$ and the remaining r_k are written analogously. Therefore the module $\mathrm{Hom}(I, A)$ is isomorphic to the submodule of A^6 consisting of vectors a such that $(r_k, a) = 0$ for $k = 1, \ldots, 12$. This is easy to describe: it is generated over A by the images of the 12 vectors of the form $\alpha_k = (\alpha_{12}, \alpha_{34}, \alpha_{13}, \alpha_{24}, \alpha_{14}, \alpha_{23})$ in which there is either z_i or z_j in the (i,j)th

place and zeros elsewhere. The image of $\mathrm{Der}(H_4)$ in $\mathrm{Hom}(I, A)$ is generated by the 4 vectors $\left\{\frac{\partial f_{ij}}{\partial z_k}\right\}$ for $k = 1, 2, 3, 4$. It is not difficult to see from this that the 12 vectors a_k generate $T^1(X_0)$ over $\mathbb{C}$, but among their images only 8 are linearly independent, i.e., $\dim T^1(X_0) = 8$. Choosing these 8 vectors in a suitable fashion, we define linear terms in the equations for the space X of a versal deformation of X_0:

$$F_{12}(z, u, v) = z_1 z_2 + u_1 z_2 + u_2 z_1 + u_2 v_1,$$
$$F_{34}(z, u, v) = z_3 z_4 + u_3 z_4 + u_4 z_3 + u_4 v_3,$$
$$F_{13}(z, u, v) = z_1 z_3 + v_1 z_3 + u_3 v_1,$$
$$F_{24}(z, u, v) = z_2 z_4 + v_2 z_4 + u_4 v_2,$$
$$F_{14}(z, u, v) = z_1 z_4 + v_4 z_1 + u_1 v_4,$$
$$F_{23}(z, u, v) = z_2 z_3 + v_3 z_2 + u_2 v_3,$$

where u_i and v_i for $i, j = 1, 2, 3, 4$ are coordinates for the base of this deformation. The terms of degree zero in z are determined uniquely by the requirement that the mapping $X \to S$ be flat. According to this requirement, for example, the relation $z_3 f_{12} - z_1 f_{23} = 0$ must extend to a relation among the F_{ij} which must necessarily have the form $(z_3 + v_3)F_{12} - (z_1 + u_1)F_{23} - u_2 F_{13} = 0$. Thus the terms of degree zero in F_{12} and F_{23} are determined uniquely, etc. Equating to zero the constant terms in this and similar relations, we obtain equations which define the base S of a versal deformation

$$\begin{aligned} u_1 v_1 - u_1 v_2 &= 0, & u_4 v_3 - u_3 v_4 &= 0, \\ u_1 v_3 - v_1(v_3 - u_3) &= 0, & u_2 v_4 - v_2(v_4 - u_4) &= 0, \\ u_2 v_3 - v_2(v_3 - u_3) &= 0, & u_1 v_4 - v_1(v_4 - u_4) &= 0, \end{aligned} \tag{3.7.1}$$

The left sides of these equations are the components of the first obstruction $T^1(X_0) \to T^2(X_0)$. The higher obstructions are equal to zero; consequently, the base S is a five-dimensional cone over some projective manifold $PS \subset \mathbb{CP}_7$. In order to describe it, we introduce the more convenient coordinates $\tilde{u}_3 = v_3 - u_3$, $\tilde{u}_4 = v_4 - u_4$. Then the equations in (3.7.1) can be expressed in the form

$$v_1 : u_1 = v_2 : u_2 = v_3 : \tilde{u}_3 = v_4 : \tilde{u}_4.$$

The common value of these four ratios is a point in $\mathbb{CP}_1$, and the vector (v_1, v_2, v_3, v_4) or the vector $(u_1, u_2, \tilde{u}_3, \tilde{u}_4)$ proportional to it defines a point in $\mathbb{CP}_3$. It follows that $PS \cong \mathbb{CP}_1 \times \mathbb{CP}_3$.

Pinkham [75] constructed a series of examples of one-dimensional germs with singular points in which all fibers of the versal deformation have singular points (nonsmoothable curves). In particular such a germ is the union of 13 lines passing through the origin in $\mathbb{C}^7$ in general position.

Example 2. Let $V_0 \subset \mathbb{CP}_3$ be the two-dimensional complex space defined by the equation $F = 0$, where F is a polynomial of degree $m \geqslant 5$ having only binary rational singular points. We consider a minimal resolution of it $\pi: X_0 \to V_0$, i.e., a resolution which replaces each singular point v_i by $k_i = \dim T^1(V_0, v_i)$ curves each isomorphic to $\mathbb{CP}_1$. Let $X \to R$ and $V \to S$ be minimal versal deformations of X_0 and V_0 respectively. We note that according to Chapter 3, Section 4, the base S is the germ of a manifold.

Theorem ([18]). *The germ R is a reduced complete intersection with*

$$\dim R = \dim S; \tag{3.7.2}$$

$$\dim T(R) - \dim R = \dim H^1(\mathbb{CP}_3, \mathcal{J}(m)), \tag{3.7.3}$$

where $\mathcal{J}(m) = \mathcal{J} \otimes \mathcal{O}(m)$ and $\mathcal{J}$ is the sheaf of ideals in the structure sheaf $\mathcal{O} = \mathcal{O}_{\mathbb{CP}_3}$ generated by the derivatives $\dfrac{\partial F}{\partial z_i}$ for $i = 0, 1, 2, 3$.

The left side of (3.7.3) is the minimal number of holomorphic equations needed to define the base R in $T(R)$ (the number of obstructions). The right side can be different from zero, for example, in the case when $m = 5$, $F(z) = z_0^5 - l_1 l_2 l_3 l_4 l_5$, where the l_i for $i = 1, \ldots, 5$ are linear functions of z_1, z_2, z_3 chosen in general position. In this case V has 10 singular points of type A_4, and calculations similar to those performed in Chapter 3, Section 4 give the equation $\dim H^1(\mathbb{CP}_3, \mathcal{J}(5)) = 8$. The obstructions to a versal deformation of X_0 can be interpreted as follows. According to Chapter 3, Section 4, the space $H^0(\mathbb{CP}_3, \mathcal{O}(m))$ maps epimorphically onto $T^1(V_0)$. On the other hand, we have $\mathcal{O}/\mathcal{J} = \mathcal{T}^1(V_0)$. We consider the commutative diagram

$$\begin{array}{ccccccccc} & & H^0(\mathbb{CP}_3, \mathcal{O}(m)) & \longrightarrow & H^0(\mathbb{CP}_3, \mathcal{O}/\mathcal{J}) & \longrightarrow & H^1(\mathbb{CP}_3, \mathcal{J}(m)) & \longrightarrow & 0 \\ & & \downarrow & & \| & & & & \\ 0 \longrightarrow & H^1(V_0, \mathcal{T}^0(V_0)) & \longrightarrow T^1(V_0) & \longrightarrow & \Gamma(V_0, \mathcal{T}^1(V_0)) & \xrightarrow{d_2} & H^2(V_0, \mathcal{T}^0(V_0)) & & \end{array}$$

with exact rows. The lower part of it is a fragment of (2.3.2). It follows from the exactness of the rows that we have an isomorphism $H^1(\mathbb{CP}_3, \mathcal{J}(m)) \cong \operatorname{Im} d_2$ and the equation

$$\dim T^1(V_0) = \dim H^1(V_0, \mathcal{T}^0(V_0)) + \dim \operatorname{Ker} d_2.$$

From this using (3.7.2) and (3.7.3), we obtain

$$\dim T(R) = \dim T^1(X_0) = \dim H^1(V_0, \mathcal{T}^0(V_0)) + \dim \Gamma(V_0, \mathcal{T}^1(V_0)),$$

$$\dim T(R) - \dim R = \dim \operatorname{Im} d_2. \tag{3.7.4}$$

The first equation can be understood as follows: the space $T(R)$ of all deformations of X_0 over the double point is equal to the sum of the subspace $H^1(X_0, \pi^*(\mathcal{T}^0(V_0)))$ and its direct complement Γ, whose dimension is equal to the dimension of the space $\Gamma(V_0, \mathscr{T}^1(V_0))$. The first summand corresponds to those deformations of X_0 which do not deform the one-dimensional subspaces $\pi^{-1}(v_i)$, where v_i are the singular points of V_0 (since the elements of $\pi^*(\mathscr{T}^0(V_0))$ are fields equal to zero on these subspaces). According to the theorem of Brieskorn-Tyurina (see Chapter 3, Section 6), the versal deformation of each singular germ (V_0, v_i) is covered by a deformation of the germ $(X_0, \pi^{-1}(v_i))$ which resolves this singular germ. In the subspace Γ with $\dim \Gamma = \sum \dim T^1(V_0, v_i)$ all such mutually independent deformations of the germs $(X_0, \pi^{-1}(v_i))$ are realized. However, in a versal deformation of V_0 the individual singular points cannot be deformed independently of each other if $d_2 \neq 0$, since the linear equation $d_2 s = 0$ for $s \in \Gamma(V_0, \mathscr{T}^1(V_0))$ defines those sets of local deformations of singular points which are realized by deformations of the entire V_0 over the double point. In deformations of X_0 over the double point such relations cannot arise, but under an extension of a deformation of X_0 from the first infinitesimal neighborhood of zero in $T^1(X_0)$ to all of $T^1(X_0)$ obstructions do appear, the number by (3.7.4) coinciding with the number of linear relations for the deformations of the singular points of V_0. Thus on passing from a singular surface of V_0 to its resolution X_0 the linear relations among the deformations of singular points of V_0 turn into nonlinear obstructions. The nonlinearity of this transition is related to the fact that a nonlinear mapping of the base φ figures in the construction of the resolution of deformations of the points A_4 (Chapter 3, Section 6).

Other examples of compact two-dimensional manifolds X_0 with obstructions to a versal deformation were found by Mumford and Kas.

Chapter 4. Deformations of Other Objects in Analytic Geometry

1. Functorial Language

For a uniform treatment of various problems in deformation theory the use of a language due to Grothendieck is convenient. Let F be a contravariant functor from some category G into the category of sets Ens. This functor is called representable and $S \in G$ a representing object if there is an isomorphism of functors $F \simeq \hom_G(\cdot, S)$. This means that for any object $R \in G$ there is a one-to-one correspondence between the sets $F(R) \simeq \hom(R, S)$ so that for each morphism $\varphi: P \to R$ in the category G the diagram

$$\begin{array}{ccc} F(R) & \simeq & \hom(R,S) \\ {\scriptstyle F(\varphi)}\downarrow & & \downarrow{\scriptstyle \varphi^*} \\ F(P) & \simeq & \hom(P,S). \end{array}$$

commutes. The mapping φ^* in the diagram is defined by the rule for composition of morphisms $h \mapsto h \circ \varphi$. In order to establish such an isomorphism of functors, it is necessary and sufficient to give the object S and an element $\xi \in F(S)$ corresponding to the identity morphism $S \to S$. Then by the representability of the functor, for any object $R \in G$ and any element $\eta \in F(R)$ there exists a unique morphism $\varphi : R \to S$ such that $\eta = F(\varphi)\xi$.

Let G be the category of germs of complex spaces and let there be given an object S in G and an element $\xi \in F(S)$. To every morphism $\varphi : R \to S$ there corresponds the element $\eta = F(\varphi)\xi \in F(R)$. Thus there is defined the mapping of sets

$$\hom(R, S) \to F(R). \tag{4.1.1}$$

In particular, if D is the double point, the set $\hom(D, S)$ is isomorphic to the tangent space $T(S)$ to the germ S (Chapter 1, Section 2); consequently, this set has the structure of a vector space. The set $F(D)$ is called the tangent space to the functor. The pair (S, ξ) is called the envelope of the functor F if the mapping (4.1.1) is always an epimorphism and in the case $R = D$ it is an isomorphism. If (S', ξ') is another envelope, then there exists an isomorphism of germs $\varphi : S \to S'$ so that $F(\varphi)\xi' = \xi$.

In deformation theory the choice of the functor F depends on the type of the objects undergoing deformation. If a complex space X is being deformed, then we consider the functor $F = F_X$, whose value on a germ S is the set of isomorphism classes of deformations of X with base S. An isomorphism class of deformations is any maximal collection of deformations of X over S in which all elements are pairwise isomorphic in the sense of Chapter 1, Section 4. If $\varphi : R \to S$ is a morphism of germs, then the mapping $F_X(\varphi) : F_X(S) \to F_X(R)$ sends a class of deformations f to the class which contains its inverse image $\varphi^*(F)$. The representability of the functor F_X means that there exists a universal deformation of X. Indeed, the representing object S is a base for such a deformation, and the image of the identity morphism of the germ S in $F(S)$ is itself a universal deformation. If the functor F_X has an envelope (S, ξ), then ξ is the class of a minimal versal deformation of X with base S. Conversely, such a deformation is an envelope for the functor F_X.

Schlessinger [85] developes this language for application to the theory of infinitesimal deformations.

2. Deformations of Vector Bundles and Coherent Sheaves

2.1. *A vector bundle on a complex space S* is a mapping of complex spaces $p : E \to X$ which possesses the following properties: 1) each fiber $E_x = p^{-1}(x)$ has the structure of a vector space over the field $\mathbb{C}$, 2) for every point $x \in X$ there is a neighborhood U and an isomorphism of complex spaces $\varphi : p^{-1}(U) \to U \times \mathbb{C}^n$ so that the diagram

$$\begin{array}{ccc} p^{-1}(U) & \xrightarrow{\varphi} & U \times \mathbb{C}^n \\ & {\scriptstyle p}\searrow \quad \swarrow{\scriptstyle \mathrm{pr}} & \\ & U & \end{array},$$

commutes, where pr is the natural projection from a product onto its factor, and for each point $y \in U$ the isomorphism $E_y \rightrightarrows \mathbb{C}^n$ generated by φ is linear. A bundle p is isomorphic to a bundle $p': E' \to X$ if there exists an isomorphism of complex spaces $a: E \to E'$ so that $p'a = p$ and for each $x \in X$ the isomorphism $E_x \to E'_x$ generated by it is linear. Let $f: Y \to X$ be a mapping of complex spaces and $p: E \to X$ a vector bundle. The space $Y \times_X E$ possesses the natural structure of a vector bundle with base Y. This bundle is called the inverse image of p.

Let $p: E \to X$ be a vector bundle on a complex space X. A *deformation of* p with base $S \in G$ is a vector bundle $\tilde{p}$ with base $X \times S$ together with an isomorphism of bundles $i: \tilde{p}|X \rightrightarrows p$. Here $\tilde{p}|X$ denotes the inverse image of $\tilde{p}$ under the imbedding $X \simeq X \times \{*\} \hookrightarrow X \times S$ ($*$ denotes the distinguished point of the germ S). An isomorphism between two deformations (p', i') and (p'', i'') with a common base is an isomorphism of the bundles $p' \cong p''$ which is consistent with i' and i''. The concepts of universal and minimal versal deformation for bundles correspond to the representability and envelope of the corresponding functor F_p. They can be introduced as follows: $F_p(S)$ for any germ S is the set of isomorphism classes of deformations of p with base S. If $\varphi: R \to S$ is a morphism of germs, then the mapping $F(\varphi)$ sends the class of the deformation of p to the class of its inverse image. For any vector bundle p on a compact complex space X the functor F_p has an envelope. The tangent space $F_p(D)$ to this functor is $H^1(X, \operatorname{ad} p)$, where $\operatorname{ad} p$ is the adjoint bundle; its fiber is the Lie algebra of the linear group of the bundle p. We note several articles in which deformations of vector bundles are studied: [24], [28], [33].

2.2. Deformations of Coherent Sheaves. Let $\mathcal{F}$ be a coherent $\mathcal{O}_X$-sheaf on a complex space X and S a germ of a complex space. A *deformation of the sheaf* $\mathcal{F}$ with base S is a coherent sheaf $\tilde{\mathcal{F}}$ on $X \times S$ together with an isomorphism $\tilde{\mathcal{F}}|X \cong \mathcal{F}$ provided that the projection $\mathrm{pr}: \operatorname{supp} \tilde{\mathcal{F}} \to S$ is a proper mapping (for some choice of representative for the germ S) and that the sheaf $\tilde{\mathcal{F}}$ is flat over S. As in Section 2.1, we may define a functor $F_{\mathcal{F}}$ on the category G; its value on a germ S is the set of isomorphism classes of deformations of F with base S. The functor F_p defined in Section 2.1 is a particular case. Indeed, if $p: E \to X$ is a vector bundle, then we can construct the sheaf $\tilde{E}$ of germs of holomorphic sections of it. The sheaf $\tilde{E}$ has an obvious $\mathcal{O}_X$-structure and is locally free. Conversely, if $\mathcal{F}$ is a locally free $\mathcal{O}_X$-sheaf, then there exists a vector bundle F so that the $\mathcal{O}_X$-sheaves $\tilde{F}$ and $\mathcal{F}$ are isomorphic. A deformation of the bundle E in the sense of Section 2.1 generates a deformation of the sheaf $\tilde{E}$ in the sense of Section 2.2, and conversely.

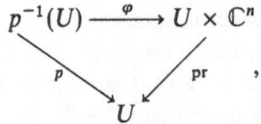

On the other hand, to each subspace $Y \subset X$ there corresponds the coherent sheaf $\mathcal{O}_Y = \mathcal{O}_X/\mathcal{I}$, where $\mathcal{I}$ is the sheaf of ideals defining the subspace. A deformation of the compact subspace Y is by definition a deformation of the coherent $\mathcal{O}_X$-sheaf $\mathcal{O}_Y$.

The study of families of subspaces of a given algebraic variety was one of the first problems in deformation theory. Such families (systems of cycles) were studied and used by the Italian school of algebraic geometry. The modern approach to deformation theory uses both singular and nonsingular bases for deformations on an equal footing. The importance of including in the theory bases with nilpotent elements such as the double point is clear in a well-known example of Zappa (see [65]) in which the space of smooth compact curves lying on a smooth projective surface has an isolated point. Using contemporary language, we would say that there are nilpotent elements in the structure algebra $\mathcal{O}_S$ of this point. Thus the tangent space to the corresponding functor has positive dimension. That the point S is isolated implies that there are nontrivial obstructions to the extension of deformations of the given curve from the first infinitesimal neighborhood to the tangent space to S.

Douady [25] showed that for any complex space X the collection K of compact subspaces can be given the structure of a complex space in such a way that K becomes the base for a universal deformation of these subspaces, even in the class of deformations whose bases are arbitrary complex spaces rather than germs. This fact is a particular case of the following more general result. For any coherent sheaf $\mathscr{E}$ on X and arbitrary complex space S we consider the coherent sheaf $\mathscr{E}_S = \tilde{\mathscr{E}} \otimes \mathcal{O}_{X \times S}$ on $X \times S$, where $\tilde{\mathscr{E}}$ denotes the inverse image of $\mathscr{E}$ under the projection $X \times S \to X$. We consider sheaves of the form $\mathscr{E}_S/\mathscr{F}$ for $\mathscr{F}$ a coherent subsheaf of $\mathscr{E}_S$, where it is assumed that the sheaf $\mathscr{E}_S/\mathscr{F}$ is flat over S and its support projects properly into S. Let $F_{\mathscr{E}}(S)$ be the set of isomorphism classes of sheaves of this form. It is shown in [25] that for any $\mathscr{E}$ the functor $F_{\mathscr{E}}$ is representable. A complex space M which represents this functor is called a Hilbert modular space. Its points are in one-to-one correspondence with coherent subsheaves $\mathscr{F} \subset \mathscr{E}$ such that the support of the quotient sheaf $\mathscr{E}/\mathscr{F}$ is compact. Each germ (M, m) of this space represents a corresponding functor $F_{\mathscr{E},\mathscr{F}}$ defined on the category of germs of complex spaces. The tangent space to the germ (M, m), which is isomorphic to the tangent space to this functor, is also isomorphic to $\mathrm{Hom}_{\mathcal{O}_X}(\mathscr{F}, \mathscr{E}/\mathscr{F})$, where $\mathscr{F}$ is the subsheaf corresponding to the distinguished point m. The method of Banach complex spaces, which was worked out by Douady, has also been applied in other problems in deformation theory.

For any coherent sheaf $\mathscr{F}$ on a compact complex space X there exists a minimal versal deformation (it is not always universal). This fact was established in work of Flenner [92]. For the sheaf $\mathscr{F}$ we may construct the sheaf of analytic algebras $\mathcal{O}_Y = \mathcal{O}_X \oplus \mathscr{F}$ in which the multiplication is defined by the rule $(a, f) \cdot (a', f') = (aa', af' + a'f)$. The pair $Y = (X, \mathcal{O}_Y)$ is a compact complex space. Each deformation of the sheaf $\mathscr{F}$ corresponds to a deformation of the complex space Y. A versal deformation of $\mathscr{F}$ can be constructed using a versal deformation of the space G. Trautmann [93] gave an explicit construction for a versal deformation of a coherent sheaf with isolated singularities (i.e., one with points in the support at which the sheaf is not locally free).

2.3. In this brief survey we shall not touch on other related questions: the deformation of invertible sheaves, principal bundles, etc. ([28]). In the next sections we shall consider the deformation theory of relative complex spaces and mappings of complex spaces. These theories all possess a number of important common features. The general constructions and their properties discussed in Chapter 2 for the theory of deformations of complex spaces carry over to the related theories mentioned here. In particular, for each of these theories one can define a tangent cohomology with the structure of a graded Lie meta-algebra, introduce the concepts of the differential of a deformation, obstructions, modular deformations, establish a semicontinuity theorem and a theorem concerning the variation of tangent cohomology, etc.

In particular, in the deformation theory of vector bundles the tangent cohomology is $H^*(X, \operatorname{ad} p)$, where $\operatorname{ad} p$ is the adjoint bundle to the principal bundle associated to p. The principal bundle is the bundle with fiber $G = \mathrm{Gl}(n, \mathbb{C})$ and transition functions the same as those for p, and the adjoint bundle has as fibers the Lie algebra $\mathfrak{g}$ of the group G on which the group acts by the adjoint representation. In particular, an infinitesimal automorphism of p is an element of $H^0(X, \operatorname{ad} p)$, the differential of a deformation of p (or more generally the distinguishing elements of different deformations) takes values in $H^1(X, \operatorname{ad} p)$, and the obstructions to the extension of deformations in $H^2(X, \operatorname{ad} p)$. The structure of a graded Lie meta-algebra in $H^*(X, \operatorname{ad} p)$ is generated by the bracket in $\mathfrak{g}$ and cohomological multiplication.

The tangent cohomology $T^*(\mathscr{F}|X)$ for an arbitrary coherent sheaf on X is described in a somewhat more complicated manner. There is a spectral sequence converging to $T^*(\mathscr{F}|X)$ similar to that which was described in Chapter 2, Section 3. Its second term is equal to $E_2^{pq} = H^p(X, \operatorname{Ext}^q_{\mathcal{O}_X}(\mathscr{F}, \mathscr{F}))$. The structure of a graded Lie algebra in $T^*(\mathscr{F}|X)$ is inherited from E_2, where a similar structure is generated by the commutator of the Ioneda multiplication in the sheaf $\operatorname{Ext}^*(\mathscr{F}, \mathscr{F})$.

3. Deformations of Relative Spaces

By a *relative space* we mean a complex space X together with a mapping f into a fixed complex space Y. This situation is sometimes denoted by the symbol X/Y. A *deformation of a relative space* is a deformation of X together with a deformation of the mapping f into the unvarying space Y. More precisely, a deformation of X/Y with base S is a commutative diagram of mappings of complex spaces

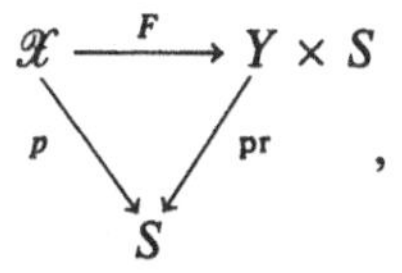

where p is a flat mapping, together with an isomorphism $i: X \simeq p^{-1}(*)$ such that $F \circ i = f$. The latter equation implies that the upper square in the diagram

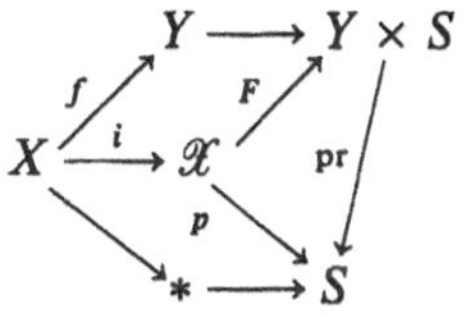

commutes. An isomorphism of deformations and the inverse image of a deformation for a mapping of the base are defined in the natural way.

In the particular case in which Y is a point, these definitions return us to the concept of a deformation of an (absolute) complex space. Another particular case occurs when f is an imbedding. A deformation of this kind of relative space X/Y is a deformation of the subspace X in a fixed space Y.

To the deformation theory of a relative space X/Y there corresponds a relative tangent cohomology $T^*(X/Y)$ which can be defined in the followng manner. We select a polyhedral covering $\mathscr{P}$ of the mapping f and a resolution of it $\mathscr{R}$ on this covering (Chapter 2, Section 2). This resolution and also the sheaf $\mathcal{O}_X$ have the structure of $\mathcal{O}_Y$-modules. An $\mathcal{O}_Y$-derivation $\delta: \mathscr{R} \to \mathcal{O}_X|\mathscr{P}$ is a mapping of the corresponding functors, defined on the nerve of the covering $\mathscr{P}$, which commutes with the multiplication on the elements of the sheaf $\mathcal{O}_Y$. The degree of this derivation is the degree of δ as a mapping of graded objects in which the sheaf $\mathcal{O}_X|\mathscr{P}$ is given the grading zero. The collection of all such derivations forms a graded vector space $T^*(\mathscr{R}, \mathcal{O}_X)$, with the differential generated by the differential of the resolution $\mathscr{R}$. By definition

$$T^*(X/Y) = H^*(T^*(\mathscr{R}, \mathcal{O}_X)).$$

In particular, $T^0(X/Y)$ is the space of vertical vector fields on X/Y, i.e., fields on X which at each point are tangent to the fiber of the mapping f. The space $T^1(X/Y)$ is isomorphic to the set of isomorphism classes of deformations of X/Y over the double point. The relative tangent cohomology is associated to the absolute exact sequence

$$0 \to T^0(X/Y) \to T^0(X) \xrightarrow{p} T^0(Y, \mathcal{O}_X) \xrightarrow{\delta} T^1(X/Y) \xrightarrow{j} T^1(X) \to T^1(Y, \mathcal{O}_X) \to \cdots, \tag{4.3.1}$$

where $T^*(Y, \mathcal{O}_X)$ is the tangent cohomology of the complex space Y with values in the $\mathcal{O}_Y$-sheaf $\mathcal{O}_X$ [70]. In particular, $T^0(Y, \mathcal{O}_X)$ is the space of derivations of the $f^*(\mathcal{O}_Y)$-module $f^*(\mathcal{O}_Y) \to \mathcal{O}_X$. The first arrows in this sequence admit a simple description. The mapping p sends a field v to its composition with the morphism of sheaves $\varphi: f^*(\mathcal{O}_Y) \to \mathcal{O}_X$. The mapping j sends a class of deformations of the relative space X/Y to the class of the corresponding deformation of X. We describe the mapping δ. It sends any derivation $t \in T^0(Y, \mathcal{O}_X)$ to a deformation of X/Y in which X is not deformed, i.e., to a deformation of the form $F: X \times D \to Y \times D$. The underlying mapping for F is the same as for f, and the morphism of sheaves $\Phi: f^*(\mathcal{O}_{Y \times D}) \to \mathcal{O}_{X \times D}$ is defined thus: we have $\mathcal{O}_{Y \times D} \cong \mathcal{O}_Y + \varepsilon\mathcal{O}_Y$, where $\varepsilon^2 = 0$, and a similar isomorphism for the sheaf $\mathcal{O}_{X \times D}$. Then $\Phi(a + \varepsilon b) = \varphi(a) + \varepsilon(t(a) + \varphi(b))$ for any $a, b \in \mathcal{O}_{Y, f(x)}$ and any $x \in X$. The pair (F, Φ) is a deformation of the mapping (f, φ); the class of this deformation is by definition $\delta(t)$.

For any morphism of germs $f: (\mathbb{C}^n, 0) \to (\mathbb{C}^m, 0)$ we have

$$T^1(X/Y) \cong H_n^m / df \cdot H_n^n, \tag{4.3.2}$$

where H_n is the algebra of convergent series in n variables, i.e., the algebra dual to the germ $(\mathbb{C}^n, 0)$, and df is the mapping $H_n^n \to H_n^m$ defined by the Jacobian matrix of f. This formula follows from (4.3.1) if we recall that $T^1(X) = 0$.

In the general case a spectral sequence with second term $E_2^{pq} = H^p(X, \mathcal{T}^q(X/Y))$ converges to $\mathcal{T}^*(X/Y)$, where $\mathcal{T}^*(X/Y)$ is the graded sheaf of relative tangent cohomology in which each of the sheaves $\mathcal{T}^q(X/Y)$ is coherent. In particular, if $f: X \to Y$ is an imbedding, then $\mathcal{T}^0(X/Y) = 0$ and $\mathcal{T}^1(X/Y) \cong \mathcal{H}om_{\mathcal{O}_Y}(\mathcal{I}, \mathcal{O}_X)$. From this we obtain

$$T^1(X/Y) \cong \mathrm{Hom}_{\mathcal{O}_V}(\mathcal{T}, \mathcal{O}_X), \qquad T^0(X/Y) = 0.$$

The first relation coincides with the formula for the tangent space to a Hilbert modular space in the theorem of Douady (Chapter 4, Section 2) for the case $\mathscr{E} = \mathcal{O}_X$, and the second explains why this space is modular also in the sense of Chapter 2, Section 7, i.e., a base for a modular deformation.

A more detailed exposition of the deformation theory of curves on surfaces can be found in the book of Mumford [65].

Horikawa in [46] considered deformations of X/Y, where Y is a manifold and X is a compact manifold, and the rank of the mapping $f: X \to Y$ at every point is equal to $\dim X$. He constructed a versal deformation and generalized other results of Kodaira and Spencer. In this case $\mathcal{T}^i(X/Y) = 0$ for $i \neq 1$ and $\mathcal{T}^1(X/Y)$ is calculated from the formula (4.3.2). The tangent space to the base of a minimal versal deformation is $T^1(X/Y) \cong \Gamma(X, \mathcal{T}^1(X/Y))$, and the obstructions take values in $T^2(X/Y) \cong H^1(X, \mathcal{T}^1(X/Y))$. This deformation is universal since $T^0(X/Y) = 0$.

Kosarew [54] constructed a versal deformation for an imbedding of a compact complex space into a noncompact manifold.

4. Stable Submanifolds

The next definition is due to Kodaira [50]. Let W be a complex manifold; a submanifold V of W is called *stable* if for every family of manifolds $p: \mathcal{W} \to B$, where B is a smooth germ with distinguished point b such that $p^{-1}(b) \cong W$, there is a submanifold $\mathscr{V} \subset \mathscr{W}$ which is proper and regular over B so that $\mathscr{V} \cap W = V$. (The regularity of $p: \mathscr{V} \to B$ means that at every point of $\mathscr{V}$ the rank of this mapping is equal to $\dim B$.) Briefly, V is stable in W if for every deformation of W with a smooth base there are in its fibers near to W compact submanifolds forming a deformation of V. Kodaira showed that V is stable in W if $H^1(V, \mathscr{N}) = 0$, where $\mathscr{N}$ is the normal sheaf to V. If in addition $H^0(V, \mathscr{N}) = 0$, then the family $\mathscr{V}$ is defined uniquely. But if $H^0(V, \mathscr{N}) \not\equiv 0$ then it is possible to construct a family of submanifolds $\mathscr{V}$ depending on a complementary parameter t which run over a neighborhood of zero in $H^0(V, \mathscr{N})$.

Penrose exploits a deformation of the space of twistors in order to obtain solutions for the self-dual Einstein-Hilbert equation (see, for example, [73]). The space of projective twistors $\mathbb{PT}$ is the space $\mathbb{CP}_3$ or an open subset of it U. Lines in it, i.e., subspaces which are isomorphic to $\mathbb{CP}_1$, are in one-to-one correspondence with points of compact complex Minkowski space $\mathbb{CM} = \mathrm{Gr}(2,4)$ or its subset V corresponding to those lines in $\mathbb{PT}$ which belong to U. In order to obtain instead of V a nonflat space with a holomorphic metric satisfying the Einstein-Hilbert equation, Penrose deforms the space U together with certain differential forms on it used to construct this metric. By the theorem of Kodaira each line in U is stable since $\mathscr{N} \cong (\mathscr{O}(1))^2$; consequently, it is imbedded in a family of submanifolds of the family deforming U. Since $\mathbb{CP}_1$ is rigid, each nearby fiber of the first family is again a line in the fiber U_s of a deformation of the domain U. Thus U_s contains, as does U, a four-parameter family of lines. According to Penrose this family, viewed as a complex manifold, is a twisted Minkowski space, and the incidence of these lines defines a conformal class of metrics. However since $H^0(\mathbb{CP}_1, \mathscr{N}) \neq 0$ there is no canonical correspondence between points of flat and points of twisted Minkowski space.

5. Deformations of Holomorphic Mappings

5.1. Let $f: X \to Y$ be a mapping of complex spaces or germs of complex spaces. A *deformation* of this mapping over a base $(S, *)$ is a triple consisting of a deformation $(\mathscr{X}, i, p)$ of the space X (i is an isomorphism from X to $p^{-1}(*)$), a deformation $(\mathscr{Y}, j, q)$ of the space Y, and a mapping $F: \mathscr{X} \to \mathscr{Y}$ such that $qF = p$ and $Fi = jf$. These equations are express the commutativity of the following diagram

$$\begin{array}{ccccc} X & \xrightarrow{i} & \mathscr{X} & & \\ & & & \searrow^{p} & \\ {\scriptstyle f}\downarrow & & {\scriptstyle F}\downarrow & & S. \\ & & & \nearrow_{q} & \\ Y & \xrightarrow{j} & \mathscr{Y} & & \end{array} \qquad (4.5.1)$$

Suppose that we are given another deformation of f with the same base S, whose components we denote with primes. This deformation is said to be isomorphic to the deformation (4.5.1) if isomorphisms $a:\mathscr{X}\to\mathscr{X}'$ and $b:\mathscr{Y}\to\mathscr{Y}'$ are defined so that a is an isomorphism of deformations of X, b is an isomorphism of deformations of Y, and the diagram

$$\begin{array}{ccc} \mathscr{X} & \xrightarrow{a} & \mathscr{X}' \\ {\scriptstyle F}\downarrow & & \downarrow{\scriptstyle F'} \\ \mathscr{Y} & \xrightarrow{b} & \mathscr{Y}' \end{array}$$

commutes. All the conditions together imply that the three-dimensional diagram which includes (4.5.1) and the analogous diagram with primes commute.

We compare these definitions with those given in Chapter 4, Section 3. A deformation of a relative space X/Y is described by the diagram (4.5.1) if we set $\mathscr{Y}=Y\times S$, define j via the imbedding $Y\times\{*\}\hookrightarrow Y\times S$, and take q to be the projection onto the first factor. Briefly then, deformations of the relative space X/Y are those deformations of the mapping $f:X\to Y$ for which Y is not deformed.

Suppose that we are given a deformation (4.5.1) and a morphism of germs $T\to S$. The inverse image of this deformation is defined in the natural manner. It contains the inverse image of the deformation of X, the inverse image of the deformation of Y, and also the mapping $F\times{}_ST:\mathscr{X}\times{}_ST\to\mathscr{Y}\times{}_ST$. Thus using the language of Chapter 4, Section 1, we may talk about versal deformations of the mapping f, etc.

5.2. We describe the tangent cohomology of the mapping f. The zero-dimensional component of it $T^0(f)$ is the collection of pairs (v_X, v_Y), where v_X (v_Y) is a tangent field on X (on Y), related by the mapping $v_X(\varphi(\alpha))=\varphi(v_Y(\alpha))$ for $\alpha\in\mathcal{O}_{Y,f(x)}$, where $\varphi:f^*(\mathcal{O}_Y)\to\mathcal{O}_X$ is the mapping of sheaves of algebras corresponding to f. In geometric language this equation means that v_X is an infinitely small motion on X sending each fiber of the mapping f to another fiber. In order to construct the tangent cohomology completely, it is necessary to proceed as follows: first we select coverings P and Q for the spaces X and Y in the sense of Chapter 2, Section 2, so that there exists a mapping $\lambda:A\to B$ of their index sets for which $f(P_\alpha)\subset Q_{\lambda(\alpha)}$ for all $\alpha\in A$. If $\mathscr{M}$ and $\mathscr{N}$ are the nerves of these coverings, then λ defines a functor $\Lambda:\mathscr{M}\to\mathscr{N}$. Further, let $\mathscr{R}_X$ and $\mathscr{R}_Y$ be resolutions on these coverings. Then by a mapping of resolutions $\mathscr{R}_f:\mathscr{R}_Y\to\mathscr{R}_X$ we mean the mapping of functors $\mathscr{R}_{Y_0}\Lambda\to\mathscr{R}_X$ defined on $\mathscr{M}$ which satisfies the conditions:

1) For any $A\in\mathscr{M}$ the morphism $\mathscr{R}_f(A):\mathscr{R}_Y(\Lambda(A))\to\mathscr{R}_X(A)$ is a mapping of resolutions of model complex spaces which lifts the mapping of sheaves $\mathcal{O}_Y|\mathscr{Q}_{\Lambda(A)}\to\mathcal{O}_X|P_A$ generated by f.

2) This mapping immerses the set of generators of the resolution $\mathscr{R}_Y(\Lambda(A))$ into the set of generators for $\mathscr{R}_X(A)$, including also generators of degree zero.

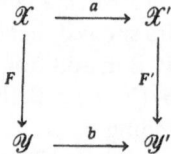

Assume that such a mapping $\mathscr{R}_f$ has been constructed. We consider the complex $\mathrm{Der}(\mathscr{R}_f)$ whose elements are pairs (t_X, t_Y) of derivations of $\mathscr{R}_X$ (resp., $\mathscr{R}_Y$) of the same degree which are related by the equation $t_X \cdot \mathscr{R}_f = \mathscr{R}_f \Lambda^*(t_Y)$, where $\Lambda^*(t_Y)$ denotes the transformation of the functor $\mathscr{R}_Y \cdot \Lambda$. The differential in this complex is given by the formula $d(t_X, t_Y) = ([t_X, s_X], [t_Y, s_Y])$ using the differentials in R_X and R_Y. The cohomology of the complex $\mathrm{Der}(R_f)$ is by definition the *tangent cohomology* of f; it is denoted by $T^*(f) = \bigoplus_0^\infty T^n(f)$. A more general construction was given in [32].

The complex $\mathrm{Der}(\mathscr{R}_f)$ maps to the complex $\mathrm{Der}(\mathscr{R}_Y)$ by the rule $(t_X, t_Y) \mapsto t_Y$. The cohomology of $\mathrm{Der}(\mathscr{R}_Y)$ is by definition $T^*(Y)$. The kernel of this mapping is the subcomplex whose cohomology is isomorphic to $T^*(X/Y)$. Thus there arises the exact sequence

$$\begin{aligned} 0 \to T^0(X/Y) \to T^0(f) \to T^0(Y) \to T^1(X/Y) \to T^1(f) \\ \to T^1(Y) \to T^2(X/Y) \to T^2(f) \to T^2(Y) \to \cdots . \end{aligned} \tag{4.5.2}$$

The first row can be easily understood. The space $T^0(X/Y)$ is the collection of vertical tangent fields on X; it is clearly a subset of the set $T^0(f)$ of fields which send a fiber of f to a fiber; each field in $T^0(f)$ generates a field on Y, i.e., an element of $T^0(Y)$. Further, each field on Y in composition with the mapping f defines an infinitesimal deformation of f which does not change Y, i.e., an element of $T^1(X/Y)$. The latter space, according to what was said in Section 5.1, maps to the space $T^1(f)$, which in turn maps to $T^1(Y)$, since every deformation of f contains a deformation of the space Y.

It is easy to localize these constructions by taking instead of a resolution of a complex space some Tyurina resolution for germs of a complex space. As a result we obtain the definition for the tangent cohomology of a morphism of germs. Each mapping of complex spaces $f: X \to Y$ defines at every point $x \in X$ a morphism of germs $f_x: (X, x) \to (Y, f(x))$ which is called the germ of the mapping f. Let $T^*(f_x)$ be its tangent cohomology. The union of $T^*(f_x)$ for all points of X is a sheaf which is denoted by $\mathscr{T}^*(f)$. It has the structure of an $\mathscr{O}_Y$-sheaf (but not an $\mathscr{O}_X$-sheaf).

5.3. Any mapping $f: X \to Y$ of compact complex spaces has a minimal versal deformation [32]. This fact is a consequence of the existence of such deformations for X and Y. The tangent space to the base of this deformation is isomorphic to $T^1(f)$. Flenner established also the existence of a versal deformation for any morphism of germs of f under two conditions: $\dim T^1(f) < \infty$ and $T^2(X/Y)$ is a finitely generated $\mathscr{O}_{Y,*}$-module. Bingener using the methods of [69] has established the existence of a versal deformation of the germ of an arbitrary proper map $f: X \to Y$ of complex spaces over any finite subset $F \subset Y$ provided that the $\mathscr{O}_Y$-module $T^1(f)$ has discrete support. This and some generalisations of the result mentioned in Chapter 2, Section 6 are obtained in [111].

5.4. We calculate the tangent cohomology of a mapping of germs $f : (\mathbb{C}^n, 0) \to (\mathbb{C}^m, 0)$. The corresponding mapping of analytic algebras $\varphi : H_m \to H_n$ sends a power series $a(y_1, \ldots, y_m)$ to the series $a(f_1(x), \ldots, f_m(x))$, where $f_1, \ldots, f_m$ are the components of f. The space $T^0(f)$ consists of germs of fields v on $(\mathbb{C}^n, 0)$ such that $v(f_i) \in \varphi(H_m)$ for $i = 1, \ldots, m$. To calculate $T^1(f)$ we employ (4.3.2) and (4.5.2). Since $T^1(Y) = 0$ and $T^0(Y) \cong H_m^m$, we obtain

$$T^1(f) \cong H_n^m/(df \cdot H_n^n + (\varphi(H_m))^m). \tag{4.5.3}$$

From (4.3.1) and (4.5.2) we find that $T^i(f) = 0$ for $i > 1$.

5.5. As an example to illustrate these constructions, we consider again the theorem of Kodaira on stability (Chapter 4, Section 4). Let X be a compact manifold and f an imbedding of X into a manifold Y. The assertion of Kodaira's theorem is that under an appropriate condition every deformation of Y is covered by a deformation of the mapping f (a small deformation of an imbedding is again an imbedding). For deformations over the double point this condition can be discerned in (4.5.2). Indeed, the image of $T^1(f)$ in $T^1(Y)$ is the set (of classes) of those deformations of Y over the double point which are covered by deformations of f. Therefore it follows from the exactness of (4.5.2) that the condition $T^2(X/Y) = 0$ is sufficient for this coverability. We calculate $T^2(X/Y)$ using the spectral sequence in Chapter 4, Section 3. We see that from the exactness of the sequence of sheaves

$$0 \to \mathcal{T}^0(X/Y) \to \mathcal{T}^0(X) \to \mathcal{T}^0(Y, \mathcal{O}_X) \to \mathcal{T}^1(X/Y) \to \mathcal{T}^1(X) \to \cdots$$

it follows that $\mathcal{T}^0(X/Y) = \mathcal{T}^2(X/Y) = 0$ and $\mathcal{T}^1(X/Y)$ is the normal sheaf $\mathcal{N}_X$ to the image of the imbedding f. It then follows that $T^2(X/Y) \cong H^1(X, \mathcal{N}_X)$; this explains the meaning of Kodaira's condition $H^1(X, \mathcal{N}_X) = 0$. A similar more complete discussion shows that under this condition any infinitesimal deformation of Y (i.e., a deformation with zero-dimensional base) is covered by a deformation of f. The theorem of Kodaira asserts that such a covering exists for a deformation whose base is any nonsingular germ. The sequence (4.5.2) also explains the role of the condition $H^0(X, \mathcal{N}_X) = 0$. Since $T^1(X/Y) \cong H^0(X, \mathcal{N}_X)$, then under this condition the mapping $T^1(f) \to T^1(Y)$ is monomorphic, and consequently the covering indicated above is unique.

6. The Stability of Germs of Smooth Mappings

The stability theory of germs of smooth mappings, which was developed by Thom, Malgrange, Mather, and V.I. Arnol'd at first independently of complex analytic geometry, has attained gradually many common features with it (see Volume 6). Tougeron and Mather [63] investigated the property of being finitely defined for a germ of a smooth mapping $f: X \to Y$ with respect to various groups of transformations acting on the germ $X \times Y$. By definition, the germ f is k-defined with respect to a group G (k is a natural number) if any germ g having the same k-jet at the distinguished point (the segment of the Taylor series up to terms of kth order) lies in an orbit of f. A germ f is finitely defined if it is k-defined for some k. A criterion for being k-defined is the finite-dimensionality of a certain vector space, which can be realized as a transversal to the orbit of G passing through f. In the case of G equal to the largest group of contact transformations this space can be written as $T^1(X_0)$, where $X_0 = f^{-1}(*)$, if for the definition of this space formula (3.1.1) is used with H_n replaced by the algebra of germs of smooth functions on X. In the case of the group of coordinate changes on X the transversal can be written as $T^1(X/Y)$ if we use formula (4.3.2) in the same way, and in the case of the group of coordinate changes on X and Y as $T^1(f)$ using formula (4.5.3).

According to Mather [63] the stability of a mapping is equivalent to its infinitesimal stability, which means the vanishing of the corresponding transversal. Here the direct analogy with the deformation theory of germs of holomorphic mappings is clear, in which the concept of rigidity corresponds with that of stability, and the criterion for rigidity is the vanishing of the corresponding space T^1.

The next logical step was to construct a versal deformation for finitely defined germs. Such deformations appeared in work of Wassermann [103] and Martinet [62] under the name of "unfolding" or "deploiment" of a germ. In the work of Martinet explicit mention is made of the direct analogy with versal deformations of germs of holomorphic mappings. A singular germ X of the level hypersurface of a smooth mapping has a versal unfolding if and only if the space $T^1(X_0)$ is finite dimensional with notation as above. Moreover, the criterion for the versality of an unfolding with base S, which is given in [62], can be formulated thus: the differential of this unfolding $T(S) \to T^1(X_0)$ (understood in the sense of Chapter 2, Section 4) is an epimorphism.

The parallelism between the real and complex analytic theories of deformations of germs becomes more complete if we direct attention to the methods. In the complex case any arguments employ the preparation lemma of Weierstrass. In the real case proofs are based on the highly nontrivial analog of this lemma—the well-known theorem of Malgrange.

7. Deformations of Holomorphic Foliations

A holomorphic foliation on a real smooth manifold, sometimes called a transversally holomorphic foliation, is a structure more general than that of a complex manifold. The definition, which is due to Haefliger, consists in the following:

an open covering $X = \bigcup U_\alpha$ is given, and for each index α a smooth covering $f_\alpha: U_\alpha \to \mathbb{C}^q$, i.e., a mapping having maximal rank $2q$ at each point. One assumes that for each pair of indices α, β there exists a holomorphic mapping $\varphi_{\alpha\beta}: f_\alpha(U_\alpha \cap U_\beta) \to f_\beta(U_\beta \cap U_\alpha)$ such that $\varphi_{\beta\alpha} \cdot f_\alpha = f_\beta$. Thus the existence of a foliation $\mathscr{F}$ defines a local decomposition of X into the union of fibers $f_\alpha^{-1}(z_\alpha)$, parametrized by points $z_\alpha \in \mathbb{C}^q$; and on passing to another chart f_β the parameters of the fibers transform via the holomorphic mappings $\varphi_{\beta\alpha}$. Consequently, each point $x \in X$ has a neighborhood $\mathscr{V}$ so that the quotient space $\mathscr{V}/\mathscr{F} \cap \mathscr{V}$ has the structure of a complex manifold. In the case when in each chart of the foliation $2q = \dim X$, the mappings f_α are local diffeomorphisms; consequently, a holomorphic foliation on X defines the structure of a complex manifold.

To describe deformations of holomorphic foliations, there are two approaches which are similar to the methods used in the deformation theory of complex manifolds. The first of these is the deformation of the distribution on X associated to the given foliation. It is analogous to the method of deforming a quasi-complex structure used in the work of Kodaira-Nirenberg-Spencer [53] and Kuranishi [56] and consists in the following. Let $L \subset TX$ be a distribution tangent to F. The normal bundle $Q = TX/L$, because F is holomorphic, admits a canonical decomposition $Q^{\mathbb{C}} = Q^{(1,0)} \oplus Q^{(0,1)}$, where $Q^{\mathbb{C}}$ is its complexification. Let E denote the kernel of the projection $T^{\mathbb{C}}X \to Q^{(1,0)}$. To every holomorphic foliation $\mathscr{F}'$ near to $\mathscr{F}$ we may associate the mapping of distributions $\varphi: E \to Q^{(1,0)}$, whose graph E_φ is related to $\mathscr{F}'$ in the same way as E is to $\mathscr{F}$. The integrability condition can be written as $[E_\varphi, E_\varphi] \subset E_\varphi$.

The other approach consists in deforming the gluing mappings $\varphi_{\beta\alpha}$ without changing the charts f_α, which is similar to deforming the gluing mappings defining the structure of a complex manifold (Chapter 2, Section 4). An infinitesimal deformation of $\varphi_{\beta\alpha}$ is generated by a local vector field on X which sends each fiber of $\mathscr{F}$ into another fiber in such a way that there arises on the parameter space a local holomorphic field. Let $\Theta_{\mathscr{F}}$ denote the sheaf of germs of such fields on X, and let Θ_0 be the subsheaf formed by fields which are tangent to F. The quotient sheaf $\Theta = \Theta_{\mathscr{F}}/\Theta_0$ here plays the same role which was played by the tangent sheaf $\mathscr{T}^0$ in Chapter 2, Section 4. If the manifold X is compact, then the cohomology of Θ is finite dimensional. The reason is that $H^*(X, \Theta)$ can be realized as the cohomology of a certain complex of elliptic differential operators. The basic result in the theory, which is similar to the theorem of Kuranishi, states that all holomorphic foliations on X near to the given foliation $\mathscr{F}$ can be parametrized by some closed analytic subset $S \subset U$, where U is a neighborhood of zero in $H^1(X, \Theta)$ ([34], [29]). Moreover, over the germ $(S, 0)$ there is a holomorphic family of holomorphic foliations on X which contains the foliation $\mathscr{F}$ over the distinguished point; and this family is versal in the category of all similar families. As in the theory discussed in Chapter 2, the set S is the preimage of zero under a nonlinear mapping $\Phi: U \to H^2(X, \Theta)$.

8. Deformations of CR-Manifolds

By a *CR-manifold* we mean a smooth real manifold M of dimension N on which there is given a p-dimensional complex distribution $E \subset T^{\mathbb{C}}M$ satisfying the conditions: 1) $[E, E] \subset E$ (involutivity) and 2) $E \cap \bar{E} = 0$ (absence of real vectors). In what follows we shall always assume that $N = 2p + 1$. The basic example of a CR-manifold is the smooth boundary of a domain X in a complex analytic manifold Y. In this case $\bar{E} = T^{\mathbb{C}}M \cap T^{(0,1)}$, where $T^{(0,1)}$ is the subbundle of $T^{\mathbb{C}}Y$ generated by the Cauchy Riemann fields $\partial/\partial\bar{z}_1, \ldots, \partial/\partial\bar{z}_{p+1}$ in each chart. Thus sections of E are tangent fields on $M = \partial X$ defining the tangential Cauchy-Riemann system. The term CR-manifold itself comes from this example. We must note, however, that not every CR-manifold can be imbedded even locally into a complex manifold in such a way that $\bar{E}$ is generated by the Cauchy-Riemann fields tangent to the image of this imbedding.

In the last two decades the theory of CR-manifolds has been developed mainly within the framework of function theory and differential equations. The basic objects of this theory are CR-functions, i.e., solutions of the tangential Cauchy-Riemann system, and the complex of differential operators associated with this system. The main difference between this theory and that of complex manifolds lies in the fact that in the case of CR-manifolds the Cauchy-Riemann system and its associated complex is not elliptic. As a consequence of this fact the geometry of a CR-manifold turns out to have a strong influence on the cohomology of this complex. The geometry of such a manifold is characterized by the *Levi form*, which is defined in the following manner. Let σ be a smooth section of the bundle TM which generates the one-dimensional bundle $T^{\mathbb{C}}M/(E + \bar{E})$. Let u be any section of E and $\bar{v}$ a section of $\bar{E}$. We consider their commutator and set

$$L(u, v)\sigma = i[u, \bar{v}] \mod E + \bar{E}.$$

It is easy to see that the value of the function $L(u, v)$ at a point of M depends only on the values of the fields u and v at this point. Consequently $L(u, v)$ is a Hermitian form on E; it is called the Levi form of the CR-manifold M, oriented by the choice of σ. If M is the boundary of a domain $X \subset\subset Y$, then it is standard to select as σ the exterior normal. A CR-manifold is called strictly pseudoconvex if its Levi form is strictly positive definite. The domain X is called strictly pseudoconvex if its boundary $M = \partial X$ is. Boutet de Monvel has shown that every compact strictly pseudoconvex CR-manifold of dimension $2p + 1$ not smaller than 5 can be imbedded locally into $\mathbb{C}^{p+1}$ by CR-functions, i.e., it is locally the boundary of a strictly pseudoconvex domain.

The deformation theory of strictly pseudoconvex CR-manifolds is closely related to that of the corresponding domains. In this sense the theory goes back to Poincaré, who studied the deformations of the ball in $\mathbb{C}^2$. Work of Segre, E. Cartan, and Tanaka is also relevant here. In an article of Chern and Moser [19] normal forms are given for germs of strictly pseudoconvex hypersurfaces in

$\mathbb{C}^n$ relative to the group of holomorphic coordinate changes. It follows from results in this article that the orbits of this group cannot be parametrized using a finite number of parameters. Wells [106] established a similar result for germs of strictly pseudoconvex domains $X \subset \mathbb{C}^2$ at a fixed boundary point x_0. He showed that for a sufficiently small ball U centered at x_0 if $\tilde{X}$ is a domain near X in the class C^∞ the intersections $\tilde{X} \cap U$ define a set of isomorphism classes of complex manifolds depending on an infinite number of parameters. It follows that the isomorphism classes of germs of strictly pseudoconvex CR-manifolds of dimension 3 (or more) also depend on an infinite number of parameters and therefore cannot have a finite-dimensional versal deformation. Further results in this direction can be found in [17] (see also Volume 7).

The formalism of the deformation theory of CR-manifolds is connected with the complex of tangential Cauchy-Riemann operators mentioned above. It is analogous to the formalism of the deformation theory of holomorphic foliations, although in Chapter 4, Section 7 the symbol E had another meaning. The difference is that in the present case the complex can have infinite dimensional and non-Hausdorff cohomology, which is a reflection of the appearance of an infinite number of parameters in the cases described above. However, if it is assumed that M is compact and the Levi form at each point has an appropriate number of eigenvalues of the appropriate sign, then by a theorem from complex analysis the cohomology of this complex in the appropriate dimensions turns out to be finite dimensional. This allows us to pass from the formal theory to existence theorems.

Kiremidjian [48] and Hamilton [44] have studied the relation between deformations of a complex manifold $X \subset\subset Y$ with boundary M and deformations of the CR-manifold M. In particular, Hamilton has established that if M is strictly pseudoconvex and $H^1(X, \mathcal{T}^0) = 0$, then any small deformation of the CR-structure on M is generated by a small deformation of the imbedding $M \to Y$, i.e., by a small deformation of the domain X in Y.

By a theorem of Rossi [108] any compact strictly pseudoconvex CR-manifold M is the boundary of some normal Stein space X with perhaps a finite number of singular points. In an article of Yau [108] the cohomology of the *Kohn-Rossi complex* on M is related to invariants of these singular points. Let $\mathscr{A}^{p,q}$ denote the sheaf of germs of smooth (p,q)-forms on $\bar{X}$ and $\mathscr{C}^{p,q}$ its subsheaf consisting of forms of type $\bar{\partial} h \wedge \Psi$, where $\Psi \in \mathscr{A}^{p,q}$ and $h \in \mathscr{A}^{0,0}$ is a function equal to zero on M. The sheaf $\mathscr{B}^{p,q} = \mathscr{A}^{p,q}/\mathscr{C}^{p,q}|M$ is called the sheaf of tangent forms on M. The operator $\bar{\partial}$ acts on $\mathscr{C}^{p,q}$, and consequently it lifts to the sheaf $\mathscr{B}^{p,q}$. The space of sections $\mathscr{B}^{p,q} = \Gamma(M, \mathscr{B}^{p,q})$ forms a complex

$$0 \to B^{p,0} \xrightarrow{\bar{\partial}_M} B^{p,1} \to \cdots \xrightarrow{\bar{\partial}_M} B^{p,n} \to 0.$$

Its cohomology $H^{p,q}(M)$ is called the Kohn-Rossi cohomology of the manifold M. Yau has shown that in the case $n \geqslant 3$ for any p and q the equation

$$\dim H^{p,q}(M) = \sum_i b_{x_i}^{p,q+1},$$

holds, where the sum of taken over all singular points $x_i \in X$. The number $b_x^{p,q}$ is the dimension of the local cohomology $H^q_{\{x\}}(X, \Omega^p(X))$. In the case when each singular germ (X, x) is the germ of a hypersurface $Z \subset \mathbb{C}^{n+1}$, Yau calculated $b^{p,q}$ using the duality between local cohomology and the space $\mathrm{Ext}^{n+1-q}(\Omega^p(Z), \mathcal{O}_{\mathbb{C}^{n+1},x})$. In this case $H^{p,q}(M) = 0$ for $p + q \neq n, n-1, 1 \leqslant q \leqslant n-2$, and

$$\dim H^{p,q}(M) = \sum_i \dim T^1(X, x_i),$$

if $p + q = n, n-1, 1 \leqslant q \leqslant n-2$. In other words, the dimension of $H^{p,q}(M)$ is equal to the dimension of the base of a versal deformation of the union of the singular germs of the space X; however, this equation does not correspond to an isomorphism but to duality. Using these results, Yau gave a solution to the complex Plateau problem: a compact oriented strictly pseudoconvex CR-manifold M of dimension $2n-1$ for $n \geqslant 3$ imbedded in a Stein space is the boundary of some complex submanifold if and only if the Kohn-Rossi cohomology $H^{p,q}(M)$ is equal to zero for $1 \leqslant q \leqslant n-2$.

Chapter 5. The Geometry of Deformations and Periods of Differential Forms

1. The Structure of the Base of a Deformation

1.1. The Discriminant. Let X_0 be a compact complex space or a germ of a complex space with a unique singular point, and let $f: X \to S$ be a minimal versal deformation of X_0. Let X and S denote representatives for the corresponding germs, chosen so that in the first case the mapping f is proper and flat, and in the second case f is also flat and the restriction of f to the set $\mathrm{Sing}(f)$ of its singular points is proper. Let $\mathrm{Sing}(f)$ denote the union of $\mathrm{Sing}(X_s)$ relative to all points $s \in S$; this subset of X is closed and analytic. Therefore its image $D = f(\mathrm{Sing}(f))$ is a closed analytic subset of the base S by a theorem of Remmert (Chapter 1, Section 3.2). It is called the *discriminant subset* or simply the *discriminant*. This subset may coincide with S; in this case the space or germ X_0 is called nonsmoothable. A number of examples of nonsmoothable one-dimensional germs are known. Pinkham [75], in particular, has shown that the germ of the union of 13 lines passing through the origin in $\mathbb{C}^8$ in general position is nonsmoothable. In [76] there is an example of a nonsmoothable compact complex space all of whose singularities are normal self intersections.

In the case when X_0 is the germ of a complete intersection, the discriminant is the germ of a hypersurface $d = 0$, where d is a generator for the sheaf of ideals $\mathcal{O}_S$ canonically associated with $\mathrm{Sing}(f)$ [80]. If X_0 is the germ of a hypersurface, its versal deformation is constructed in the manner described in Chapter 1,

Section 1.1, and the variable s_1 corresponds to the monomial $e_1 = 1$, then the function $d(s_1,0,\ldots,0)$ has at the point $s_1 = 0$ a root of multiplicity equal to the Milnor number μ (Chapter 5, Section 3). According to Wirthmüller [107] a germ X_0 of a complete intersection is uniquely determined by the germ of the discriminant of its versal deformation, but only the type of X_0 (see Chapter 5, Section 2) is determined if X_0 is the germ of a hypersurface.

1.2. Stratification. We shall assume in addition that the representative S for the base of a versal deformation of f is chosen so that it is versal for each of its fibers X_s. The germ of this deformation on the fiber X_s is generated from the minimal versal deformation $Y \to (T, *)$ of the space X_s by some mapping $\Psi: (S, s) \to (T, *)$. We consider the germ of a complex space $N_s = \Psi^{-1}(*)$. By the versality of f the rank of the mapping Ψ is equal to dim $T_*(T)$; consequently, the tangent space $T(N_s)$ is equal to the kernel of the differential $D_s f: T_s(S) \to T^1(X_s)$ of the deformation f at the point s. Among all germs contained in the germ (S, s) over which the deformation f is trivial, the germ N_s is maximal; in the case when X_0 is the germ of a complex space, by trivial we mean the triviality of the germ at each singular point X_s. A *stratum* of S is a connected subset S_α whose germ at each point s coincides with N_s, where S_α is maximal among the sets possessing these properties. We note some properties of the strata:

1) the entire base S is a union of strata;
2) the closure of a stratum is a union of strata;
3) any stratum is the difference of two closed analytic sets;
4) a point is a stratum if and only if it belongs to a modular subspace $M \subset S$.

The discriminant is always a union of strata. If the closure of a stratum S_α contains the distinguished point $* \in S$, we call this stratum attractive, and the isomorphism class of spaces containing all the fibers X_s for $s \in S$ a disintegration of the space X_0. In the case when X_0 is a germ, its disintegration is defined to be the collection of singular germs of fibers of the deformation f over some attractive stratum. Thus any complex space Y, resp., any collection of singular germs, which is isomorphic to some fiber of the restriction of the deformation f to any neighborhood of the distinguished point in S, is a disintegration of X_0. The relation is transitive, i.e., if Y is a disintegration of X_0 and Z is a disintegration of Y, then Z is also a disintegration of X_0. This follows from the fact that f is also a versal deformation for Y.

2. The Classification of Germs

Germs of complex spaces can be classified according to the degree of complication of the decomposition of the base of their versal deformation into strata. If we are concerned with germs of hypersurfaces, then it is useful to combine them into equivalence classes called *types*. The germ of a hypersurface Y_0 is called a Morse suspension of a germ X_0 if it is isomorphic to the germ of a hypersurface of the form $f(x_1,\ldots,x_n) + x_{n+1}^2 + \cdots + x_N^2 = 0$, where $f(x_1,\ldots,x_n) = 0$ is an equation of the germ X_0. Two germs of hypersurfaces are said to be equivalent if they are isomorphic to Morse suspensions of the same germ. For equivalent germs X_0 and Y_0 there is a natural isomorphism $T^1(X_0) \cong T^1(Y_0)$ and a consistent isomorphism of the minimal versal deformations which preserves the decompositions of their bases into strata.

The type of a germ of a hypersurface is called simple if in the base of its minimal versal deformation there are only a finite number of strata, which are all necessarily attractive. In other words, the type of a germ of simple if the germ disintegrates in only a finite number of ways. For nonsimple types in the base of a deformation there are holomorphic families depending essentially on some number of complex parameters. The maximal number of parameters in such families can depend on the representative of the versal deformation, but it stabilizes with a shrinking of the base. The stable maximum of this number is called the modality of the type for germs of hypersurfaces. This terminology almost coincides with that used by V.I. Arnol'd [3], [4] in the classification of germs of holomorphic functions $f: \mathbb{C}^n \to \mathbb{C}$. The difference lies in the fact that functions are classified according to the orbits of the group of local coordinate changes in $\mathbb{C}^n$ (i.e., as relative germs $\mathbb{C}^n/\mathbb{C}$), while the classification of germs of hypersurfaces $\{f = 0\}$ is the enumeration of orbits of the contact group. However, the enumeration of simple germs in both cases reduces to the same result: a germ X_0, resp., a germ f, is simple if and only if it is equivalent to one of the singular germs of Klein surfaces, resp., to the germ of a function defining this surface (see Chapter 3, Section 6). Thus the types of simple germs form series A_k for $k \geqslant 1$; D_k for $k \geqslant 4$; and E_k for $k = 6, 7, 8$. The class of simple germs coincides also with the class of elliptic germs, i.e., germs for which the intersection form in the Milnor bundle (Chapter 5, Section 3) in definite in sign provided that $\dim X_0$ is even [95].

Those germs having in the base of a versal deformation a modular subspace of positive dimension and only a finite number of other strata can also be classified. It follows from [3] that each of them is equivalent to a germ in one of the following three modular families:

$$x^3 + y^3 + z^3 + \lambda xyz, \qquad \lambda^3 + 27 \neq 0; \qquad \mu = 8;$$

$$x^4 + y^4 + \lambda x^2y^2, \qquad \lambda^2 \neq 4; \qquad \mu = 9;$$

$$x^3 + y^6 + \lambda x^2y^2, \qquad 4\lambda^3 + 27 \neq 0; \qquad \mu = 10.$$

The collection of these germs coincides with the class of parabolic germs, i.e., those whose intersection form is semidefinite. Various information concerning elliptic and parabolic germs can be found in [5] and [81]; in particular, their decays are described as is the topology of the complement of the discriminant.

The classification carried out by V.I. Arnol'd [3], [4] covers all germs of functions f with $\mu \leqslant 16$, where μ (the Milnor number) is equal to the dimension of the space $T^1(\mathbb{C}^n/\mathbb{C}) \cong H^n/df \cdot H^n_n$, and also all germs with modality not exceeding

2 (all real forms of these germs are also enumerated). Any deformation of a germ of a function f is simultaneously a deformation of the germ X_0 of its zero level surface, however, the dimension τ of the base of a minimal deformation of X_0 as calculated by formula (3.1.1) can differ from but can never exceed μ. The difference $\delta = \mu - \tau$ is the dimension of the subspace in the base of a minimal versal deformation of f along which the germ of the zero level hypersurface of the deforming function remains isomorphic to X_0. This number is equal to zero if and only if the germ X_0 can be shrunk holomorphically to its singular point; this property is equivalent to the property that in an appropriate coordinate system the function f is a weighted homogeneous polynomial (K. Saito). Within Arnol'd classification the number δ does not exceed the modality of the germ.

This classification was completed by Giusti [35], who described all simple germs of complete intersections with a unique singular point, and by A.G. Aleksandrov [2], who supplemented Giusti's classification with a list of the weighted homogeneous curves of modality 1. All these classifications are contained in a single diagram of disintegrations in which various empirical regularities are visible. In particular, modular families of germs separate this diagram into connected parts in each of which the modality of the germs is constant.

3. The Milnor Bundle

Let X_0 be the germ of a hypersurface $f = 0$ in $\mathbb{C}^{n+1}$ with distinguished point $z = 0$, having no singularities except for this point. If $U \subset \mathbb{C}^{n+1}$ is a sufficiently small ball centered at this point, then the sphere ∂U intersects X_0 transversally and also all neighboring level manifolds $X_s = f^{-1}(s)$ if $|s| < \varepsilon$ for $\varepsilon > 0$ a sufficiently small number. In this situation according to a theorem of Milnor [64], the open manifold $X_s \cap U$ has the homotopy type of a bouquet of n-dimensional spheres. The number $\mu = \mu(X_0)$ of spheres in this union is called the *Milnor number* of the germs X_0 and f (it does not depend on the choice of U). This number coincides with the dimension of the space $T^1(\mathbb{C}^{n+1}/\mathbb{C})$ (see [67]). The union of the manifolds $X_s \cap U$ over the punctured disc $S' = \{s | 0 < |s| < \varepsilon\}$ forms the so-called *Milnor bundle*. It is locally trivial as a bundle of smooth real manifolds. Suppose that we fix a point $s_0 \in S'$ and a cycle $\gamma_0 \subset X_{s_0}$ corresponding to some element $c_0 \in H_n(X_{s_0}, \mathbb{Z})$. In view of the local triviality of the Milnor bundle, we may define a continuous family of cycles $\gamma(s) \subset X_s$ over some neighborhood V of the point s_0, which contains the cycle $\gamma_0 = \gamma(s_0)$. This family defines a section over V of the homological bundle $\mathscr{H}_n = \bigcup_{S'} H_n(X_s, \mathbb{Z})$ which depends only on the class c_0. This construction defines a connection on the homological bundle, that connection for which the indicated section is horizontal.

The cohomological bundle $\mathscr{H}^n = \bigcup_{s'} H^n(X_s, \mathbb{C})$ is dual to the homological one. There is defined in it a flat connection dual to the one described above. It is called the *local Gauss-Manin connection.* A section h of the cohomological bundle over a domain $V \subset S'$ is horizontal if for any horizontal section c of the homological bundle the function $h(c)$ is constant. It is known that this connection has at the critical value $s = 0$ a regular singularity (see the article [41] and the references there). The property of regularity can be characterized in the following manner. If the cohomological bundle is supplied with a norm, then every horizontal section grows as $s \to 0$ no faster than $C|s|^{-q}|\ln s|^p$ for some constants p and q. We choose some path in S' beginning and ending at the point s_0 and going round the center of S' counterclockwise. Any element of the fiber of $\mathscr{H}_n$ and also any element of the fiber of $\mathscr{H}^n$ over the point s_0 can be continued around this path as a horizontal section. As a result we obtain the classical monodromy operator M in $H_n(X_{s_0}, \mathbb{Z})$ and its adjoint operator M^* in $H^n(X_{s_0}, \mathbb{C})$. By a theorem of Landman [16] the operator M is unipotent, more precisely, $(M^N - E)^{n+1} = 0$, where N is some natural number. Further information can be found in [86] and [90].

In the case of the germ of a hypersurface with a nonisolated singular point the neighboring nonsingular level hypersurfaces also form a locally trivial smooth bundle, but they may have nonzero homology in dimensions q for $0 < q < n$. Nonetheless the monodromy operator as before is unipotent. (Lê Dûng Tráng [58]).

Let $X_0 \subset \mathbb{C}^m$ be the germ of a complete intersection with a unique singular point and $X \to S$ its minimal versal deformation, where X is a model subspace of $\mathbb{C}^m \times S$. Again let U be a small ball in $\mathbb{C}^m$ centered at the singular point of X_0. If the base S of the deformation is sufficiently small, then the fibers X_s for points $s \in S \backslash D$, where D is the discriminant of the deformation, also form a locally trivial differentiable bundle with fiber homotopically equivalent to a union of μ n-dimensional spheres, where $n = \dim X_0$ (Hamm [45]). For a weighted homogeneous complete intersection the number μ is equal to the dimension τ of the base S. In the general case $\tau \leqslant \mu$ ([41], [60]). The role of the monodromy operator M^* is played by the group homomorphism

$$\pi_1(S\backslash D, s_0) \to \mathrm{Aut}(H^n(X_{s_0}, \mathbb{C})), \tag{5.3.1}$$

which is called the monodromy of the Picard-Lefshetz deformation. A description of it is known in the case when X_0 is the germ of a hypersurface of one of the types A, D, or E (Chapter 5, Section 2). In this case the pair (S, D) is homeomorphic to the pair $(V/W, \Sigma/W)$, where V is a suitable neighborhood of zero in $\mathbb{C}^\tau$ on which the Weyl group W of the same type (i.e., A, D or E) acts by mirror reflections (cf., Chapter 3, Section 6), and Σ is the union of the singular orbits of W, i.e., orbits with number of points smaller than the maximal. The space $\mathbb{C}^\tau$ contains a universal covering space of $(V\backslash\Sigma)/W$ and at the same time is isomorphic to a fiber of the cohomological bundle over the distinguished point

s_0. To every loop in $S \backslash D$ with beginning and end at s_0 there corresponds a loop in $(V \backslash \Sigma)/W$, and consequently a path in $\mathbb{C}^\tau$ whose end is obtained from the beginning by the action of some element $w \in W$. The action of w is also the image of the class of this path under the homomorphism (5.3.1) ([15], [3]). For further results see [114].

4. Picard-Fuchs Equations in the Milnor Bundle

Again let X_0 be the germ of a complete intersection with a unique singular point and $f: X \to S$ a minimal versal deformation of X_0 in which X is a closed subspace of $U \times S$, where $U \subset \mathbb{C}^m$ is a ball centered at the singular point of X_0. Let D be the discriminant of the deformation, $S' = S \backslash D$, $X' = X \backslash f^{-1}(D)$, and $f': X' \to S'$ the restriction of the mapping f. We shall assume that a representative for the base S is chosen sufficiently small that f' is the Milnor bundle (Chapter 5, Section 3). Further, let $\Omega_{S'}$ denote the sheaf of holomorphic differential forms on S', and $\Omega_{X'}$ the sheaf of holomorphic forms on X', endowed with the differential d. The mapping f' generates the inverse image mapping $f^*: \Omega_{S'} \to \Omega_{X'}$. We consider the subsheaf $f^*(\Omega_{S'}^{\prime}) \wedge \Omega_{X'}$ in $\Omega_{X'}$. It coincides with the sheaf of those forms on X' which vanish when restricted to each fiber of f. The quotient sheaf

$$\Omega_{X'/S'} = \Omega_{X'}/f^*(\Omega_{S'}^1) \wedge \Omega_{X'}$$

is called the sheaf of relative differential forms on f'. The operator d induces a differential in this sheaf. A form $\omega \in \Omega_{X'}$ is called relatively closed if $d\omega \in f^*(\Omega_{S'}^1) \wedge \Omega_{X'}$. The cycles in the sheaf $\Omega_{X'/S'}$ are generated by the relatively closed forms and the boundaries by the exact ones.

We consider the sheaf $R^0 f'(\Omega_{X'/S'})$ defined on S'. Its fiber at the point $s \in S'$ can be described as the quotient of the space of holomorphic forms, each of which is defined on $f^{-1}(V)$, where V is some ball centered at s, relative to the space of forms $\sum df_i \wedge v_i$, where the v_i are holomorphic forms on $f^{-1}(V)$ and $s_i = f_i$ for $i = 1, \ldots, \tau$ are the equations of the mapping f. This description follows from the fact that $f^{-1}(V)$ is a Stein manifold (because of the convexity of U), as does the vanishing of the higher direct images $R^k f'(\Omega_{X'/S'})$. In the sheaf $R^0 f'(\Omega_{X'/S'})$ there is a differential induced from the sheaf $\Omega_{X'/S'}$). Consequently the cohomology sheaf

$$\mathscr{H}^*_{\mathrm{DR}}(X'/S') = H^*(R^0 f'(\Omega_{X'/S'})),$$

is defined, which is called the relative de Rham cohomology sheaf of the mapping f'. This sheaf has the natural structure of an $\mathcal{O}_{S'}$-sheaf. It follows from Serre's theorem on the cohomology of a Stein manifold (see Article I) that $\mathscr{H}^*_{\mathrm{DR}}(X'/S')$ is isomorphic to the sheaf of germs of holomorphic sections of the cohomological bundle $\mathscr{H}^*$, described in Chapter 5, Section 3.

To the local Gauss-Manin connection (Chapter 5, Section 3) there corresponds the differential operator

$$\nabla : \mathscr{H}^n_{\mathrm{DR}}(X'/S') \to \mathscr{H}^n_{\mathrm{DR}}(X'/S') \otimes_{\mathcal{O}_{S'}} \Omega^1_{S'},$$

whose kernel coincides with the sheaf of horizontal sections of the cohomological bundle. As a connection operator, it must satisfy the equation $\nabla(\varphi\sigma) = d\varphi \otimes \sigma + \varphi\nabla(\sigma)$ for any element $\varphi \in \mathcal{O}_{S'}$. We give a more explicit description of this operator. We represent a class of relative de Rham cohomology by a holomorphic form ω defined on the manifold $f^{-1}(V)$, where V is some ball in S'. Let $\gamma(s)$ be a continuous family over V of cycles in the Milnor bundle. We consider the integrals

$$a(s) = \int_{\gamma(s)} \omega.$$

The function a is holomorphic in V. In order to find its differential, we write $d\omega = \sum df_i \wedge \omega_i$, where ω_i are holomorphic forms on $f^{-1}(V)$. It is easy to verify using Stokes' formula that

$$da = \sum b_i\, ds_i, \qquad b_i(s) = \int_{\gamma(s)} \omega_i;$$

consequently for the constancy of the function a it is necessary and sufficient that $b_i \equiv 0$ for $i = 1, \ldots, \tau$. It follows that the connection operator has the form

$$\nabla(\mathrm{cl}(\omega)) = \sum_i \mathrm{cl}(\omega_i) \otimes ds_i.$$

Suppose that we have chosen holomorphic forms $\omega_1, \ldots, \omega_\mu$ on X' whose restrictions to each fiber of f' generate its cohomology in dimension n. These forms define sections $\bar{\omega}_1, \ldots, \bar{\omega}_\mu$ of the bundle $\mathscr{H}^n$ and form a basis for it. By what was said above, $d\omega_i = \sum df_i \wedge \omega_i^j$, where ω_i^j are holomorphic forms on X'. Each of them defines a holomorphic section of the cohomological bundle, which can be written in the form $\sum \Theta_i^{jk}\bar{\omega}_k$, where Θ_i^{jk} are holomorphic functions on S'. Denoting by $\bar{\omega}$ the column $(\bar{\omega}_1, \ldots, \bar{\omega}_\mu)$, we rewrite these relations in the matrix form $d\bar{\omega} = \Theta\bar{\omega}$, where $\Theta = \{\Theta_i^k = \sum \Theta_i^{jk}\, ds_j\}$ is a matrix with elements in $\Omega^1_{S'}$. This matrix is called the connection form for ∇. If the sections $\bar{\omega}_i$ are integrated relative to some continuous family of cycles $\sigma(s)$ and the integrals are denoted by $a_i(s)$, we obtain the system of differential equations of first order $da = \Theta a$, which is called the *local Picard-Fuchs system*.

The Gauss-Manin connection was studied in articles of Brieskorn [16], Greuel [41], and K. Saito [82].

Example. Let X_0 be the hypersurface in $\mathbb{C}^{1+\nu}$ defined by the equation $f(x, y) \equiv x^{k+1} + y_1^2 + \cdots + y_\nu^2 = 0$. The singular point $x = 0$, $y = 0$ has type A_k. Its minimal versal deformation is defined over the base $S = \mathbb{C}^k$ by the equation $F(x, y, s) = 0$, where

$$F(x, y, s) = x^{k+1} + s_{k-1}x^{k-1} + \cdots + s_1 x + s_0 + y_1^2 + \cdots + y_\nu^2.$$

In this case there is no necessity to intersect the hypersurface X defined by this equation by a ball, since the projection $f : X \to S$ is a locally trivial smooth bundle over the complement of the discriminant D. It is defined by the equation $d(s) = 0$, where d is the usual discriminant of the polynomial $F(x, 0, s)$. The forms

$$\omega_i = \frac{x^i\, dx \wedge dy_1 \wedge \cdots \wedge dy_\nu}{dF}, \qquad i = 0, 1, \ldots,$$

are holomorphic on $X' = X \setminus f^{-1}(D)$ and constitute a basis in the cohomological bundle if we restrict to the first k of them. We write the Picard-Fuchs system for their integrals a_i. This system can be written more simply if we consider a redundant number of integrals $a_0, \ldots, a_{k+1}$. This system contains the equations

$$\frac{\partial a_i}{\partial s_j} = \frac{\partial a_\alpha}{\partial s_\beta}, \qquad i + j = \alpha + \beta, \qquad 0 \leqslant j, \qquad \beta \leqslant k - 1.$$

So we may introduce functions

$$b_l = \frac{\partial a_i}{\partial s_j}, \qquad i + j = l, \qquad l = 0, \ldots, 2k.$$

Let b denote the column formed from these functions. The remaining part of the Picard system has the form $\Delta \cdot b = a'$, where Δ is the matrix of order $2k + 1$:

$$\Delta = \begin{vmatrix}
s_0 & s_1 & \cdots & s_{k-1} & 0 & 1 & 0 & \cdots & 0 \\
0 & s_0 & s_1 & \cdots & s_{k-1} & 0 & 1 & 0 \cdots & 0 \\
\vdots & \ddots & \ddots & & & \ddots & \ddots & \ddots & \vdots \\
0 & \cdots & 0 & s_0 & s_1 & \cdots & s_{k-1} & 0 & 1 \\
s_1 & 2s_2 & \cdots & (k-1)s_{k-1} & 0 & k+1 & 0 & \cdots & 0 \\
0 & s_1 & 2s_2 & \cdots & (k-1)s_{k-1} & 0 & k+1 & 0 \cdots & 0 \\
\vdots & \ddots & \ddots & & & \ddots & \ddots & \ddots & 0 \\
0 & \cdots & 0 & s_1 & 2s_2 & \cdots & (k-1)s_{k-1} & 0 & k+1
\end{vmatrix}$$

and a' is the column $((\nu/2 - 1)a_0, \ldots, (\nu/2 - 1)a_{k-1}, 0, -a_0, -2a_1, \ldots, -ka_{k-1})$.

We note that the determinant of the matrix Δ is the discriminant d of the polynomial $F(x, 0, s)$. It follows that on $S \setminus D$ the Picard-Fuchs system of equations is a system of complete differentials; that however is a property of this system for any germ X_0 of a complete intersection with a unique singular point. In other words, the characteristic set $M \subset T^*(S)$ of the Picard-Fuchs system lies over the discriminant set D. A more detailed analysis shows that for germs X_0 of types A, D, or E the set M is the cotangent bundle to the stratification of S described in Chapter 5, Section 1.

5. Deformations of Singular Points of Real Hypersurfaces

Let f be a real infinitely differentiable function in a domain $U \subset \mathbb{R}^n$ and $x = 0$ a critical point of f with critical value zero. The family of noncritical level hypersurfaces $f^{-1}(s)$ in a neighborhood of zero is the analog of the Milnor bundle. In fact, if the multiplicity of this critical point is finite (see below), then there exists a ball U' centered at this point and a number $\varepsilon > 0$ so that the manifolds with boundary $X_s = f^{-1}(s) \cap U'$ form a locally trivial bundle over the open set $\{0 < |s| < \varepsilon\}$. Thus all the manifolds X_s are mutually diffeomorphic if $0 < s < \varepsilon$ and also if $-\varepsilon < s < 0$ (they can also be empty). The multiplicity of the critical point $x = 0$ is the dimension of the quotient algebra

$$\mathbb{R}[[x]]/J(f), \tag{5.5.1}$$

where $\mathbb{R}[[x]]$ is the algebra of formal power series in $x_1, \ldots, x_n$ and $J(f)$ is the ideal generated by the Maclaurin series of the functions $\partial f/\partial x_1, \ldots, \partial f/\partial x_n$.

If the Maclaurin series of the function f converges, then it can be continued to a holomorphic function in a complex neighborhood of the point $z = 0$, and we can associate to this extension a Milnor bundle (Chapter 5, Section 3). The Milnor number μ of this bundle is equal to the multiplicity m. For the real analog of the Milnor bundle, which was described above, there is also a relation between the topology of the fibers and the algebra (5.5.1), however, this relation is more complicated.

In the case of $m = 1$ the critical point $x = 0$ of the function f is called a Morse point. According to a theorem of Morse, the function f is a quadratic form of rank n in a suitable smooth coordinate system defined in a neighborhood of the critical point:

$$f(x(y)) = y_1^2 + \cdots + y_k^2 - y_{k+1}^2 - \cdots - y_n^2.$$

The homotopy type of the level set as $X_s \to X_0$ changes by attaching cells of dimension k if $s > 0$ and cells of dimension $n - k$ if $s < 0$. If we denote by χ the reduced Euler characteristic of the manifold, i.e., the usual Euler characteristic minus 1, then $\chi(X_0) = 0$, since the germ of a critical level hypersurface retracts to the critical point. Therefore $\chi(X_s) = (-1)^k$ if $s > 0$ and $\chi(X_s) = (-1)^{n-k}$ if $s < 0$. It follows that

$$\chi(X_{-s}) = (-1)^n \chi(X_s). \tag{5.5.2}$$

According to [68], equation (5.5.2) is valid for any function f having a critical point at zero of finite multiplicity m. Here $|\chi(X_s)| = m - 2p$, where p is some nonnegative integer. This last relation is the real analog of the formula for the Milnor number, according to which $|\chi(Z_s)| = m$, where Z_s is a level hypersurface for the holomorphic extension of f.

The reason underlying this analogy is the fact that both formulas are related to properties of the mapping ∇f, whose coordinates are the partial derivatives of f. If the function f is holomorphic, then the local degree of the mapping $\nabla f:(\mathbb{C}^n,0)\to(\mathbb{C}^n,0)$ is equal to the number of spheres in the bouquet which is homotopically equivalent to Z_s. In the real case the local degree of the mapping $\nabla f:(\mathbb{R}^n,0)\to(\mathbb{R}^n,0)$ is equal to $-\chi(X_s)$ for $s<0$.

In papers of Khimshiashvili [47] and Eisenbud and Levin [30] (see [4]) an algebraic formula is obtained for the local degree $\deg F$ for any smooth mapping $F:(\mathbb{R}^n,0)\to(\mathbb{R}^n,0)$ with finite multiplicity. To such a mapping there is associated the algebra $Q=\mathbb{R}[[x]]/(F)$, where (F) is the ideal in the algebra $\mathbb{R}[[x]]$ generated by the Maclaurin series of the components of the mapping F. The dimension of Q as a vector space over the field $\mathbb{R}$ is called the multiplicity of the mapping F (at the distinguished point $x=0$). The determinant of the Jacobian matrix of the mapping F generates the ideal (J) in the algebra Q. This ideal always has dimension 1. Let N denote the collection of all ideals I in the algebra Q satisfying the relation $I^2=0$. This collection is partially ordered by inclusion. Let I be a maximal element and $\operatorname{ann} I$ its annihilator, i.e., the ideal in Q formed by all elements a such that $aI=0$. It follows from the equation $I^2=0$ that $I\subset\operatorname{ann} I$. We choose an arbitrary element $a\in\operatorname{ann} I\setminus I$; one can show that its square belongs to (J) and does not equal zero. The formula of Khimshiashvili-Eisenbud-Levin takes the following form:

$$\deg F=\operatorname{sgn}\frac{a^2}{J}\cdot\dim(\operatorname{ann} I/I), \tag{5.5.3}$$

where the first factor on the right is the sign of the number λ for which $a^2=\lambda J$. The second factor can be replaced by the quantity $\dim Q-2\dim I$, which is equal to it. If $\operatorname{ann} I=I$, then $\deg F=0$.

In particular, if $F=\nabla f$, then the algebra Q coincides with (5.5.1); consequently, the multiplicity of F coincides with the multiplicity of the critical point $x=0$ of the function f. In this case the left side of (5.5.3) is equal to the number $-\chi(X_s)$ for $s<0$, and in the right side J is the image of the function $\operatorname{Hess} f$ in the algebra (5.5.1), where $\operatorname{Hess} f$ is the determinant of the matrix of second derivatives of the function f at the point $x=0$.

For functions f of two variables a closer relation is known between the topologies of the fibers of the real and complex Milnor bundles. As Guseĭn-Zade and A'Campo showed (see [5]), the intersection matrix of the cycles of the complex Milnor bundle can be found by considering a real curve $\tilde{f}=0$, where $\tilde{f}$ is an appropriate deformation of the function f sufficiently near to it. We assume that the critical point $x_1=x_2=0$ of the function f has finite multiplicity m and that the deformation $\tilde{f}$ is chosen so that $\tilde{f}$ has in some disc U exactly m critical points so that they all, with the exception of the maximum and minimum, lie on the curve $\tilde{f}=0$. This curve divides U into a finite number of domains. A count of the incidence of these domains and the critical points allows us to write the intersection matrix of Z_s.

6. Periods of Differential Forms and the Torelli Theorems

6.1. Let X be a complex manifold, S a connected complex manifold, and $f: X \to S$ a proper holomorphic mapping whose rank at each point is equal to $\dim S$. Its fibers X_s are compact manifolds which form a locally trivial family of smooth manifolds. We consider the homological bundle $\mathscr{H}_*$ on S, whose fibers are the $\mathbb{Z}$-Modules $H_*(X_s, \mathbb{Z})$. It has a canonical connection; a section of $\mathscr{H}_*$ over a set $V \subset S$ is flat if it is generated by a continuous family of cycles $\gamma(s) \subset X_s$ for $s \in V$. A holomorphic form ω defined on X is called relatively closed if the form $d\omega$ vanishes on each fiber of the mapping f. If $\gamma(s)$ for $s \in V$ is a continuous family of cycles in the fibers of f over V, then the function $\int_{\gamma(s)} \omega$ (cf., Chapter 5, Section 4) is defined. It is holomorphic and depends only on the horizontal section of the bundle $\mathscr{H}_*$ which is generated by this family of cycles and on the section of the cohomological bundle $\mathscr{H}^* = \bigcup_s H^*(X_s, \mathbb{C})$ generated by the form ω. Functions of this kind in various cases can provide natural coordinates on the moduli space of compact manifolds, i.e., on the base of a universal deformation.

Let us assume that over S continuous families of cycles $\gamma_1(s), \ldots, \gamma_b(s)$ have been chosen generating a basis in the homological bundle modulo its torsion, and also relatively closed forms $\omega_1, \ldots, \omega_g$ generating the sheaf $H^*_{\mathrm{DR}}(X/S)$ (Chapter 5, Section 4). Then the matrix-valued function $\Omega = \{\Omega_{ij}\}$, where

$$\Omega_{ij}(s) = \int_{\gamma_j(s)} \omega_i,$$

is called the period matrix of the family f. It is uniquely determined up to multiplication by an invertible holomorphic matrix A on the left (the choice of another basis for relative cohomology) and an integer-valued matrix $\Lambda \subset SL(b, \mathbb{Z})$ on the right: $\tilde{\Omega} = A\Omega\Lambda$. In order to remove the arbitrariness in the choice of the differential forms, we consider instead of these matrices the g-dimensional subspace of b-dimensional space spanned by the rows of this matrix. Thus for a fixed homology basis there is a holomorphic mapping $I: S \to G$, where G is the Grassmann manifold. If this mapping is an imbedding, then the period matrix can give the desired coordinates. These coordinates are the functions $\pi_k(\Omega)$, where π_k are the Plücker coordinates in the Grassmann manifold. Here the collection of quantities $\pi_k(\Omega)$ is viewed as a point in the projective space $\mathbb{C}\mathbb{P}_N$. The assertion that I is an imbedding in a neighborhood of any point $s \in S$ is called the *local Torelli theorem*. The *global Torelli theorem* is the assertion that I is an imbedding. We note that it follows from the local Torelli theorem that f is a modular deformation for each of its fibers X_s (Chapter 2, Section 7). This means that the restriction of f to a neighborhood of X_s is isomorphic to the restriction of the minimal versal deformation of X_s to some subspace M' of the modular subspace M of the base of this deformation. This assertion follows from the fact that the period mapping I is constant on each stratum of the stratification described in Chapter 5, Section 1.

6.2. Each Riemann surface of genus 1 can be obtained as a quotient of the complex plane $\mathbb{C}$ by a lattice Z generated by the integer 1 and some complex number λ belonging to the upper halfplane $\mathbb{C}^+$. The collection of all such surfaces forms a family $X \to \mathbb{C}^+$, which is a universal deformation for each of its fibers. In each fiber there are cycles $\gamma_1(\lambda)\ \gamma_2(\lambda)$ which are the images of the segments $[0, 1]$ and $[0, \lambda]$ under the quotient mapping. They form a basis for the one-dimensional homology. The space of holomorphic forms of first degree on the fiber is one-dimensional and is generated by a form which on the covering space $\mathbb{C}$ is denoted by dz. The period matrix in this case is the column vector formed by the quantities $\int_0^1 dz = 1$ and $\int_0^\lambda dz = \lambda$. An affine coordinate is given by the ratio $\lambda : 1 = \lambda$, which coincides with a coordinate in the base of the family. Consequently, the global Torelli theorem holds in this case. We note that in order to obtain from the base of this family a moduli space M_1 for Riemann surfaces of genus 1 it is necessary to take quotient of $\mathbb{C}^+$ relative to the action of the group $SL(2, \mathbb{Z})$ corresponding to base changes in homology. It acts on $\mathbb{C}^+$ by integral Möbius transformations.

6.3. The analog of this construction for Riemann surfaces of genus $g > 1$ is the Teichmüller space T_g [10]. It is a complex manifold of dimension $3g - 3$ which is homeomorphic to a ball. There is defined a holomorphic family $f: X \to T_g$ whose fibers are all Riemann surfaces of genus g with distinguished canonical cuts. By a canonical cut we mean a choice of basis $c_1, \ldots, c_{2g}$ in $H_1(X, \mathbb{Z})$ with intersection matrix

$$Q = \begin{pmatrix} 0 & E_g \\ -E_g & 0 \end{pmatrix},$$

where E_g is the identity matrix of order g. Let $\omega_1, \ldots, \omega_g$ be holomorphic forms on $f^{-1}(V)$, where V is an open subset of X, which generate the space of holomorphic forms on each fiber X_s for $s \in V$. We consider the period matrix of these forms relative to the corresponding family of canonical cuts. It has dimension $g \times 2g$ and can be written in block form as $\Omega = (A, B)$. This matrix satisfies the Riemann bilinear relations

$$\Omega Q \Omega' = 0, \qquad i\Omega Q \bar{\Omega}' > 0,$$

where prime denotes transposition of the matrix and the bar—complex conjugation. If we pass to the equivalent period matrix $\tilde{\Omega} = A^{-1}\Omega = (E_g, \Sigma)$, then these relations mean that Σ is a symmetric matrix with positive imaginary part, i.e., a point in the upper Siegel halfplane H_g. By what has been said, the holomorphic mapping $I: V \to H_g$ is defined. It does not depend on the choice of the forms $\omega_1, \ldots, \omega_g$; consequently, it extends to a holomorphic mapping $I: T_g \to H_g$ which is called the period mapping. The local Torelli theorem for Riemann surfaces of genus g is the assertion that the rank of the period mapping is equal to $3g - 3$. This assertion is true if $g = 2$ and also in the case $g > 2$ at those points $t \in T_g$ for which the fiber X_t is not a hyperelliptic curve. In the remaining cases the local Torelli theorem does not hold. This is related to the fact that in these cases by Noether's theorem the space of quadratic differentials on X_t (which is dual by Serre to the space $H^1(X_t, \mathscr{T}^0)$) is generated by holomorphic forms.

The moduli space M_g for Riemann surfaces of genus g is the quotient space of T_g relative to the action of the group Γ_g of transformations of the one-dimensional homology group of the surface which preserve the intersection form. This group is isomorphic to the group $G = Sp(2g, Q)$ of automorphisms of the $2g$-dimensional integral lattice which preserve the form Q. The period mapping lifts to a holomorphic mapping $M_g \to H_g/G$ in which the group G acts in the Siegel upper halfplane by the generalized linear fractional transformations $\Sigma \mapsto (a + \Sigma c)^{-1}(b + \Sigma d)$.

6.4. For multi-dimensional complex manifolds the Torelli theorem has been established in various special cases. The most complete result is known for surfaces of type $K3$. I.R. Shafarevich and I.I. Pyatetskiĭ-Shapiro [78] established the global Torelli theorem for algebraic surfaces of type $K3$, and it has now been proven for arbitrary surfaces of type $K3$. For three-dimensional cubic manifolds $X \subset \mathbb{CP}_4$ the global Torelli theorem was obtained by A.N. Tyurin [94] and Clemens and Griffiths [20].

Griffiths showed in [42] that the local Torelli theorem is valid for all manifolds with trivial canonical bundle (see also [74]) and also for all hypersurfaces in $\mathbb{CP}_n$ with the exception of the cubic surfaces. The cubic surfaces, i.e., surfaces in $\mathbb{CP}_3$ defined by equations of third degree, form a universal family with a four-dimensional base, but on them there are no nonzero holomorphic forms. Recently Griffiths, Donagi and Green (1983–1984) have obtained the global Torelli theorem for hypersurfaces of sufficiently large degree.

6.5. Let $f: \mathbb{C}^n \to \mathbb{C}$ be the germ of a holomorphic function with a unique critical point and $X \to S$ a minimal versal deformation of f. For any holomorphic form ω of degree $n - 1$ on X we may define a period mapping $S \backslash D \to \mathbb{C}^\mu$ by integrating this form relative to a continuous family of cycles in the Milnor bundle. Every single-valued branch of this mapping is holomorphic. In papers of Looijenga [59] and A.N. Varchenko and A.B. Givental' [100] it has been shown that for "almost any" form ω this mapping is a locally biholomorphic. Consequences of this fact are studied in [100].

7. The Variation of Hodge Structures

7.1. In a sequence of papers (see the summary [42], Part 3) Griffiths has studied the period mapping in a more complete and invariant form. The basic geometric object is a family of nonsingular compact algebraic manifolds $X \to S$ over the punctured disc $S = \{s : 0 < |s| < \varepsilon\}$. For any $k > 0$ the cohomology of the fibers $H^k(X_s, \mathbb{C})$ forms a locally trivial bundle over S with a canonical flat connection. If this bundle is lifted to the universal covering space $\tilde{S} \to S$, then this connection defines in the lifting a natural trivialization; consequently, we can write it in the form $H \times \tilde{S}$, where H is the fiber of the cohomological bundle over a fixed point s. The complex analytic structure of the fiber X_s of the family generates the *Hodge filtration*

$$0 \subset F_s^n \subset \cdots \subset F_s^{q+1} \subset F_s^q \subset \cdots \subset H, \tag{5.7.1}$$

where $n = \dim X_s$,

$$F_s^q = H^{n,k-n}(X_s) \oplus \cdots \oplus H^{q,k-q}(X_s); \qquad H^{r,*}(X_s) = H^*(X_s, \Omega^r),$$

and Ω^r is the sheaf of germs of holomorphic forms of order r on X_s. We note that the period mapping considered in Chapter 5, Section 6 takes into account only the one term F_s^n of this filtration. If the family is trivial, then the Hodge filtration does not depend on s; in the general case this filtration is a varying flag of subspaces in the fixed space H. The dimensions of these subspaces are the same for all points $s \in S$. Thus a mapping from S into the direct product of the corresponding Grassmann manifolds is defined. This mapping is holomorphic, and the Gauss-Manin connection operator acts from F^q to F^{q-1}:

$$\nabla(F^q) \subset F^{q-1} \otimes \Omega_S^1. \tag{5.7.2}$$

The latter property follows from the fact that the connection operator can be represented as the composition of the Kodaira-Spencer mapping $T_s(S) \to H^1(X_s, \mathcal{T}^0)$ and the direct sum of the mappings

$$H^1(X_s, \mathcal{T}^0) \otimes H^q(X_s, \Omega^r) \to H^{q+1}(X_s, \Omega^{r-1}),$$

which are induced by the interior product of sheaves $\mathcal{T}^0 \otimes \Omega^r \to \Omega^{r-1}$. The Hodge structure on a fiber is the direct sum decomposition

$$H = \bigoplus_{p+q=k} H_s^{p,q},$$

where $H_s^{p,q} = F_s^p \cap \bar{F}_s^q = H^q(X_s, \Omega^p)$, is called the Hodge subspace. Griffiths' theory includes also the polarization of Hodge structures. To introduce it, we first imbed the algebraic family into the space $\mathbb{CP}_N \times S$ for a suitable N. The Kähler structure in the projective space $\mathbb{CP}_N$ is defined by a closed $(1, 1)$-form ω. The restriction of this form to X defines the structure of a Kähler manifold on each fiber X_s. Using this, a bilinear form is introduced on cohomology:

$$B(\alpha, \beta) = \int_{X_s} \omega^l \wedge \alpha \wedge \beta, \qquad \alpha, \beta \in H, \qquad l = n - k.$$

It possesses the following properties:

$$B(F^p, F^{k-p+1}) = 0, \qquad p = 0, 1, \ldots \tag{5.7.3}$$

$$i^{p-q} B(v, \bar{v}) > 0, \qquad v \in H^{p,q}, \qquad v \neq 0. \tag{5.7.4}$$

7.2. The codification of the properties of Hodge structures arising in the situation described in Section 7.1 leads to the concept of an abstract Hodge structure with a polarization. This is a finite-dimensional complex space H, which is the complexification of a real space $H_{\mathbb{R}}$, which in turn is the hull of an integral lattice $H_{\mathbb{Z}}$. We define a number k, called the weight of the Hodge structure, and also integers $h^{p,q} \geqslant 0$ which satisfy the relations $\sum h^{p,q} = \dim H$, $h^{q,p} = h^{p,q}$, and $h^{p,q} > 0$ only in case $p + q = k$. Further, we define the filtration $0 \subset \cdots \subset F^{q+1} \subset F^q \subset \cdots \subset H$ formed by subspaces satisfying the relations $\dim F^p/F^{p+1} = h^{p,k-p}$ for all p. A polarization of this structure is a bilinear form B on H which takes rational values on $H_{\mathbb{Z}}$ and satisfies the relations (5.7.3) and (5.7.4), where $H^{p,q} = F^p \cap \bar{F}^q$, and the bar denotes complex conjugation in H using the decomposition $H = H_{\mathbb{R}} + iH_{\mathbb{R}}$. From these relations it follows, in particular, that $\dim H^{p,q} = h^{p,q}$ and the direct sum of these subspaces is equal to H.

The next step in Griffiths theory is to consider the classifying space for Hodge structures. We follow here the exposition of Schmid [86]. Let $\mathscr{F}$ denote the set of all filtrations in H with the conditions mentioned above on dimensions. It is a submanifold of the product of Grassmann manifolds. Further, let B be a polarization, $\check{\mathscr{D}} \subset \mathscr{F}$ the subset formed by the filtrations satisfying (5.7.3), and $\mathscr{D} \subset \check{\mathscr{D}}$ the part of it on which (5.7.4) holds. $\check{\mathscr{D}}$ is a closed submanifold of F, and $\mathscr{D}$ is an open subset of $\check{\mathscr{D}}$. This subset is by definition the classifying space of Hodge structures of weight k with Hodge numbers $h^{p,q}$ polarized by the form B. Let G denote the subgroup of the linear group $GL(H)$ consisting of the transformations which preserve the form B. This group acts transitively on $\check{\mathscr{D}}$ and its real subgroup $G_{\mathbb{R}}$ acts transitively on $\mathscr{D}$.

The most important property of the classifying space for the Griffiths-Schmid theory is that on it there is a distinguished invariant distribution, i.e., a subbundle of the tangent bundle. This distribution, which is called horizontal, can be described as follows. Every tangent field v on $\check{\mathscr{D}}$ defines an operator on the space of sections of the trivial bundle with fiber H. In it are the tautological subbundles $\mathscr{F}^p$ for $p = \cdots -1, 0, 1, \ldots$; the fiber of $\mathscr{F}^p$ over a point $d \in \check{\mathscr{D}}$ is the pth term of the filtration corresponding to this point. The field v is called horizontal if for each p it acts from $\mathscr{F}^p$ into $\mathscr{F}^{p-1}$. The key analytic fact is formulated as follows:

Lemma. *There is on $\check{\mathscr{D}}$ a G-invariant Hermitian metric whose curvature relative to the holomorphic horizontal sections is negative.*

This lemma opens the door for applying hyperbolic analysis.

7.3. We return to the family of algebraic manifolds $X \to S$ which we considered in Section 7.1. By what was said, to each point s in the universal covering space $\tilde{S}$ of the base there corresponds a polarized Hodge structure in $H = H^k(X_{s_0}, \mathbb{C})$ of weight k with Hodge numbers $h^{p,q} = \dim H^q(X_s, \Omega^p)$. Consequently a mapping is defined from $\tilde{S}$ into the corresponding classifying space $\mathscr{D}$. This mapping is holomorphic and horizontal in view of (5.7.2). Therefore from the Lemma in

Section 7.2 and also from a lemma of Ahlfors (see Volume 9) we have the important result [86]: the mapping $\Phi : \tilde{S} \to \mathcal{D}$ is uniformly continuous if $\tilde{S}$ is given the standard Poincaré metric. As Borel noticed, it follows that for the monodromy operator M acting in H all the eigenvalues are roots of unity.

The next step in the theory is to "twist" using the monodromy operator the mapping Φ so that it descends to S. If the covering mapping $\tilde{S} \to S$ is defined by the formula $z \mapsto \exp(2\pi i z)$, then by the definition of the monodromy operator we have $\Phi(z + 1) = M\Phi(z)$. We choose a natural number N so that the Nth powers of the eigenvalues of M are equal to 1. In this case the operator $M^N - E$ is nilpotent; consequently, we can define a logarithm of the operator M by the formula

$$L = \frac{1}{N} \ln M^N = \frac{1}{N} \sum_{k \geqslant 1} \frac{(-1)^{k-1}}{k} (M^N - E)^k.$$

The operator L is an element of the Lie algebra of the group $G_{\mathbb{R}}$ and the exponential function $\exp(-zL)$ is a holomorphic function on $\tilde{S}$ with values in the group G. Therefore the holomorphic mapping

$$\Psi(z) = \exp(-zL)\Phi(z) : \tilde{S} \to \check{\mathcal{D}},$$

is defined and has period N. It is uniformly continuous for the same reason as Φ, and by the periodicity it can be lowered from the halfplane $\tilde{S}$ to the disc. It follows that the limit

$$a = \lim_{\operatorname{Im} z \to \infty} \Psi(z) \in \check{\mathcal{D}}.$$

exists. It is called the *limiting Hodge structure.* The theorem on nilpotent orbits [86] describes the speed of convergence to this limit. As a consequence Schmid has obtained the following refinement of Landman's theorem on monodromy: $(M_u - E)^l = 0$, where M_u is the unipotent part of M and l is the largest natural number such that the inequality $h^{r,k-r} \neq 0$ holds for some p and all r satisfying $p \leqslant r < p + l$. But perhaps the most important consequences of the Griffiths–Schmid theory are those related to the limiting mixed Hodge structure.

7.4. The next definitions are due to Deligne [22]. Let the symbols $H_{\mathbb{Z}} \subset H$ have the same meaning as in Section 7.2. We choose a Hodge filtration $\{F^q\}$ in H. We are also given an increasing filtration

$$0 \subset W_0 \subset \cdots \subset W_{m-1} \subset W_m \subset \cdots \subset W_{2k} = H,$$

which is called a weight filtration. It must be defined over $\mathbb{Q}$, i.e., it must be generated by its intersections with $H_{\mathbb{Q}} = H_{\mathbb{Z}} \otimes \mathbb{Q}$ ($\mathbb{Q}$ denotes the field of rational numbers). A Hodge filtration generates a decreasing filtration on each quotient space $\mathrm{Gr}_m = W_m/W_{m-1}$, namely

$$F^q(\mathrm{Gr}_m) = F^q \cap W_m / F^q \cap W_{m-1}.$$

We assume that this generated filtration defines a Hodge structure of weight m

on Gr_m for any m. In this case the collection of the Hodge and weight filtrations in H is called a *mixed Hodge structure.*

Let (H, W_m, F^q) and (H', W'_m, F'^q) be two mixed Hodge structures. A morphism of type (r,r) of these structures, where r is an integer, is by definition a linear mapping $A: H \to H'$ such that $A(W_k) \subset W'_{k+2r}$ and $A(F^q) \subset F'^{q-r}$. The next property partly explains the importance of the concept of mixed Hodge structure: for any k and q the relations $A(W_k) = A(H) \cap W'_{k+2r}$ and $A(F^q) = A(H) \cap F'^{q-r}$ hold.

We again return to the family $X \to S$ considered in Section 7.1. Deligne introduced a weight filtration into the cohomology $H = H^k(X_{s_0}, \mathbb{C})$ associated to this family using the nilpotent operator L. This filtration can be described as follows. We have $L^{k+1} = 0$, and we set $W_0 = \mathrm{Im}\, L^k$ and $W_{2k-1} = \mathrm{Ker}\, L^k$. Replacing the space $H = W_{2k}$ by W_{2k-1}/W_0, we consider on it the well defined operator L', generated by L. We have $L'^k = 0$, and we set: W_1 = the preimage of $\mathrm{Im}\, L'^{k-1}$ and W_{2k-2} = the preimage of $\mathrm{Ker}\, L'^{k-1}$ under the mapping $W_{2k-1} \to W_{2k-1}/W_0$. Arguing by induction, we construct the weight filtration $\{W_m\}$. It follows from the construction that for any l the operator

$$L^l : \mathrm{Gr}_{k+l} \to \mathrm{Gr}_{k-l} \tag{5.7.5}$$

is an isomorphism. One of the basic results of the theory is the *theorem of Schmid–Deligne*: the Deligne weight filtration $\{W_m\}$ and the limiting Hodge structure a (see Section 7.3) form a mixed Hodge structure on H. The operator L on H is a morphism of these structures of type $(-1, -1)$ and the Kähler operator $\Lambda : H^k(X_s, \mathbb{C}) \to H^{k+2}(X_s, \mathbb{C})$ is a morphism of type (1, 1).

8. Mixed Hodge Structures and Characteristic Numbers

Let $f: \mathbb{C}^{n+1} \to \mathbb{C}$ be a holomorphic function for which the point $z = 0$ is an isolated critical point. Steenbrink [90] has introduced a mixed Hodge structure into the cohomology of the Milnor bundle associated to the germ of this function at the critical point. Using a construction of Brieskorn, we can construct a family $X \to T$ of projective hypersurfaces in $\mathbb{CP}_{n+1}$ over the disc $T = \{|t| < \varepsilon\}$ so that all fibers X_t for $t \neq 0$ of this family are nonsingular, the fiber X_0 has a unique singular point x_0, and the germ of this family at the point x_0 is isomorphic to the germ of the function f. Let U be a sufficiently small ball in $\mathbb{CP}_{n+1}$ centered at the point x_0. It intersects transversally the fibers X_t for all t with $|t| < \varepsilon'$; consequently, the manifolds $U_t = X_t \cap U$ form a Milnor bundle for this germ. Since the fiber X_0 is homeomorphic to $X_t/\bar{U}_t$, there is the exact cohomology sequence

$$0 \to H^n(X_0, x_0) \to H^n(X_{t_0}) \to H^n(\bar{U}_{t_0}) \to H^{n+1}(X_0) \to H^{n+1}(X_{t_0}) \to 0$$

with coefficients in $\mathbb{C}$, where t_0 is a fixed point. Steenbrink introduced a mixed

Hodge structure into $H^n(\bar{U}_{t_0})$ in such a way that all mappings in this sequence turn out to be morphisms of mixed Hodge structures. There is defined in $H^*(X_0)$ the mixed Hodge structure of Deligne, which is defined for singular manifolds in [22] and in $H^*(X_{t_0})$ a structure coinciding with the limiting structure of Schmid, described in Chapter 5, Section 7. Steenbrink's construction of a mixed Hodge structure is similar to that used by Deligne and employs a resolution of the singularity of X_0. However, in the definition of the weight filtration there is a small variation from the construction described in Chapter 5, Section 7. The space $H^n(\bar{U}_{t_0})$ decomposes into the direct sum of eigenspaces for M. On each subspace H_λ corresponding to an eigenvalue $\lambda \neq 1$, a weight filtration is introduced so that the mapping (5.7.5) will be an isomorphism for $k = n$. On the eigenspace with eigenvalue $\lambda = 1$ the mapping (5.7.5) must be an isomorphism if n is replaced by $n + 1$.

Steenbrink associates to the critical point of the function f a collection of characteristic pairs $(u_1, w_1), \ldots, (u_\mu, w_\mu)$ defined as follows. For every eigenvalue λ of the operator M there is a rational number l so that $\exp(2\pi i l) = \lambda$ and $q \leqslant l < q + 1$, if λ is an eigenvalue of M on the subspace $H^{p,q}$. The corresponding pair is $(l, p + q)$ if $\lambda \neq 1$ and $(q, p + q - 1)$ if $\lambda = 1$. Here $H^{p,q}$ is the Hodge space corresponding to the Hodge filtration on Gr_{p+q}. The set of characteristic pairs is an important invariant of the critical point. If the set is known, one can find, in particular, the rank and signature of the intersection form on the fibers of the Milnor bundle [90].

A.N. Varchenko [78] has constructed a mixed Hodge structure on the Milnor bundle using a different method. The weight filtration in his method is chosen in the same way as in Steenbrink's. To introduce the Hodge filtration he considers the integral $\int_{\gamma(t)} \frac{\omega}{df}$ relative to a continuous family of cycles $\gamma(t) \subset f^{-1}(t)$, where ω is an arbitrary holomorphic form of order $n + 1$ defined in a neighborhood of the critical point of the function f. If $t = 0$ is a critical value, then as is well known, this integral has for small $|t|$ the convergent expansion

$$\sum_{\lambda \in \Lambda} \sum_{\alpha \in L(\lambda)} \sum_{k=0}^{n} a_{\alpha,k} t^\alpha (\ln t)^k, \tag{5.8.1}$$

where Λ is the set of eigenvalues of the monodromy operator and $L(\lambda)$ is the set of rational numbers α such that $\alpha > -1$ and $\exp(2\pi i\alpha) = \lambda$. If we fix a branch of the logarithm $\ln t$, then the coefficients $a_{\alpha,k}$ are uniquely determined and are linear functionals from the homology class of the cycle $\gamma(t_0)$, where t_0 is a fixed point in the base of the Milnor bundle. A.N. Varchenko defines the weight of the form ω/df to be the minimal number $\alpha = \alpha(\omega)$ such that some element $a_{\alpha,k} \in H^n(X_{t_0})$ is different from zero. The sum of terms (5.8.1) with $\alpha = \alpha(\omega)$ he calls the principal part of this form, viewed as a section of the cohomological bundle. The filtration $\{F^q\}$ in the fibers of this bundle is generated by the principal parts of all forms having weight $\alpha(\omega) \leqslant n - 1 - q$. This filtration, in general, is different from the Hodge filtration defined by the method of Deligne–Steenbrink, which was described in Chapter 5, Sections 7 and 8; however, as A.N. Varchenko has shown, both filtrations induce on Gr^W the same Hodge structure. It follows that the weight filtration and Varchenko's filtration $\{F^q\}$ generate a mixed Hodge structure on the Milnor bundle.

From this result follows, in particular, another description of Steenbrink's characteristic pairs: their first components u_i are the numbers $\alpha(\omega) + 1$, where $\alpha(\omega)$ is the weight of the forms. Under a deformation of the germ of the function f, the spectrum is lower semicontinuous [99], [91] so, in particular, the minimum value in the spectrum is a lower semicontinuous function. Steenbrink [71] has shown that for any real half-open interval $(\sigma, \sigma + 1]$ the number of elements in the spectrum of the germ f lying in this interval is not smaller that the number of elements in the spectrum of any critical point of any function sufficiently near to f.

In papers of M. Saito [83] and also Scherk and Steenbrink [84] there are constructions of the Hodge–Steenbrink filtration which do not use a resolution of the singularities. For the most complete exposition of the theory see [92], [110].

Bibliography*

1. Ahlfors, L.V.: The complex analytic structure of the space of closed Riemann surfaces. Princeton, Math. Ser. 24, 45–66 (1960). Zbl. 100.289
2. Aleksandrov, A.G.: Normal forms of one-dimensional quasi-homogeneous complete intersections. Mat. Sb., Nov. Ser., 117(159), 3–31 (1982); English transl.: Math. USSR, Sb. 45, 1–30 (1983). Zbl. 508.14001
3. Arnol'd, V.I.: A remark on the method of stationary phase and Coxeter numbers (Russian) Usp. Mat. Nauk 28, No. 5 17–44 (1973). Zbl. 285.40002
4. Arnol'd, V.I., Varchenko, A.N., Guseĭn-Zade, S.M.: Singularities of differentiable mappings. Classification of critical points, caustics, and wave fronts. Nauka, Moscow (1982). Zbl. 513.58001; English transl.: Monographs in Math. 82, Birkhäuser Boston–Basel–Stuttgart, Boston (1985). Zbl. 554.58001
5. Arnol'd, V.I., Varchenko, A.N., Guseĭn-Zade, S.M.: Singularities of differentiable mappings. Monodromy and asymptotics of integrals. (Russian), Nauka, Moscow (1984). Zbl. 545.58001
6. Artin, M.: On isolated rational singularities of surfaces. Am. J. Math. 88, 129–136 (1966). Zbl. 142.186
7. Artin, M.: Algebraic construction of Brieskorn's resolutions. J. Algebra 29, 330–348 (1974). Zbl. 292.14013
8. Bănică, C., Putinar, M., Schumacher, G.: Variation der globalen Ext in Deformationen kompakter komplexer Räume, Math. Ann. 250, 135–155 (1980). Zbl. 438.32007
9. Bănică, C., Stănăşilă, O.: Méthodes algébriques dans la théorie globale des espaces complexes. Gauthier-Villars, Paris (1977). Zbl. 349.32006

*For the convenience of the reader, references to reviews in Zentralblatt für Mathematik (Zbl.), compiled using the MATH database, and Jahrbuch über die Fortschritte der Mathematik (Jrb.) have, as far as possible, been included in this bibliography.

10. Bers, L.: Quasiconformal mappings and Teichmüller's theorem. Princeton Math. Ser. 24, 89–119 (1960). Zbl. 100.289
11. Bers, L.: Finite dimensional Teichmüller spaces and generalisations. Proc. Symp. Pure Math. 39, Part 1, 115–156 (1983). Zbl. 559.32003
12. Borcea, C.: Some remarks on deformations of Hopf manifolds. Rev. Roum. Math. Pures Appl. 26, 1287–1294 (1981). Zbl. 543.32010
13. Borcea, C.: Smooth global complete intersections in certain compact homogeneous complex manifolds, J. Reine Angew. Math. 344, 65–70 (1983). Zbl. 511.14027
14. Brieskorn, E.: Die Auflösung der rationalen Singularitäten holomorpher Abbildungen. Math. Ann. 178, 255–270 (1968). Zbl. 159.377
15. Brieskorn, E.: Singular elements of semi-simple algebraic groups. Actes Congres Intern. Math. 1970, Gauthier–Villars, Paris, Part 2, 279–284 (1971). Zbl. 223.22012
16. Brieskorn, E.: Die Monodromie der isolierten Singularitäten von Hyperflächen, Manuscr. Math. 2, 103–161 (1970). Zbl. 186.261
17. Burns, D., Shnider, S., Wells, R.O.: Deformations of strictly pseudoconvex domains, Invent. Math. 46, 237–253 (1978). Zbl. 412.32022
18. Burns, D., Wahl, J. Local contributions to global deformations of surfaces. Invent. Math. 26, 67–88 (1974). Zbl. 288.14010
19. Chern, S.S., Moser, J.K.: Real hypersurfaces in complex manifolds. Acta Math. 133, 219–271 (1975). Zbl. 302.32015
20. Clemens, C.H., Griffiths, Ph.A.: The intermediate Jacobian of the cubic threefold. Ann. Math., II. Ser. 95, 281–356 (1972). Zbl. 214,483
21. Dabrowski, K.: Moduli spaces for Hopf surfaces. Math. Ann. 259, 201–225 (1982). Zbl. 497.32017
22. Deligne, P.: Theorie de Hodge. II, III. Inst. Haut Etud. Sci., Publ. Math. 40, 5–57 (1972); 44, 5–77 (1975). Zbl. 237.14003
23. Donin, I.F.: Complete families of deformations of germs of complex spaces. Mat. Sb., Nov. Ser, 89(131), 390–399 (1972); English transl.: Math. USSR, Sb. 18, 397–406 (1972). Zbl. 255.32011
24. Donin, I.F.: Construction of a versal family of deformations for holomorphic bundles over a compact complex space. Mat. Sb., Nov. Ser. 94(136), 430–443 (1974); English transl.: Math. USSR, Sb. 23, 405–416 (1975). Zbl. 325.32008
25. Douady, A.: Le problème des modules pour les sous-espaces analytiques compacts d'un espace analytique donné. Ann. Inst. Fourier 16, No. 1 1–95 (1966). Zbl. 146.311
26. Douady, A.: Flatness and privilege. Enseign. Math., II. Ser. 14, 47–74 (1968). Zbl. 183.351
27. Douady, A.: Le problème des modules locaux pour les espaces ℂ-analytiques compacts. Ann. Sci. Ec. Norm. Supér., IV., Ser. 7, 569–602, (1975). Zbl. 313.32036
28. Douady, A.: Verdier, J.-L.: Séminaire de géometrie analytique.[Astérisque, No. 16–17 (1974).] Paris: Société Mathématique de France (1976), Zbl. 334.00011
29. Duchamp, T., Kalka, M.: Deformation theory for holomorphic foliations. J. Diff. Geom. 14, 317–337 (1979). Zbl. 451.57015
30. Eisenbud, D., Levine, H.I.: An algebraic formula for the degree of C^∞-map germ. Sur une inégalité à la Minkowski pur les multiplicités. Ann. Math., II. Ser. 106, 19–44 (1977). Zbl. 398.57020
31. Elkik, R.: Singularités rationelles et déformations. Invent. Math. 47, 139–147 (1978). Zbl. 363.14002
32. Flenner, H.: Über Deformationen holomorphen Abbildungen, Habilitationsschrift, Osnabrück, 1–142 (1978).
33. Forster, O., Knorr, K.: Über die Deformationen von Vektorraumbündeln auf kompakten komplexen Raümen. Math. Ann. 209, 291–346 (1974). Zbl. 272.32004
34. Girbau, J., Haefliger, A., Sundararaman, D.: On deformations of transversely holomorphic foliations. J. Reine Angew. Math. 345, 122–147 (1983). Zbl. 538.32015
35. Giusti, M.: Sur les singularités isolées d'intersection completes quasi-homogènes. Ann. Inst. Fourier. 27, No. 3 163–192 (1978). Zbl. 353.14003

36. Godement, R.: Topologie algébrique et théorie des faisceaux. Hermann, Paris (1958). Zbl. 80.162
37. Grauert, H.: Ein Theorem der analytischen Garbentheorie und Modulraüme komplexer Strukturen. Inst. Hautes Etud. Sci., Publ. Math., No. 5, 5–64 (1960). Zbl. 100.80
38. Grauert, H.: Über die Deformation isolierter Singularitäten analytischer Mengen. Invent. Math. 15, 171–198 (1972). Zbl. 237.32011
39. Grauert, H.: Der Satz von Kuranishi für kompakte komplexe Raüme. Invent. Math. 25, 107–142 (1974). Zbl. 286.32015
40. Grauert, H., Kerner, H.: Deformationen von Singularitäten komplexer Raüme. Math. Ann. 153, 236–260 (1964). Zbl. 118.304
41. Greuel, G.-M.: Der Gauss-Manin-Zusammenhang isolierter Singularitäten von vollstandigen Durchschnitten. Math. Ann. 214, 235–266 (1975). Zbl. 285.14002
42. Griffiths, Ph.A.: Periods of integrals on algebraic manifolds. I, II, III. Am. J. Math. 90, 568–626. Zbl. 169.523, 805–865 (1968). Zbl. 183.255; Periods of integrals on algebraic manifolds: Summary of main results and discussion of open problems. Bull. Am. Math. Soc. 76, 228–296 (1970). Zbl. 214.198
43. Griffiths, Ph.A., Schmid, W.: Locally homogeneous complex manifolds. Acta Math. 123, 253–302 (1970). Zbl. 209.257
44. Hamilton, R.S.: Deformation of complex structures on manifolds with boundary II: Families of non-coercive boundary value problems. J. Differ. Geom. 14, 409–473 (1979). Zbl. 512.32015
45. Hamm, H.: Lokale topologische Eigenschaften komplexer Raüme. Math. Ann. 191, 235–252 (1971). Zbl. 214.228
46. Horikawa, E.: On deformations of holomorphic maps I, II. J. Math. Soc. Japan 25, 372–396 (1973). Zbl. 254.32022; 26, 647–667 (1974). Zbl. 286.32014
47. Khimshiashvili, G.N.: On the local degree of a smooth mapping. Soobshch. Akad. Nauk Gruz. SSR 85, 309–312 (1977). Zbl. 346.55008
48. Kiremidjian, G.: On complex structures with a fixed induced CR-structure. Ann. Math. II. Ser. 109, 87–119 (1979). Zbl. 415.32007
49. Klein, F.: Vorlesungen über das Ikosa eder und die Auflösungen der Gleichungen vom fünften Grade. Teubner, Leipzig (1884). Jrb. 16, 61
50. Kodaira, K.: On stability of compact submanifolds of complex manifolds. Am. J. Math. 85, 79–94 (1963). Zbl. 173.331
51. Kodaira, K., Spencer, D.C.: On deformations of complex analytic structures I, II. Ann. Math., II. Ser. 67, 328–401, 403–466 (1958). Zbl. 128.169
52. Kodaira, K., Spencer, D.C.: Stability theorems for complex structures. Ann. Math., II. Ser. 71, 43–76 (1960). Zbl. 128.169
53. Kodaira, K., Spencer, D.C., Nirenberg, L.: On the existence of deformations of complex analytic structures. Ann. Math., II. Ser. 68, 450–459 (1958). Zbl. 88.380
54. Kosarew, S.: Das Modulproblem für holomorphe Einbettungen mit konkaver Umgebungsstruktur. Math. Z. 180, 307–329 (1982). Zbl. 502.32016
55. Kuranishi, M.: On the locally complete families of complex analytic structures. Ann. Math., II. Ser. 75, 536–577 (1962). Zbl. 106.153
56. Laudal, O.A., Pfister, G.: The local moduli problem applications to the classification of isolated hypersurface singularities. Preprint series. Mat. Inst. Univ. Oslo, No. 11 (1983). Zbl. 560.01202
57. Laufer, H.: Deformations of resolutions of two-dimensional singularities. Rice Univ. Studies 59, No. 1, 53–96 (1973). Zbl. 281.32009
58. Lê Dũng Tráng, The geometry of the monodromy theorem. Tata Inst. Fund. Res., Stud. Math. 8, 157–173 (1978). Zbl. 434.32010
59. Looijenga, E.: A period mapping of certain semiuniversal deformations. Compos. Math. 30, 299–316 (1975). Zbl. 312.14006
60. Looijenga, E., Steenbrink, J. (H.M.): Milnor number and Tjurina number of complete intersection. Math. Ann. 271, 121–124 (1985). Zbl. 539.14002
61. Malgrange, B.: Analytic spaces. Enseign. Math., II. Sér. 14, 1–28 (1968). Zbl. 165.405

62. Martinet, J.: Deploiement versels des applications différentiables et classification des applications stables, (Springer) Lect. Notes Math., 535, 1–44 (1976). Zbl. 362.58004
63. Mather, J.N.: Stability of C^∞-mappings. II, III. Ann. Math., II. Ser. 89, 254–291 (1969). Zbl. 177.260. Inst. Hautes Etud. Sci. Publ. Math. 127–156 (1969). Zbl. 159.250
64. Milnor, J.: Singular points of complex hypersurfaces, Princeton Univ. Press, Princeton (1968). Zbl. 184.484
65. Mumford, D.: Lectures on curves on an algebraic surface. Princeton Univ. Press, Princeton (1966). Zbl. 187.427
66. Namba, M.: Automorphism groups of Hopf surfaces. Tôhoku Math. J., II. Ser. 26, 133–157 (1974). Zbl. 283.32023
67. Palamodov, V.P.: On the multiplicity of a holomorphic mapping. Funkts. Anal. Prilozh. 1, Nr. 3 54–65 (1967). English transl.: Funct. Anal. Appl. 1, 218–226 (1968). Zbl. 164.92
68. Palamodov, V.P.: Remarks about differentiable mappings of finite multiplicity. Funkts. Anal. Prilozh. 6, No. 2. 52–61 (1972); English transl.: Funct. Anal. Appl. 6, 128–135 (1972). Zbl. 264.58004
69. Palamodov, V.P.: Deformations of complex spaces. Usp. Mat. Nauk 31, Nr. 3 129–194 (1976); Zbl. 332.32013. English transl.: Russ. Math. Surv. 31, No. 3 129–197 (1976). Zbl. 347.32009
70. Palamodov, V.P.: The tangent complex of an analytic space. (Russian) Tr. Semin. Im. I.G. Petrovskogo No. 4, 173–226 (1978). Zbl. 416.32008
71. Palamodov, V.P.: Moduli in versal deformations of complex spaces. Varietes analytiques compactes. (Springer) Lect. Notes Math. 683, 74–115 (1978). Zbl. 387.32007
72. Palamodov, V.P.: Deformations of Hopf manifolds and the Poincaré–Dulac theorem. Funkts. Anal. Prilozh. 17, No. 4 7–16 (1983); English transl.: Funct. Anal. Appl. 17, 252–259 (1983). Zbl. 561.32009
73. Penrose, R.: Nonlinear gravitons and curved twistor theory. Gen. Relativ. Gravitation 7, 31–52 (1976). Zbl. 354.53025
74. Petrovski, I.G., Nikolskiĭ, S.M., ed.: Algebraic surfaces. Seminar of I.R. Shafarevich. Tr. Mat. Inst. Steklova 75 (1965); English transl.: Proc. Steklov Inst. Math. 75 (1967). Zbl. 154,210
75. Pinkham, H.C.: Deformations of algebraic varieties with G_m-action. Astérisque 20, 1–131 (1974). Zbl. 304.14006
76. Pinkham, H.C., Persson, U.: Some examples of nonsmoothable varieties with normal crossings. Duke Math. J. 50, 477–486 (1983). Zbl. 529.14007
77. Popp, H.: Moduli theory and classification theory of algebraic varieties. (Springer) Lect. Notes Math. 620 (1977). Zbl. 359.14005
78. Pyatetskiĭ-Shapiro, I.I., Shafarevich, I.R.: The Torelli theorem for algebraic surfaces of type $K3$. Izv. Akad. Nauk SSSR, Ser. Mat. 35, 530–572 (1971); English transl.: Math. USSR, Izv. 5, 547–588 (1971). Zbl. 219.14021
79. Quillen, D.: On the (co)homology of commutative rings. Proc. Symp. Pure Math. 17, 65–87 (1970). Zbl. 234.18010
80. Saito, K.: Calcul algébrique de la monodromie. Astérisque, No. 7–8, 195–211 (1974). Zbl. 294.14005
81. Saito, K.: Einfach-elliptische Singularitäten. Invent. Math. 23, 289–325 (1974). Zbl. 296.14019
82. Saito, K.: Primitive forms for a universal unfolding of a function with an isolated critical point. J. Fac. Sci., Univ. Tokyo, Sec. IA 28, 775–792 (1981). Zbl. 523.32015
83. Saito, M.: Gauss–Manin systems and mixed Hodge structure. Proc. Jap. Acad. Ser. A. 58, 29–32 (1982). Zbl. 516.32012
84. Scherk, J., Steenbrink, J.H.M.: On the mixed structure on the cohomology of the Milnor fibre. Math. Ann. 271, 641–665 (1985). Zbl. 618.14002
85. Schlessinger, M.: Functors of Artin rings. Trans. Am. Math. Soc. 130, 208–222 (1968). Zbl. 167.495
86. Schmid, W.: Variation of Hodge structure: the singularities of the period mapping. Invent. Math. 22, 211–319 (1973). Zbl. 278.14003

87. Seminaire H. Cartan, 1960/61. Familles d'espaces complexes et fondements de la geometrie analytique, 1962. Ecole Norm. Sup. Paris (1962). Zbl. 124.241
88. Serre, J.-P.: Algèbre locale. Multiplicités. (Springer) Lect. Notes Math., 11, (1965). Zbl. 142.286
89. Slodowy, P.: Simple singularities and simple algebraic groups. (Springer) Lect. Notes Math. 815 (1980). Zbl. 441.14002
90. Steenbrink, J.H.M.: Mixed Hodge structure on the vanishing cohomology. Nordic Summer School, Symp. Math., Oslo, 1976, 525–563 (1977). Zbl. 373.14007
91. Steenbrink, J.H.M.: Semicontinuity of the singularity spectrum. Invent. Math. 79, 557–565 (1985). Zbl. 568.14021
92. Steenbrink, J.H.M., Zucker, S.: Variation of mixed Hodge structure I., Invent. Math. 80, 489–542 (1985). Zbl. 626.14007
93. Trautmann, G.: Deformations and moduli of coherent analytic sheaves with finite singularities. (Springer) Lect. Notes Math. 670, 233–302 (1978). Zbl. 398.32013
94. Tyurin, A.N.: The geometry of the Fano surface of a nonsingular cubic $F \subset \mathbb{P}^4$ and Torelli theorems for Fano surfaces and cubics. Izv. Akad. Nauk SSSR, Ser. Mat. 35, 498–529 (1971); English transl.: Math. USSR, Izv. 5, 517–546 (1971). Zbl. 215,82
95. Tyurina, G.N.: The topological properties of isolated singularities of complex spaces of codimension one. Izv. Akad. Nauk SSSR 32, 605–620 (1968); English transl.: Math. USSR, Izv. 2, 557–572 (1968). Zbl. 176,509
96. Tyurina, G.N.: Locally semi-universal flat deformations of isolated singularities of complex spaces. Izv. Akad. Nauk SSSR, Ser. Mat., 33, 1026–1058 (1969); English transl.: Math. USSR, Izv. 3, 976–1000 (1969). Zbl. 196,97
97. Tyurina, G.N. Resolution of singularities of flat deformations of binary rational points. Funkt. Anal. Prilozh. 4, No. 1, 77–83 (1970); English transl.: Funct. Anal. Appl. 4, 68–73 (1970). Zbl. 221.32008
98. Varchenko, A.N.: The asymptotics of holomorphic forms determine a mixed Hodge structure. Dokl. Akad. Nauk SSSR 225, 1035–1038, (1980); English transl.: Sov. Math., Dokl. 22, 772–775 (1980). Zbl. 516.14007
99. Varchenko, A.N.: Semicontinuity of the complex singularity index. Funkts. Anal. Prilozh. 17, No. 4 77–78 (1983); English transl.: Funct. Anal. Appl. 17, 307–308 (1983). Zbl. 536.32005
100. Varchenko, A.N., Givental', A.B.: Mappings of periods and the intersection form. Funkts. Anal. Prilozh. 16, No. 2 7–20 (1982); English transl.: Funct. Anal. Appl. 16, 83–93 (1982). Zbl. 497.32008
101. Wahl., J.: Equisingular deformations of normal surface singularities I., Ann. Math., II. Ser. 104, 325–356 (1976). Zbl. 358.14007
102. Wahl., J.: Simultaneous resolution and discriminantal loci. Duke Math. J. 46, 341–375 (1979). Zbl. 472.14002
103. Wassermann, G.: Stability of unfoldings. (Springer) Lect. Notes Math. 393 (1974). Zbl. 288.57017
104. Wehler, J.: Versal deformation of Hopf surfaces, J. Reine Angew. Math. 328 22–32 (1981). Zbl. 459.32009
105. Wehler, J.: Deformations of complete intersection with singularities. Math. Z. 179, 473–491 (1982). Zbl. 473.14021
106. Wells, R.O.: Deformations of strongly pseudoconvex domains in $\mathbb{C}^2$. Proc. Symp. Pure Math. 30, Part 2, 125–128 (1977). Zbl. 357.32013
107. Wirthmüller, K.: Singularities determined by their discriminants. Math. Ann. 252, 237–245 (1980). Zbl. 425.32003
108. Yau, S.S.-T.: Kohn-Rossi cohomology and its application to the Plateau problem I. Ann. Math., II. Ser. 113, 67–110 (1981). Zbl. 464.32012
109. Yau, S.S.-T.: Deformations and equitopological deformations of strongly pseudoconvex manifolds. Nagoya Math. J. 82, 113–129 (1981). Zbl. 443.14019
110. Zucker, S.: Variation of mixed Hodge structure II. Invert. Math. 80, 543–565 (1985). Zbl. 615.14003

Additional bibliography

111. Bingener J.; Lokale Modulräume in der analytischen Geometrie. Bd 1, 2. Aspekte der Mathematik, Bde. D2, D3, Braunschweig/Wiesbaden Friedr. Vieweg & Sohn (1987)
112. Forster O., Knorr K.: Konstruktion verseller Familien kompakter komplexer Räume. (Springer) Lect. Notes Math. 705 (1979). Zbl. 408.32004
113. Illusie L.: Complex Contangent et déformation I, II. (Springer) Lect. Notes Math. 239, 283 (1971, 1972). Zbl. 224. 13014 Zbl. 238.13017
114. Looijenga E.: Homogeneous spaces associated to certain semi-universal deformations. Proc. Intern. Congress Math., 1978, Vol. 2, 529–536 Helsinki (1980). Zbl. 464.32004
115. Pourcin G.: Deformation de singularité isolées. Astérisque 16, 161–173 (1974). Zbl. 292.32014
116. Retakh V.S.: Massey operations in Lie super algebras and deformations of complex-analytic algebras. Funkts. Anal. Prilozh. (Russian) 11, No. 4, 88–89 (1977); Engl. transl. Funct. Anal. Appl. 11, 319–321 (1978). Zbl. 383.17005
117. Rim D.S.: Formal deformation theory. [In: Groupes des Monodromie en Géométrie Algébrique (SGA 71). Exp VI] (Springer) Sem. Geom. algébrique Bois-Maire, 1967–1969, SGA 7 I, Nr. 6, Lect. Notes Math. 288, 32–132 (1972). Zbl. 246.14001

IV. Homogeneous Complex Manifolds

D.N. Akhiezer

Translated from the Russian
by J. Nunemacher

Contents

Introduction

The subject of this article is the set of complex manifolds X whose group of automorphisms (of biholomorphic transformations) acts transitively on X. The list of one-dimensional complex manifolds having this property was surely known already to Poincaré. It consists of the complex plane $\mathbb{C}$, the punctured plane $\mathbb{C}^* = \mathbb{C}\backslash\{0\}$, the unit disc in $\mathbb{C}$, the Riemann sphere, and one-dimensional complex tori (elliptic curves). In each of these cases it is easy to calculate the automorphism group and to see that it is a Lie group of transformations of the manifold X.

In the sequel we shall assume that there is defined on the complex manifold X a differentiable and transitive action of a (real) Lie group G and that each element of G acts as a biholomorphic transformation. In this case the complex manifold X is called a *homogeneous complex manifold* of the Lie group G.

This assumption is essential already in the two-dimensional case. There exist two-dimensional complex manifolds which do not admit a transitive Lie group of transformations but on which the automorphism group is nonetheless transitive [57]. For two-dimensional homogeneous complex manifolds of Lie groups as well as in the one-dimensional case the classification problem has been solved; however, a complete list was obtained only recently [105], [46], [81].

We shall consider two important classes of complex manifolds for which the automorphism groups are Lie groups. With them are associated two different directions of investigation in the theory of homogeneous complex manifolds.

The first class consists of the bounded domains in $\mathbb{C}^n$. In a bounded domain we can introduce a special Hermitian metric, the so-called Bergman metric (see Chapter 1). A significant property of the Bergman metric is that it is invariant with respect to the automorphism group of the domain. From this and from a general theorem in Riemannian geometry (see [62], Volume 1, Chapter 6) it follows that this group is a Lie group. The first proof of this fact is due to H. Cartan (1935). We note that if bounded holomorphic functions separate points of a complex manifold X, then no connected complex Lie group can act holomorphically and nontrivially on X. This follows easily from the Liouville Theorem. Thus a homogeneous bounded domain in $\mathbb{C}^n$ is a homogeneous complex manifold of a real Lie group.

For $n = 2$ there are two homogeneous bounded domains—the complex ball and the bicylinder. Poincaré (1907) proved that they are not isomorphic, and E. Cartan [24] proved that there are no other homogeneous bounded domains in $\mathbb{C}^2$. For the classification of homogeneous bounded domain see Chapters 1 and 8.

The second class is made up of compact complex manifolds. In this case the automorphism group is a complex Lie group of transformations (the Bochner-Montgomery Theorem, 1947, see [62]). The reason for its finite-dimensionality lies in a general property of compact complex manifolds: the space of holomorphic sections of a holomorphic vector bundle over a compact manifold is finite-dimensional. In particular, this pertains to the space of holomorphic vector fields with the operation of Lie bracket, which is the Lie algebra of the Lie group of automorphisms.

For $n = 2$ there are precisely the following compact homogeneous complex manifolds [105]: the complex projective plane $\mathbb{P}^2$, the product of complex projective lines (Riemann spheres) $\mathbb{P}^1 \times \mathbb{P}^1$, two-dimensional complex tori, products of the form $\mathbb{P}^1 \times Y$, where Y is an elliptic curve, and the so-called homogeneous

Hopf surfaces (see Example 4.6). The classification of compact homogeneous complex manifolds is discussed in Chapter 4.

The systematic study of homogeneous complex manifolds began with the work of E. Cartan in the theory of Riemannian and, in particular, Hermitian symmetric spaces. One of the most significant results of E. Cartan is the duality discovered by him between Riemannian symmetric spaces of compact and noncompact types. In the Hermitian symmetric case spaces of noncompact type can be realized as homogeneous bounded domains of special type in $\mathbb{C}^n$. The duality allows us to study them simultaneously with certain compact homogeneous complex manifolds, which turn out to be projective algebraic. Unfortunately in the general theory a similar intrinsic connection between compact manifolds and bounded domains does not exist, although several results in this direction are known (see Chapter 1 and [113]). Homogeneous manifolds of complex Lie groups, in particular, compact homogeneous complex manifolds, and homogeneous complex manifolds of real Lie groups, in particular, homogeneous bounded domains in $\mathbb{C}^n$, in recent articles have been studied separately from one another, and the present survey reflects this dichotomy.

The contents of the article are as follows. In Section 1 we give the classification of Hermitian symmetric spaces. A generalization of Hermitian symmetric spaces of compact type is the so-called flag manifolds, which are considered in Section 2. They are interesting, in particular, for their applications to the theory of group representations (see Section 3). Section 4 is devoted to the general classification of compact homogeneous complex manifolds. In Sections 5 and 6 we discuss some results from the theory of analytic functions on homogeneous manifolds of complex Lie groups. In Section 7 we consider almost homogeneous spaces of complex Lie groups, i.e., complex spaces on which a complex Lie group acts with an open orbit. In Section 8 we return to homogeneous bounded domains and real transformation groups. Finally in Section 9 we survey results concerning homogeneous Kähler manifolds.

We shall continually employ the following notation and terminology. Let X be a smooth manifold and G a Lie group acting differentiably on X. The closed subgroup $U = \{g \in G | go = o\}$, where $o \in X$, is called the stationary subgroup of the point o. If the action is transitive, i.e., if X is a homogeneous manifold, then there is a natural bijection between the quotient space G/U and the manifold X in which $gU \mapsto go$. There is the standard structure of a smooth manifold on G/U according to which this bijection becomes a diffeomorphism. We shall say that the homogeneous manifold X is written in Klein form $X = G/U$. The equality denotes the above-mentioned diffeomorphism. If X is a complex manifold, G a complex Lie group, and the action $G \times X \to X$ is holomorphic, then there is an invariant complex structure on G/U and the equality $X = G/U$ indicates an isomorphism of complex manifolds.

We denote by e the identity element of the group G, by G' its commutator, and by $N_G(U)$ the normalizer in G of a subgroup $U \subset G$. For a topological group G the symbol G_0 denotes the connected component of the identity. Lie groups will be denoted by upper case Latin letters and their Lie algebras by the corresponding lower case Gothic letters. We denote by exp the exponential mapping $\mathfrak{g} \to G$ and by $\mathrm{Ad}_G g : \mathfrak{g} \to \mathfrak{g}$ the adjoint operator for $g \in G$. For the classical groups we shall use the notation in Helgason [44]. The space of all complex matrices of type $p \times q$ we denote by the symbol $M_{p,q}$. For $A \in M_{p,q}$ we denote by A' the transpose of the matrix and by $\bar{A}$ the complex conjugate of the matrix. The notation $A > B$ (resp., $A \geqslant B$), where A and B are square Hermitian matrices, means that the difference $A - B$ has positive (nonnegative) eigenvalues. The symbol I_n denotes the identity matrix of order n (if it is not necessary to indicate n, then we write simply I). Further, we set

$$I_{p,q} = \begin{pmatrix} -I_p & 0 \\ 0 & I_q \end{pmatrix}, \quad J_p = \begin{pmatrix} 0 & I_p \\ -I_p & 0 \end{pmatrix}.$$

The notations for the standard manifolds are as follows: $\mathbb{P}^n$ is n-dimensional complex projective space, $\mathbb{C}^n$ is complex affine space, and also the corresponding vector group, $\mathbb{C}^* = \mathbb{C} \setminus \{0\}$, $\mathbb{B}^n$ is the ball in $\mathbb{C}^n$, $\mathbb{G}_{m,n}$ is the Grassmann manifold whose points are the m-dimensional planes in $\mathbb{C}^n$, $\mathbb{Q}_n$ is the quadric in $\mathbb{P}^{n-1}$ which in homogeneous coordinates is defined by the equation $\sum_{i=1}^{n} z_i^2 = 0$, and $\mathbb{S}^n$ is the n-dimensional sphere.

All complex spaces are assumed to be reduced. We denote by $\mathrm{Aut}\, X$ the group of all automorphisms of the complex space X. If X is irreducible, then the symbol $\dim X$ denotes the complex dimension (if nothing to the contrary is explicity noted).

The author thanks E.B. Vinberg, S.G. Gindikin, and A.L. Onishchik for their remarks and suggestions concerning this survey.

Chapter 1. Hermitian Symmetric Spaces

Let X be a connected complex manifold with a Hermitian metric h. In local coordinates $h = \sum h_{\alpha\beta}\, dz_\alpha \otimes d\bar{z}_\beta$, where $h_{\beta\alpha} = \bar{h}_{\alpha\beta}$. Let $h = g - 2i\Omega$, where g and Ω are real bilinear forms. Then g is a Riemannian metric on X. All metric concepts in what follows pertain to this metric. The form Ω is nondegenerate and skew-symmetric and in local coordinates is given by $\Omega = i/2 \sum h_{\alpha\beta}\, dz_\alpha \wedge d\bar{z}_\beta$. A Hermitian metric is called *Kähler* if the form Ω is closed. A complex manifold is called *Kähler* if it admits a Kähler metric[1].

A manifold X endowed with a Hermitian metric is called a *Hermitian symmetric space* if each point $x \in X$ is an isolated fixed point for some involutive holomorphic isometry $s_x : X \to X$[2]. There can be only one such isometry. It is called the symmetry at the point x.

[1] Sometimes a Kähler manifold refers to a complex manifold with a fixed Kähler metric.
[2] A Hermitian symmetric space is a particular case of a Riemannian symmetric space (see [44], [62]).

It is easy to show that the metric of a Hermitian symmetric space is Kähler. Indeed, the form Ω is invariant with respect to the symmetry s_x. Therefore $d\Omega$ possesses the same property. Since $(ds_x)_x = -\mathrm{Id}$, the 3-form $d\Omega$ vanishes at the point x. Since x is an arbitrary point, $d\Omega = 0$.

As above let X be a complex manifold with a Hermitian metric. We denote by $G(X)$ the connected component of the identity in the group of holomorphic isometries of the manifold X. If X is a Hermitian symmetric space, then the group of holomorphic isometries, and thus also its connected component $G(X)$, acts transitively on X. Indeed, using the symmetries it is possible to extend infinitely any geodesic on X. It follows that any two points $x_1, x_2 \in X$ can be joined by a segment of a geodesic. But then $s_x(x_1) = x_2$, where x is the middle of this segment.

Example 1.1. Let $X = \mathbb{C}^n$ and $h = \sum dz_\alpha \otimes d\bar{z}_\alpha$. The automorphism $z \mapsto -z$ is the symmetry at the point 0. Since $\mathbb{C}^n$ is homogeneous with respect to the group of translations, this symmetry exists at each point, i.e., $\mathbb{C}^n$ is a Hermitian symmetric space. Here $\mathbb{C}^n$ is a flat manifold, i.e., its curvature is equal to zero. One can show that any flat homogeneous Kähler manifold is holomorphically isometric to $\mathbb{C}^n/\Gamma$, where Γ is a discrete subgroup of the vector group. In particular, any such manifold is a Hermitian symmetric space.

Now let X be the homogeneous complex manifold of a compact Lie group K. We assume that for some point $x \in X$ there exists an element $s \in K$ so that $s^2 = e$ and x is an isolated fixed point for the transformation s. Then the manifold X, viewed with any K-invariant Hermitian metric, is a Hermitian symmetric space with $K_0 \subset G(X)$.

Example 1.2. On the space of matrices $M_{n,q}$ there is defined an action of the group $\mathrm{GL}(n, \mathbb{C})$ (namely, $A \mapsto gA$, where $g \in \mathrm{GL}(n, \mathbb{C})$) and of $\mathrm{GL}(q, \mathbb{C})$ (namely, $A \mapsto Ah^{-1}$, where $h \in \mathrm{GL}(q, \mathbb{C})$). Let $q \leqslant n$ and let $M^0_{n,q}$ be the open subset of $M_{n,q}$ consisting of matrices of rank q. Then $M^0_{n,q}$ is invariant under both actions, and $M^0_{n,q}/\mathrm{GL}(q, \mathbb{C}) = \mathbb{G}_{q,n}$. More precisely, $M^0_{n,q}$ is the total space of a principal bundle with base $\mathbb{G}_{q,n}$ and with structure group $\mathrm{GL}(q, \mathbb{C})$. We denote by π the projection of this bundle. The group $\mathrm{GL}(n, \mathbb{C})$ acts on $M^0_{n,q}$, and its action drops to the base of the bundle $\mathbb{G}_{q,n}$. One verifies immediately that the subgroup $U(n) \subset \mathrm{GL}(n, \mathbb{C})$ is transitive on $\mathbb{G}_{q,n}$.

Let

$$o = \pi\left(\begin{pmatrix} 0 \\ I_q \end{pmatrix}\right) \in \mathbb{G}_{q,n}, \quad s = I_{p,q} \in U(n), \quad p + q = n.$$

Then $s^2 = e$, $s(o) = o$, and in some neighborhood of the point o the transformation s has no other fixed points. Thus the Grassmannian $\mathbb{G}_{q,n}$, viewed with any $U(n)$-invariant Hermitian metric, is a Hermitian symmetric space. It is not hard to verify that $G(\mathbb{G}_{q,n})$ coincides with the quotient of the group $U(n)$ by its center.

One can show that the curvature of the symmetric space in this example is positive ([44]). Now we turn to Hermitian symmetric spaces with negative curvature. For this we need the concept of the Bergman metric.

Let D be a bounded domain in $\mathbb{C}^n$. We denote by $L^2(D)$ the Hilbert space of complex functions on D for which $\int |f|^2 d\mu < \infty$, where $d\mu$ is Lebesgue measure on $\mathbb{C}^n$. Let $H(D)$ be the set of holomorphic functions in $L^2(D)$. It is well known that $H(D)$ is a closed subspace in $L^2(D)$. We select in $H(D)$ any orthonormal basis $\{\varphi_k\}$, where $k = 0, 1, \ldots$. Then the series

$$\sum_{k=0}^{\infty} \varphi_k(z)\overline{\varphi_k(\zeta)}$$

converges uniformly on compact subsets of $D \times D$, and its sum $K(z, \zeta)$ does not depend on the choice of basis. The function K is called the *Bergman kernel function* of the domain D. It has the following significant property: if $\varphi : D \to D'$ is an isomorphism from D onto some other bounded domain D', then

$$K(z, \zeta) = K'(\varphi(z), \varphi(\zeta)) \cdot J_\varphi(z) \cdot \overline{J_\varphi(\zeta)},$$

where J_φ is the Jacobian of the mapping φ and K' is the Bergman kernel function of the domain D'. It follows that the Hermitian form

$$h = \sum_{\alpha,\beta=1}^{n} \frac{\partial^2 \log K(z, z)}{\partial z_\alpha \partial \bar{z}_\beta} dz_\alpha \otimes d\bar{z}_\beta$$

is invariant under all automorphisms of the domain D. It is not difficult to verify that the form h is positive definite. The Hermitian metric h is called the *Bergman metric* of the bounded domain D. The form Ω satisfies $\Omega = \frac{i}{2} d'd'' \log K(z, z)$, which implies that $d\Omega = 0$, i.e., that the Bergman metric is a Kähler metric (see also Volume 9, Article III).

A bounded domain D is called *symmetric* if for any point $x \in D$ there exists an involutive automorphism $s_x : D \to D$ (the symmetry at the point x) for which s is the unique fixed point[3]. If it is already known that the domain D is homogeneous, then to prove that it is symmetric it suffices to find a symmetry s_x for some single point x. A symmetric domain with the Bergman metric is a Hermitian symmetric space with negative curvature in which $G(D) = (\operatorname{Aut} D)_0$.

The simplest examples of symmetric domains are the polycylinder and the ball. The second of these can be generalized as follows.

Example 1.3. Let $p \geqslant q > 0$ be integers. We shall show that the bounded domain $D_{p,q} = \{Z \in M_{p,q} | I_q - \bar{Z}^t Z > 0\}$ is symmetric. It is clear that the automorphism $Z \mapsto -Z$ has the unique fixed point $Z = 0$. Therefore it suffices to show that the domain $D_{p,q}$ is homogeneous.

We define an action of the group $U(p, q)$ on $D_{p,q}$. If a matrix $g \in M_{n,n}$ for $n = p + q$ is divided into blocks

$$g = \begin{pmatrix} A & B \\ C & D \end{pmatrix},$$

[3] An equivalent definition is obtained if we require that the point x be an isolated fixed point of the transformation s_x.

where $A \in M_{p,p}$, $B \in M_{p,q}$, $C \in M_{q,p}$, and $D \in M_{q,q}$, then

$$g \in U(p,q) \Leftrightarrow \bar{A}^t A - \bar{C}^t C = I_p; \quad -\bar{B}^t B + \bar{D}^t D = I_q; \quad \bar{B}^t A = \bar{D}^t C.$$

From these relations it is easy to deduce the following identity, which holds for all $Z \in M_{p,q}$:

$$(\bar{Z}^t \bar{C}^t + \bar{D}^t)(CZ + D) = (\bar{Z}^t \bar{A}^t + \bar{B}^t)(AZ + B) + (I_q - \bar{Z}^t Z).$$

If $Z \in D_{p,q}$, then the right side is a positive Hermitian metrix, which implies that $\det(CZ + D) \neq 0$. We set $W = (AZ + B)(CZ + D)^{-1}$. Then from the same identity it follows that $I_q - \bar{W}^t W = \bar{S}^t \cdot (I_q - \bar{Z}^t Z) \cdot S$, where $S = (CZ + D)^{-1}$. Therefore if $Z \in D_{p,q}$ then $W \in D_{p,q}$, so the formula

$$Z \mapsto W = (AZ + B)(CZ + D)^{-1}$$

defines an automorphism of the domain $D_{p,q}$. Thus we have proven that group $U(p,q)$ acts on $D_{p,q}$.

Let us verify the transitivity of this action. For a given $Z_0 \in D_{p,q}$ there exist two matrices $P \in \mathrm{GL}(p, \mathbb{C})$ and $Q \in \mathrm{GL}(q, \mathbb{C})$ such that

$$(\bar{P}^t P)(I_p - Z_0 \bar{Z}_0^t) = I_p; \quad (\bar{Q}^t Q)(I_q - \bar{Z}_0^t Z_0) = I_q.$$

It is easy to see that

$$\begin{pmatrix} P & -PZ_0 \\ -Q\bar{Z}_0^t & Q \end{pmatrix} \in U(p,q)$$

and the corresponding automorphism sends Z to 0.

And so $D_{p,q}$ is a symmetric domain. One can show that $G(D_{p,q})$ coincides with the quotient group of $U(p,q)$ by its center.

Returning to Example 1.2, we consider the sequence of holomorphic mappings

$$D_{p,q} \to M_{n,q}^0 \xrightarrow{\pi} \mathbb{G}_{q,n} \quad (n = p + q),$$

in which the first is defined by the formula $Z \mapsto \begin{pmatrix} Z \\ I_q \end{pmatrix}$. As a result we obtain an imbedding $D_{p,q} \hookrightarrow \mathbb{G}_{p,n}$, which, as one can easily check, commutes with the action of the group $U(p,q)$ (it acts on $\mathbb{G}_{q,n}$ as a subgroup of $\mathrm{GL}(n, \mathbb{C})$).

Hermitian symmetric spaces were classified by E. Cartan (see [24] and also [44], [112]). A Hermitian symmetric space X is said to be of *compact* (resp., *noncompact*) *type*, if the group $G(X)$ is semisimple and X is compact (resp., noncompact). In the first case the curvature is positive, and in the second negative. Hermitian symmetric spaces of compact and noncompact types are simply connected; a Hermitian symmetric space of noncompact type is homeomorphic to a cell. Any Hermitian symmetric space X, viewed as a Hermitian manifold, decomposes into a direct product $X = X^{(-)} \times X^{(0)} \times X^{(+)}$, where $X^{(-)}$ is of noncompact type, $X^{(+)}$ of compact type, and $X^{(0)}$ of Euclidean type (i.e., is flat). The space $X^{(0)}$ is holomorphically isometric to $\mathbb{C}^n/\Gamma$, where Γ is a lattice in the vector group (see Example 1.1). Let $G = G(X)$ and K the stationary subgroup of some point in X. If X is a Hermitian symmetric space of compact (resp., noncompact) type, then the group G is compact (resp., K is a maximal compact subgroup in G).

There exists a duality between Hermitian symmetric spaces of compact and noncompact types, which is a special case of the duality between Riemannian symmetric spaces. Using the language of Lie algebras one can describe this duality in the following way. Let X be a Hermitian symmetric space of one of these two types and $\mathfrak{m}$ the orthogonal complement in $\mathfrak{g}$ of the stationary subalgebra $\mathfrak{k}$ with respect to the Killing form. Then $\mathfrak{g} = \mathfrak{k} + \mathfrak{m}$. The subspace $\mathfrak{g}^\wedge = \mathfrak{k} + i\mathfrak{m}$ of the complex Lie algebra $\mathfrak{g}^C = \mathfrak{g} + i\mathfrak{g}$ is a subalgebra. There exists a unique Hermitian symmetric space $X^\wedge$ for which $G(X^\wedge)$ has Lie algebra $\mathfrak{g}^\wedge$, and the stationary subalgebra as before is equal to $\mathfrak{k}$. If X is of compact type, then $X^\wedge$ is of noncompact type and conversely, and $(X^\wedge)^\wedge = X$. The Hermitian symmetric spaces X and $X^\wedge$ are called dual in the sense of E. Cartan.

Every Hermitian symmetric space of noncompact type can be realized as a domain in the dual Hermitian symmetric space. The imbedding can be chosen invariant relative to G (Borel 1952; see also [44]). More precisely, if $X = G/K$ is a Hermitian symmetric space of noncompact type, then the dual Hermitian symmetric space can be written in the Klein form $X^\wedge = G^{\mathbb{C}}/P$, where P is a parabolic subgroup in the complex Lie group $G^{\mathbb{C}}$ containing K such that $P \cap G = K$ (the homogeneous manifold $X^\wedge$ is an example of a flag manifold, see Chapter 2). The Borel imbedding is defined by the formula $g \cdot K \mapsto g \cdot P$, and its image is the orbit of the real Lie group G on the manifold $X^\wedge$. An example is the imbedding $D_{p,q} \hookrightarrow \mathbb{G}_{q,p+q}$ (see Examples 1.2 and 1.3).

A Hermitian symmetric space of noncompact type can also be realized as a bounded symmetric domain in $\mathbb{C}^n$ (Harish–Chandra, 1956; see also [44]). Namely, the subspace $\mathfrak{m}$ complementary to $\mathfrak{k}$ can be identified with the tangent space to the manifold $X = G/K$ at the point $e \cdot K$. Since there is a complex structure on X, we have

$$\mathfrak{m}^{\mathbb{C}} = \mathfrak{m} + i\mathfrak{m} = \mathfrak{m}^+ + \mathfrak{m}^-,$$

where $\mathfrak{m}^+$ (resp., $\mathfrak{m}^-$) is identified with the space of holomorphic (resp., antiholomorphic) tangent vectors. The Lie algebra of the group P is equal to $\mathfrak{p} = \mathfrak{k}^{\mathbb{C}} + \mathfrak{m}^-$. The mapping $\xi : \mathfrak{m}^+ \to X^\wedge$ defined by the formula

$$\xi(A) = (\exp A) \cdot P \quad (A \in \mathfrak{m}^+),$$

is a regular imbedding. Its image is dense and contains the open subset X imbedded in $X^\wedge$ as indicated above. The domain $\xi^{-1}(X)$ is the required realization. It is not hard to verify that the resulting domain centered at 0 is mapped to itself under the homotheties $z \mapsto \lambda z$ for $\lambda \in \mathbb{C}$ and $0 \leqslant |\lambda| \leqslant 1$.

A Hermitian symmetric space of compact (resp., noncompact) type decomposes uniquely into a product of irreducible Hermitian symmetric spaces of compact (resp., noncompact) type. For an irreducible space the group $G(X)$ is simple. The classification of irreducible Hermitian symmetric spaces of noncompact type coincides with the classification of symmetric bounded domains, and the classification of irreducible spaces of compact type can be obtained from it by duality. In Table 1 all irreducible Hermitian symmetric spaces X of non-compact type are listed together with their dual spaces $X^\wedge$ (the space of type IV for $p = 2$ is reducible; it is included in the table to complete the series). A group in Klein form for X (resp., for $X^\wedge$) as a rule is locally isomorphic to $G(X)$ (resp., to $G(X^\wedge)$). An exception is the space of type II for $p = 2$. In the singular case E_n is a compact group, and $E_n^{(\delta)}$ is a real form of its complex hull $E_n^{\mathbb{C}}$, which is uniquely determined by the condition $\dim \mathfrak{m} - \dim \mathfrak{k} = \delta$. By a model for X we mean a realization of X as a symmetric bounded domain. The dimension of X is the complex dimension, and the rank, as usual, is the (real) dimension of a maximal abelian subspace in $\mathfrak{m}$.

The domains of types I–IV are called the classical domains. In the low dimensions there are the following isomorphisms among them (and among the corresponding dual Hermitian symmetric spaces):

$$I(p = q = 1) \simeq \mathrm{II}(p = 2) \simeq \mathrm{III}(p = 1) \simeq \mathrm{IV}(p = 1);$$

$$\mathrm{IV}(p = 2) \simeq \mathrm{I}(p = q = 1) \times I(p = q = 1);$$

$$\mathrm{II}(p = 3) \simeq \mathrm{I}(p = 3, q = 1);$$

$$\mathrm{IV}(p = 3) \simeq \mathrm{III}(p = 2);$$

$$\mathrm{IV}(p = 4) \simeq \mathrm{I}(p = q = 2);$$

$$\mathrm{IV}(p = 6) \simeq \mathrm{II}(p = 4).$$

A large number of papers are devoted to the classical domains (see [45], [93], [100]). In particular, the full automorphism groups of the classical domains have been determined as have explicit formulas for their Bergman kernels, Riemannian distances and geodesics.

From the point of view of harmonic analysis and automorphic functions an important role is played by the concept of a *boundary component* of a domain D, which was introduced by I.I. Pyatetskiĭ-Shapiro [93]. Let D be a domain in a complex manifold. A subset $F \subset \partial D$ is called a boundary component if: (a) F is analytic in a neighborhood of each of its points; (b) any holomorphic curve lying entirely in ∂D and intersecting F is completely contained in F; and (c) F is a minimal subset possessing property (b).

To study the boundaries of the classical domains, instead of the standard model for the domain X in $\mathbb{C}^n$, we may consider its Borel imbedding $X \hookrightarrow X^\wedge$, since the whole affine space containing the bounded domain is actually imbedded in $X^\wedge$. It is clear that under the imbedding $X \hookrightarrow X^\wedge$ the boundary ∂X is invariant with respect to the group $G = G(X)$, i.e., it decomposes into the orbits of this group. Thus there arises the problem of studying the orbits of a real Lie group G acting on the flag manifold $X^\wedge = G^{\mathbb{C}}/P$.

The most general results on the orbits of a real Lie group G on a flag manifold of its complex hull $G^{\mathbb{C}}$ were obtained by Wolf [113]. Wolf proved that for

Table 1

Type of Domain according to E. Cartan	X	Dimension of X	Rank of X	Model of X	$X^\wedge$
I	$SU(p,q)/S(U(p) \times U(q))$ $(p \geqslant q \geqslant 1)$	pq	$\min(p,q)$	$\{Z \in M_{p,q} \mid \bar{Z}^t Z < I_q\}$	$\mathbb{G}_{q,p+q} = SU(p+q)/S(U(p) \times U(q))$
II	$SO^*(2p)/U(p)$ $(p \geqslant 2)$	$\frac{p(p-1)}{2}$	$\left[\frac{p}{2}\right]$	$\{Z \in M_{p,p} \mid \bar{Z}^t Z < I_p, Z = -Z^t\}$	$SO(2p)/U(p)$
III	$Sp(p,\mathbb{R})/U(p)$ $(p \geqslant 1)$	$\frac{p(p+1)}{2}$	p	$\{Z \in M_{p,p} \mid \bar{Z}^t Z < I_p, Z = Z^t\}$	$Sp(p)/U(p)$
IV	$SO_0(p,2)/SO(p) \times SO(2)$ $(p \geqslant 1)$	p	$\min(2,p)$	$\{z \in \mathbb{C}^p \mid \Sigma \lvert z_i \rvert^2 < \frac{1}{2}(1 + \lvert \Sigma z_i^2 \rvert^2) < 1\}$	$\mathbb{Q}_{p+2} = SO(p+2)/SO(p) \times SO(2)$
V	$E_6^{(-14)}/SO(10) \cdot SO(2)$	16	2	See [63], [54], [29], [30]	$E_6/SO(10) \cdot SO(2)$
VI	$E_7^{(-25)}/E_6 \cdot SO(2)$	27	3		$E_7/E_6 \cdot SO(2)$

such an action the number of G-orbits is finite, and he described a decomposition of each G-orbit into classes according to the following equivalence relation. Two points in a G-orbit are called equivalent if they can be joined by a finite chain of holomorphic curves which are contained entirely in the given G-orbit.

In the Hermitian symmetric case the classes are complex submanifolds. Also, a holomorphic curve belonging to ∂X is entirely contained in one of the G-orbits. Therefore for any point $y \in \partial X$ its class in $G(y)$ is a boundary component of ∂X.

Let us consider a Hermitian symmetric space X which is irreducible and has rank r. In this case more precise results have been obtained (see Korányi–Wolf [64], [114], Takeuchi [103]). The number of the orbits of the group G on $X^\wedge$ is equal to $1/2(r+1)(r+2)$, the number of open orbits is equal to $r+1$, (one of them coincides with X), and there is exactly one closed orbit (a part of ∂X). The number of orbits in ∂X is equal to r. Every component of the boundary of the domain $X \subset X^\wedge$ is a Hermitian symmetric space (of another group). For any s with $0 \leqslant s < r$ among the components there is exactly one (up to a transformation from G) which as a symmetric space has rank s (for $s = 0$ it is a point). The types of the components are known. For the classical domains a component belongs to the same series as does the space X (see [93]). The unique closed orbit $S \subset \partial X$ is the union of zero-dimensional components. The orbit S is the *Bergman–Shilov boundary* of the domain X, i.e., any function holomorphic in X and continuous in $\bar{X}$ attains its maximum modulus on S and S is minimal with respect to this property.

Example 1.4. The boundary of a domain of type III

$$D_p = \{Z \in M_{p,p} | \bar{Z}^t Z < I_p, Z = Z^t\}$$

consists of those $Z \in M_{p,p}$ for which $\det(I_p - \bar{Z}^t Z) = 0$, $I_p - \bar{Z}^t Z \geqslant 0$, and $Z = Z^t$. A transformation of the group $G(D_p)$ has the form

$$Z \mapsto (AZ + B)(CZ + D)^{-1},$$

where the matrix $g = \begin{pmatrix} A & B \\ C & D \end{pmatrix}$ satisfies the conditions $g^t I_{p,p} \bar{g} = I_{p,p}$ and $g^t J_p g = J_p$. The group $G = G(D_p)$ has p orbits on the boundary ∂D_p, whose representatives are the matrices

$$y_s = \begin{pmatrix} I_{p-s} & 0 \\ 0 & 0 \end{pmatrix} \quad (s = 0, 1, \ldots, p-1).$$

The component of the boundary containing y_s consists of the matrices $\begin{pmatrix} I_{p-s} & 0 \\ 0 & Z \end{pmatrix}$, where $Z \in M_{s,s}$, $\bar{Z}^t Z < I_s$, and $Z = Z^t$, i.e., $Z \in D_s$. For $s = 0$ the component consists of the single point $y_0 = I_p$. The Bergman–Shilov boundary $S = G(y_0)$ can be written in Klein form as $S = U(p)/O(p)$; in particular, $\dim_{\mathbb{R}} S = (p^2 + p)/2 = \dim_{\mathbb{C}} D_p$.

Chapter 2. Flag Manifolds

Any compact homogeneous complex manifold X can be written in the Klein form $X = G/U$, where G is a connected complex Lie group and U is a closed complex subgroup. In order to define flag manifolds, it is necessary to impose certain restrictions on the stationary subgroup U. Although these restrictions are of algebraic character, we shall show later that the class of flag manifolds can be characterized also in geometric terms. This class contains many interesting homogeneous manifolds, in particular, all Hermitian symmetric spaces of compact type, for example, projective space, the Grassmann manifold, and projective quadrics.

A maximal connected solvable subgroup B of a connected complex Lie group G is called a *Borel subgroup*. A subgroup $P \subset G$ which contains some Borel subgroup is called a *parabolic subgroup*. A homogeneous complex manifold $X = G/P$ is called a *flag manifold* (of the group G) if the stationary subgroup P is a parabolic subgroup. If $X = G/P$ is a flag manifold, then without loss of generality we may assume that the group G is semisimple, since the parabolic subgroup P contains the radical of the group G.

Example 2.1. (flag manifolds of the group $SL(n, \mathbb{C})$). We fix a decomposition of the integer n:

$$n = d_1 + d_2 + \cdots + d_k, \quad d_j \in \mathbb{Z}, \quad d_j > 0.$$

A flag of type $\{d_1, d_2, .., d_k\}$ in the space $\mathbb{C}^n$ is by definition a sequence of linear subspaces

$$\{0\} = V_0 \subset V_1 \subset V_2 \subset \cdots \subset V_k = \mathbb{C}^n$$

such that $\dim V_j = d_1 + d_2 + \cdots + d_j$ for $j = 1, 2, \ldots, k$. We denote the set of all flags of type $\{d_1, d_2, \ldots, d_k\}$ by $\mathbb{F}(d_1, d_2, \ldots, d_k)$. The group $SL(n, \mathbb{C})$ acts in the natural way on the set $\mathbb{F}(d_1, d_2, \ldots, d_k)$ and this action is transitive. Let $z_1, z_2, \ldots, z_n$ be coordinates in $\mathbb{C}^n$; we set

$$V_j^0 = \{z \in \mathbb{C}^n | z_1 = z_2 = \cdots = z_{n-d_1-\cdots-d_j} = 0\} \quad (j = 1, 2, \ldots, k).$$

We denote the stationary subgroup of the resulting flag by $P(d_1, d_2, \ldots, d_k)$. It consists of unimodular matrices such that along the main diagonal there are square blocks of dimensions $d_k, d_{k-1}, \ldots, d_1$, above these blocks there are zeros, and below them the elements are arbitrary. The subgroup $P(d_1, d_2, \ldots, d_k)$ contains the subgroup of lower triangular unimodular matrices, which is a Borel subgroup in $SL(n, \mathbb{C})$. Thus $P(d_1, d_2, \ldots, d_k)$ is a parabolic subgroup. Conversely, one can show that any parabolic subgroup of $SL(n, \mathbb{C})$ is conjugate to one of the subgroups $P(d_1, d_2, \ldots, d_k)$. The natural bijective correspondence

$$\mathbb{F}(d_1, d_2, \ldots, d_k) \rightleftarrows SL(n, \mathbb{C})/P(d_1, d_2, \ldots d_k)$$

allows us to introduce into the set $\mathbb{F}(d_1, d_2, \ldots, d_k)$ the structure of a complex manifold. This manifold will be a flag manifold of the group $SL(n, \mathbb{C})$, and any flag manifold of this group can be obtained in this manner. We note that $\mathbb{G}_{m,n} = \mathbb{F}(m, n-m)$ and, in particular, $\mathbb{P}^{n-1} = \mathbb{G}_{1,n} = \mathbb{F}(1, n-1)$.

Example 2.2. Fix $m \leqslant n/2$. We denote by Y the set of all linear subspaces of dimension m in $\mathbb{C}^n$ which are completely isotropic with respect to the bilinear form $b(z, w) = \sum z_i w_i$, i.e., we set

$$Y = \{V \in \mathbb{G}_{m,n} | b(V, V) = 0\}.$$

The group $SO(n, \mathbb{C})$ acts on Y in the natural manner; for $m < n/2$ this action is transitive, and for $m = n/2$ Y is disconnected and consists of two orbits of the group $SO(n, \mathbb{C})$. We set $X = Y$ in the first case and denote by X one of the two orbits in the second case. Then X is a flag manifold of the group $SO(n, \mathbb{C})$. For $m = 1$ and $n > 2$ we have $X = \mathbb{Q}_n$.

Example 2.3. (the manifold of Borel subgroups). Let G be a semisimple complex Lie group. It is known that any two Borel subgroups B_1 and B_2 are conjugate, i.e., that $gB_1g^{-1} = B_2$ for some $g \in G$. The group G acts on the set of its Borel subgroups by the formula $B \overset{g}{\mapsto} gBg^{-1}$. By what has been said above this action is transitive. The stationary subgroup of the point (subgroup) B under this action coincides with $N_G(B)$. But it can be shown that $N_G(B) = B$, which means that the Borel subgroups are in natural bijective correspondence with the points of the flag manifold G/B.

Example 2.4. Let X be a Hermitian symmetric space of compact type. The action of the compact group $K = G(X)$ (see Section 1) extends to a holomorphic action of its complex hull $K^{\mathbb{C}}$. The manifold X is a flag manifold of the group $K^{\mathbb{C}}$.

An important property of a parabolic subgroup is its connectedness. Therefore flag manifolds are simply connected. In view of a well known theorem of Montgomery, a maximal compact subgroup $K \subset G$ is transitive on the flag manifold $X = G/P$; i.e., X can be written in the "compact" Klein form $X = K/L$, where $L = K \cap P$. It is not difficult to show that L is the centralizer of some torus in K.

Conversely, Wang [110] has shown that the quotient space of a connected compact Lie group K by the centralizer L of any torus in K possesses a K-invariant complex structure. In general this structure is not unique (see [21]). Viewed with any invariant complex structure, the manifold K/L is a flag manifold of the complex hull $G = K^{\mathbb{C}}$ of K.

Example 2.5. If X is the manifold from Example 2.2, then $K = SO(n)$ and $L = U(m) \times SO(n-2m)$. The torus which has L as its centralizer is one-dimensional and consists of all matrices of the form

$$\begin{pmatrix} \alpha I_m & \beta I_m & 0 \\ -\beta I_m & \alpha I_m & 0 \\ 0 & 0 & I_{n-2m} \end{pmatrix},$$

where $\alpha^2 + \beta^2 = 1$. For $n = 2m$ the manifold $X = K/L$ is a Hermitian symmetric space of compact type (see Table 1).

The next theorem solves completely the problem of determining the invariant complex structures on simply connected homogeneous manifolds of compact Lie groups.

Theorem 2.6. (Wang [110]). *Let K be a connected compact semisimple Lie group and J a connected closed subgroup of K. A K-invariant complex structure on the homogeneous manifold K/J exists if and only if K/J is even dimensional, and the semisimple part of J coincides with the semisimple part of the centralizer L of some torus in K. If such a torus exists, then it can be chosen so that $L \supset J$. The manifold K/L can be endowed with a K-invariant complex structure so that the natural mapping $K/J \to K/L$ is holomorphic.*

Corollary 2.7. *Any simply connected compact homogeneous complex manifold is the total space of a holomorphic fiber bundle, whose base is a flag manifold and whose fiber is a complex torus.*

A result analogous to Theorem 2.6 holds for noncompact semisimple Lie groups if the rank of a maximal compact subgroup coincides with the rank of the group [69]. In the case of an arbitrary (semisimple) Lie group the situation is more complicated. The problem of determining the invariant complex structures on the homogeneous manifolds S/J reduces to the algebraic problem of decomposing the Lie algebra $\mathfrak{s}^{\mathbb{C}} = \mathfrak{s} + \mathfrak{h}$, where $\mathfrak{h}$ is a complex subalgebra of $\mathfrak{s}^{\mathbb{C}}$ such that $\mathfrak{h} \cap \mathfrak{s} = \mathfrak{j}$. The problem has been solved only in particular cases (see [84], [70], [71]).

Example 2.8. By Theorem 2.6 there exists on any even dimensional compact semisimple Lie group K a left-invariant complex structure. Let $K = \mathrm{SU}(3)$. We consider in the group $G = K^{\mathbb{C}} = \mathrm{SL}(3, \mathbb{C})$ the subgroup

$$U_\lambda = \left\{ \begin{pmatrix} e_z & 0 & 0 \\ * & e^{\lambda z} & 0 \\ * & * & e^{-z-\lambda z} \end{pmatrix} \middle| z \in \mathbb{C} \right\},$$

where $\lambda \in \mathbb{C} \setminus \mathbb{R}$ is fixed and the stars denote arbitrary numbers. The subgroup U_λ is closed and $U_\lambda \cap K = \{e\}$. On the complex homogeneous manifold $X_\lambda = G/U_\lambda$ the group K acts simply transitively. Thus on K there is a left-invariant complex structure which depends on the parameter λ. The manifold X_λ fibers over the flag manifold $\mathbb{F}(1,1,1) = G/B$, where B is the lower triangular Borel subgroup in G. The fiber of this bundle is a one-dimensional complex torus (an elliptic curve)

$$B/U_\lambda = \mathbb{C}^* \times \mathbb{C}^*/\{(e^z, e^{\lambda z}) | z \in \mathbb{C}\} = \mathbb{C}^*/\{e^{2\pi i n \lambda} | n \in \mathbb{Z}\} = \mathbb{C}/\mathbb{Z} + \lambda\mathbb{Z}.$$

Flag manifolds can be characterized in geometric terms.

Theorem 2.9 (Wang [110]). *Let X be a simply connected compact homogeneous complex manifold. In order for X to be Kähler it is necessary and sufficient that X be a flag manifold.*

For flag manifolds a cell decomposition is known consisting of cells of even dimensions (Borel [19]). Therefore flag manifolds have no torsion, and their odd Betti numbers are equal to zero. The Poincaré polynomial of a flag manifold can be calculated using the formula of Leray-Hirsch.

A function theoretic description of flag manifolds is contained in the next theorem.

Theorem 2.10 (F.A. Berezin- I.I. Pyatetskiĭ–Shapiro [12], Goto [38]). *A compact homogeneous complex manifold X is a flag manifold if and only if the field of meromorphic functions on X is isomorphic to the field of rational functions in n independent variables, where* $n = \dim X$. *The flag manifold X is a rational projective algebraic variety and the imbedding* $X \hookrightarrow \mathbb{P}^N$ *can be chosen to be equivariant.*

The degrees of the equivariant projective imbeddings of flag manifolds are calculated in [21].

To conclude this chapter we pause to consider one geometric application of the concept of flag manifold.

Let X be a complex manifold with $\dim X = 2n + 1$. A *contact structure* on X is by definition a holomorphic distribution of tangent hyperplanes which satisfies the following nondegeneracy condition. In a neighborhood of any point of the manifold X the distribution is defined by a holomorphic 1-form ω having no zeros. The nondegeneracy condition is that $\omega \wedge (d\omega)^n \neq 0$ everywhere in this neighborhood.

Automorphisms of a contact structure are defined in the obvious fashion. If X is a contact complex manifold (i.e., a complex manifold with a given contact structure) and if X is compact, then the group of automorphisms of the contact structure is a complex Lie group of transformations of X. The manifold X is called a homogeneous contact complex manifold if this group acts transitively on X.

Theorem 2.11 (Boothby [17], [18]). *Let X be a compact homogeneous contact complex manifold. If X is simply connected, then X is a flag manifold of a simple complex Lie group G. For each group G this manifold is unique. Namely, X is the (unique) closed G-orbit in* $P(\mathfrak{g})$ *(here* $\mathfrak{g}$ *is the adjoint G-module and* $P(\mathfrak{g})$ *is the associated projective space).*

Chapter 3. Homogeneous Vector Bundles

Flag manifolds are interesting, in particular, because for them there has been constructed a theory of homogeneous vector bundles which has numerous important applications. The main result—the Borel–Weil–Bott theorem—gives a method of calculating cohomology which is applicable to a certain natural class of coherent sheaves and at the same time indicates a geometric realization for the irreducible finite-dimensional representations of complex semisimple Lie groups.

Let there be given a holomorphic vector bundle $\mathbb{E} = \mathbb{E}(Z, X, \pi, E)$. Here Z is the total space of the bundle, X is the base, $\pi : Z \to X$ is the projection, and E is a typical fiber. We assume that $X = G/U$, where G is a connected complex Lie group and U a closed complex subgroup. We agree to identify E with $\pi^{-1}(o)$, where $o = e \cdot U$.

The bundle $\mathbb{E}$ is called a (holomorphic) *homogeneous vector bundle* if there is defined a holomorphic action of the group G on the total space Z so that for every $g \in G$ the diagram

$$\begin{array}{ccc} Z & \xrightarrow{\;g\;} & Z \\ \downarrow{\scriptstyle \pi} & & \downarrow{\scriptstyle \pi} \\ X & \xrightarrow[\;g\;]{} & X \end{array}$$

commutes, and for every pair $(g, x) \in G \times X$ the induced mapping $g : \pi^{-1}(x) \to \pi^{-1}(gx)$ is an isomorphism of vector spaces.

A homogeneous vector bundle $\mathbb{E}$ defines a holomorphic linear representation of the subgroup U in the fiber $E = \pi^{-1}(o)$. Conversely, for a given representation $\varphi : U \to \mathrm{GL}(E)$ it is not difficult to construct a homogeneous vector bundle. To do this, we form the fiber product $Z = G \times_U E$, i.e., the quotient space of the direct product $G \times E$ by the action of U

$$(g, \xi) \overset{u}{\mapsto} (gu^{-1}, \varphi(u)\xi) \quad (g \in G, u \in U, \xi \in E),$$

and define a projection $\pi : Z \to X$ by setting $\pi(\overline{g, \xi}) = g \cdot U$ (here the line denotes the equivalence class). The group G acts on Z by the formula $h(\overline{g, \xi}) = (\overline{hg, \xi})$ for g, $h \in G$ and $\xi \in E$. It is not difficult to verify that $\mathbb{E} = \mathbb{E}(Z, X, \pi, E)$ is a homogeneous vector bundle in which the action of the subgroup U in the fiber E coincides with the given action. Thus we obtain the one-to-one correspondence

$$\left\{\begin{matrix}\text{holomorphic homogeneous} \\ \text{vector bundles over } G/U\end{matrix}\right\} \rightleftarrows \left\{\begin{matrix}\text{holomorphic representations} \\ \text{of the subgroup } U\end{matrix}\right\}.$$

This correspondence has various natural properties. Let $\mathbb{E}^\varphi$ be the homogeneous vector bundle corresponding to the representation $\varphi : U \to \mathrm{GL}(E)$. Then $\mathbb{E}^{\varphi_1 \oplus \varphi_2} = \mathbb{E}^{\varphi_1} \oplus \mathbb{E}^{\varphi_2}$, $\mathbb{E}^{\varphi_1 \otimes \varphi_2} = \mathbb{E}^{\varphi_1} \otimes \mathbb{E}^{\varphi_2}$, $\mathbb{E}^{\wedge^n \varphi} = \wedge^n \mathbb{E}^\varphi$, and $E^{\varphi^*} = (\mathbb{E}^\varphi)^*$. Here we use the standard notation for direct sum, tensor product, exterior power, and dual representation, and also the analogous notation for vector bundles.

Example 3.1. The holomorphic tangent space to a manifold $X = G/U$ at the point o is identified with the quotient space $\mathfrak{g}/\mathfrak{u}$. The representation $\theta : U \to \mathrm{GL}(\mathfrak{g}/\mathfrak{u})$ defined by the formula

$$\theta(u)(Y + \mathfrak{u}) = \mathrm{Ad}_G(u)(Y) + \mathfrak{u} \quad (Y \in \mathfrak{g}),$$

is thus identified with the isotropy representation. Let $\mathbb{T}$ be the holomorphic tangent bundle over G/U and let $\mathbb{K} = \wedge^n \mathbb{T}^*$ be the canonical bundle (here $n = \dim G/U$). Then $\mathbb{T} = \mathbb{E}^\theta$ and $\mathbb{K} = \mathbb{E}^\delta$, where $\delta : U \to \mathbb{C}^*$ is the one-dimensional representation

$$\delta(u) = \det \mathrm{Ad}_U(u) \cdot (\det \mathrm{Ad}_G(u))^{-1} \quad (u \in U),$$

We denote by $\mathscr{E}$ (resp., $\mathscr{E}^\varphi$) the sheaf of germs of holomorphic sections of the vector bundle $\mathbb{E}$ (resp., $\mathbb{E}^\varphi$). Since the bundle $\mathbb{E}^\varphi$ is homogeneous, in the cohomology spaces $H^k(X, \mathscr{E}^\varphi)$ there are induced representations of the group G, and the problem of calculating them arises naturally. We note that the space of global sections $H^0(X, \mathscr{E}^\varphi)$ admits the following description:

$$H^0(X, \mathscr{E}^\varphi) = \{f : G \to E \mid f \text{ is holomorphic},$$
$$f(gu) = \varphi(u)^{-1} f(g) \quad (g \in G, u \in U)\}.$$

A representation of the group G in this space is defined by the formula $(gf)(x) = f(g^{-1}x)$ for $x, g \in G$.

Example 3.2. We calculate the induced representations of the group $G = \mathrm{SL}(m + 1, \mathbb{C})$ in the cohomology spaces of linear bundles over $\mathbb{P}^m$. Let P be the parabolic subgroup consisting of the matrices

$$p = \left(\begin{array}{c|c} & 0 \\ * & \vdots \\ & 0 \\ \hline * \ldots * & x(p) \end{array}\right),$$

and let $\varphi_n : P \to \mathbb{C}^*$ be the character defined by the formula $\varphi_n(p) = x(p)^{-n}$ and $\mathbb{E}^n = \mathbb{E}^{\varphi_n}$. Any linear bundle over $\mathbb{P}^m = G/P$ is homogeneous and is isomorphic to one of the $\mathbb{E}^n$. We note that $\mathbb{K} = \mathbb{E}^{-m-1}$ (see Example 3.1). The sheaf corresponding to the bundle $\mathbb{E}^n$ is usually denoted by $\mathcal{O}(n)$. Its global sections are identified with the holomorphic functions on $\mathbb{C}^{m+1} \setminus \{0\}$ satisfying the equation

$$f(tz) = t^n f(z) \quad (z \in \mathbb{C}^{m+1}, t \in \mathbb{C}^*).$$

Using Hartogs' Theorem, we conclude that f is a homogeneous polynomial of degree n for $n \geqslant 0$ and $f = 0$ for $n < 0$. Thus

$$H^0(\mathbb{P}^m, \mathcal{O}(n)) = \begin{cases} W_n & (n \geqslant 0) \\ 0 & (n < 0). \end{cases}$$

Here W_n is the space of homogeneous polynomials of degree n on $\mathbb{C}^{m+1}$ on which the group $\mathrm{SL}(m + 1, \mathbb{C})$ acts naturally, and equality is understood in the sense of equivalence of representations.

According to a well known theorem of Serre, $H^k(\mathbb{P}^m, \mathcal{O}(n)) = 0$ if $0 < k < m$. To calculate the representation in $H^m(\mathbb{P}^m, \mathcal{O}(n))$ we can use Serre duality, which has equivariant character. Since $(\mathbb{E}^n)^* \otimes \mathbb{K} = \mathbb{E}^{-(m+n+1)}$, we obtain that

$$H^m(\mathbb{P}^m, \mathcal{O}(n)) = H^0(\mathbb{P}^m, \mathcal{O}(-m-n-1)) = \begin{cases} W_{-m-n-1} & (m+n+1 \leqslant 0) \\ 0 & (m+n+1 > 0). \end{cases}$$

The theorem to which we are now heading is a far reaching generalization of this example. In what follows the complex Lie group G will be assumed to be semisimple and the stationary subgroup, which we now denote by P, parabolic. We select a Cartan subalgebra $\mathfrak{h} \subset \mathfrak{g}$ contained in $\mathfrak{p}$. Let Δ denote the root system of the pair $(\mathfrak{g}, \mathfrak{h})$ with $\mathfrak{g}^\alpha$ the root subspace corresponding to a root $\alpha \in \Delta$. We fix an ordering in the root system Δ so that $\mathfrak{g}^{-\alpha} \in \mathfrak{p}$ if $\alpha \in \Delta^+$ (here Δ^+ is the set of positive roots). Then

$$\mathfrak{p} = \mathfrak{h} \oplus \sum_{\alpha \in \Phi} \mathfrak{g}^\alpha \oplus \sum_{\alpha \in \Delta^+} \mathfrak{g}^{-\alpha},$$

where Φ is some subset of Δ^+. We note that in the space of an irreducible finite-dimensional representation of the group P the unipotent radical of the algebra $\mathfrak{p}$, which is equal to $\sum_{\alpha \in \Delta^+ \setminus \Phi} \mathfrak{g}^{-\alpha}$, acts trivially. Therefore describing the irreducible representations of the group P reduces to describing the irreducible representations of its reductive part. We shall employ here, as usual, the concepts of highest vector and highest weight. By the highest vector of a representation of G (resp., P) we mean an eigenvector of the subgroup B^+ (resp., $P \cap B^+$).

The real subspace in $\mathfrak{h}^*$ generated (over $\mathbb{R}$) by the elements of Δ is the real form of the space $\mathfrak{h}^*$. We denote it by $\mathfrak{h}^*_{\mathbb{R}}$. In $\mathfrak{h}^*_{\mathbb{R}}$ there is a positive definite scalar product $(,)$ defined by duality via the Killing form. Let $C = \{\mu \in \mathfrak{h}^*_{\mathbb{R}} | (\mu, \alpha) \geqslant 0, \forall \alpha \in \Delta^+\}$ be a positive Weyl chamber, C^0 its interior port, W a Weyl group, and ρ the half-sum of positive roots. The form $\mu \in \mathfrak{h}^*_{\mathbb{R}}$ is called singular if $(\mu, \alpha) = 0$ for some $\alpha \in \Delta$, and regular otherwise. If μ is a regular form, then there exists a unique element $w = w_\mu \in W$ such that $w(\mu) \in C^0$. The index of an element of the Weyl group is by definition the number of positive roots which the element sends to negative ones. The index w_μ is equal to the number of roots $\alpha \in \Delta^+$ for which $(\mu, \alpha) < 0$.

Theorem 3.3 (Bott [23]). *Let $\mathbb{E}^\varphi$ be a homogeneous vector bundle over a flag manifold $X = G/P$ defined by an irreducible holomorphic representation $\varphi : P \to$ GL(E), and let $\lambda \in \mathfrak{h}^*_{\mathbb{R}}$ be the highest weight of the representation φ. If the form $\lambda + \rho$ is singular, then $H^k(X, \mathscr{E}^\varphi) = 0$ for all k. If the form $\lambda + \rho$ is regular, then we denote by l the index of the element $w = w_{\lambda+\rho} \in W$. Then $H^k(X, \mathscr{E}^\varphi) = 0$ for $k \neq l$, and $H^l(X, \mathscr{E}^\varphi)$ is an irreducible G-module with highest weight $\Lambda = w(\lambda + \rho) - \rho$.*

The assertion of the theorem for a representation in $H^0(X, \mathscr{E}^\varphi)$, where φ is a one-dimensional representation (character) was proven by Borel and A. Weil (1954). They showed that an irreducible finite-dimensional representation of a complex semisimple Lie group can be realized in $H^0(G/B^-, \mathscr{E}^\varphi)$, where φ is the character corresponding to the highest weight of the representation. For various proofs of Theorem 3.3 see articles by Kostant [65], Aribaud [9], and Demazure [26]. A generalization of Theorem 3.3 to arbitrary simply connected compact homogeneous complex manifolds is due to Griffiths [40].

Now let X be a compact complex manifold and $\mathbb{T}$ its holomorphic tangent bundle with $\mathscr{T}$ the corresponding sheaf. Sections of the sheaf $\mathscr{T}$ are holomorphic vector fields on X. The space $H^0(X, \mathscr{T})$ viewed with the operation of Lie bracket is the Lie algebra of the group $\operatorname{Aut} X$. The space of one-dimensional cohomology $H^1(X, \mathscr{T})$ plays an important role in deformation theory. Namely, there exists a versal deformation of the complex structure on X, parametrized by an analytic subset in a neighborhood of zero in the space $H^1(X, \mathscr{T})$. In particular, if $H^1(X, \mathscr{T}) = 0$, then any deformation is locally trivial, i.e., the complex structure on X is locally rigid (the Frölicher–Nijenhuis Theorem, see Article III).

We calculate the representations of the group G in the spaces $H^k(X, \mathscr{T})$, where $X = G/P$ is a flag manifold. Immediate application of Theorem 3.3 to the bundle $\mathbb{T}$ is possible only in rare cases, since the isotropy representation $P \to \mathrm{GL}(\mathfrak{g}/\mathfrak{p})$ is generally not irreducible. However, in the space $\mathfrak{g}/\mathfrak{p}$ there is a P-invariant filtration on whose sequence of quotients the unipotent radical of the algebra $\mathfrak{p}$ acts trivially. Therefore these quotients are completely reducible, and their highest weights are positive roots. One can show that in an irreducible root system there are at the most three positive roots α for which the form $\alpha + \rho$ is regular. These are the maximal root $\alpha_{\max}$, the maximal root among the short roots α_0; and the root $\beta = s_\gamma(\alpha_0 + \gamma)$, where γ is the (unique) simple root such that $\alpha_0 + \gamma$ is again a root and $s_\gamma \in W$ is the corresponding reflection (it is clear that the two latter possibilities can occur only in the root systems of types B_n, C_n, G_2, and F_4, where there exist roots of different lengths). Using Theorem 3.3 and the filtration mentioned above, one can prove the following proposition of which the first part is again due to Bott [23].

Proposition 3.4. *Let $X = G/P$ be a flag manifold and $\mathscr{T}$ the sheaf of germs of holomorphic vector fields on X. Then:*

1) $H^k(X, \mathscr{T}) = 0$ for $k \geqslant 1$.

Assume now that the group G is simple.

2) *If all roots of $\mathfrak{g}$ have the same length or if β is not a root of the subalgebra $\mathfrak{p}$, then $H^0(X, \mathscr{T}) = \mathfrak{g}$ (the adjoint G-module).*

3) *If β is a root of the subalgebra $\mathfrak{p}$, then*

$$H^0(X, \mathscr{T}) = \mathfrak{g} \oplus W^{\alpha_0},$$

where W^{α_0} is an irreducible G-module with highest weight α_0.

Corollary 3.5. *The complex structure on a flag manifold is locally rigid.*

Using assertions 2) and 3) of the theorem it is possible to calculate the group $(\operatorname{Aut} X)_0$ for flag manifolds.

Theorem 3.6 (A.L. Onishchik [83]). *Let X be a flag manifold of a simple adjoint complex Lie group G. Then the group* $(\mathrm{Aut}\, X)_0$ *is simple and as rule* $G = (\mathrm{Aut}\, X)_0$. *All exceptions are listed in* Table 2.

Table 2

Type of G	Type of $(\mathrm{Aut}\, X)_0$	X
$B_n (n \geqslant 2)$	D_{n+1}	$\mathrm{SO}(2n+2)/U(n+1)$
$C_n (n \geqslant 3)$	A_{2n-1}	$\mathbb{P}^{2n-1}$
G_2	B_3	$\mathbb{Q}_7$

Other proofs of Theorem 3.6 are given in [56] and [101] (see also [105]). The proof in [101] is based on the theorem of Borel-Weil-Bott. In [56] the full groups $\mathrm{Aut}\, X$ are also calculated.

As was already noted in the beginning of Section 3, the Borel-Weil-Bott Theorem gives geometric realizations of the irreducible finite-dimensional representations of complex semisimple Lie groups, or equivalently, realizations of irreducible unitary representations of compact semisimple Lie groups. Now let G be a connected and, in general, noncompact semisimple Lie group, and let K be a maximal compact subgroup with the rank of G equal to the rank of K. As Harish-Chandra has proven, the latter condition is necessary and sufficient for the group G to have a unitary representation with integrable square (this means that the matrix elements of the representation belong to $L^2(G)$). The irreducible unitary representations with integrable squares are called representations of the *discrete series*.

We denote by H a Cartan subgroup of the group K (it is also a Cartan subgroup of G). Let $\mathfrak{h}^{\mathbb{C}}$ be the corresponding Cartan subalgebra in $\mathfrak{g}^{\mathbb{C}}$, and let $\mu \in (h^{\mathbb{C}})^*$ be a regular linear form whose restriction to $\mathfrak{h}$ is the differential of a globally defined character $H \to \mathbb{C}^*$. According to a theorem of Harish-Chandra this form μ corresponds to a representation of the discrete series T_μ of the group G, and T_μ is unitarily equivalent to $T_{\mu'}$ if and only if $\mu = w(\mu')$ for some $w \in W$ (here we use notation which was introduced before the statement of Theorem 3.3; only $\mathfrak{g}$ (resp., $\mathfrak{h}$) is now replaced by $\mathfrak{g}^{\mathbb{C}}$ (resp., $\mathfrak{h}^{\mathbb{C}}$)).

We assume that G is imbedded in the connected complex Lie group $G^{\mathbb{C}}$ as a real form. Let B be a Borel subgroup in $G^{\mathbb{C}}$ containing $H^{\mathbb{C}}$. We choose an ordering of the root system Δ of the pair $(\mathfrak{g}^{\mathbb{C}}, \mathfrak{h}^{\mathbb{C}})$ so that $B = B^-$. Then the homogeneous space $D = G/H$ is identified with the orbit of the point $e \cdot B^-$ on the flag manifold $X = G^{\mathbb{C}}/B^-$. Thus D has an invariant complex structure.

Langlands (1966) conjectured that the representations of the discrete series T_μ are realized in the properly defined L^2-cohomology spaces of homogeneous linear bundles over D (see also [65]). This conjecture was proven in final form by Schmid [97], [98]. A related result for Hermitian symmetric spaces was obtained by Narasimhan and Okamoto [78]. We formulate the theorem of Schmid.

Let $\lambda \in (h^{\mathbb{C}})^*$ be a form corresponding to a holomorphic character $\varphi: H^{\mathbb{C}} \to \mathbb{C}^*$. We extend φ to B^- by setting $\varphi = 1$ on the unipotent radical. We again denote the extension by φ. Over the flag manifold X there is defined the homogeneous linear bundle $\mathbb{E}^{\varphi}$. We restrict it to the domain $D \subset X$ and denote by $H^k(\lambda)$ the Hilbert space of square integrable harmonic $(0, k)$-forms with values in $\mathbb{E}^{\varphi}|D$. In the space $H^k(\lambda)$ there is an induced unitary representation of the group G (see Griffiths-Schmid [41]).

We denote by Δ^c (resp., Δ^n) the set of compact (resp., noncompact) roots:

$$\Delta^c = \{\alpha \in \Delta | (\mathfrak{g}^{\mathbb{C}})^{\alpha} \subset \mathfrak{k}^{\mathbb{C}}\}, \quad \Delta^n = \{\alpha \in \Delta | (\mathfrak{g}^{\mathbb{C}})^{\alpha} \not\subset \mathfrak{k}^{\mathbb{C}}\}.$$

Theorem 3.7 (Schmid [97], [98]). *If the form $\lambda + \rho$ is singular, then $H^k(\lambda) = 0$ for all k. If the form $\lambda + \rho$ is regular, then we set*

$$l = \#\{\alpha \in \Delta^c \cap \Delta^+ | (\lambda + \rho, \alpha) < 0\} + \#\{\alpha \in \Delta^n \cap \Delta^+ | (\lambda + \rho, \alpha) > 0\}.$$

Then $H^k(\lambda) = 0$ if $k \neq l$ and in $H^l(\lambda)$ there is induced an irreducible representation which is unitarily equivalent to $T_{\lambda+\rho}$.

We note that unlike in the case of a compact group, the spaces $H^0(\lambda)$ in general are not sufficient for the realization of all representations of the discrete series.

Chapter 4. Compact Homogeneous Complex Manifolds

Recall that a complex manifold X is called *parallelizable* if its holomorphic tangent bundle is trivial, i.e., if there exist on X holomorphic vector fields $\xi_1, \ldots, \xi_n$ ($n = \dim X$) which are linearly independent at each point. If X is compact, then the Lie algebra generated by these fields has dimension n. Indeed,

$$[\xi_i, \xi_j] = \sum_{k=1}^{n} c_{ij}^k \xi_k,$$

where c_{ij}^k are holomorphic functions on X, which must be constant by the Maximum Principle. It follows that a parallelizable compact complex manifold is homogeneous, and the dimension of the transitive group of transformations is equal to n. More precisely, we have

Proposition 4.1 (Wang [111]). *A compact complex manifold X is parallelizable if and only if it can be expressed in the Klein form $X = G/\Gamma$, where G is a connected complex Lie group and Γ is a discrete uniform subgroup.*

In this chapter we show that any compact homogeneous complex manifold fibers over a flag manifold in such a way that the fiber of this bundle is a parallelizable manifold.

The *Tits bundle* of a compact homogeneous complex manifold X is by definition a locally trivial holomorphic bundle (X, Y, π), where X is the total space of the bundle, Y the base of the bundle, and $\pi: X \to Y$ the bundle projection, such that Y is a flag manifold and the following universality property holds. For any similar bundle (X, Y', π') with the base Y' being a flag manifold, the projection $\pi': X \to Y'$ can be represented in the form $\pi' = \varphi \circ \pi$, where $\pi: X \to Y$ is the projection of the Tits bundle, and $\varphi: Y \to Y'$ is a holomorphic mapping.

The next theorem asserts the existence of the Tits bundle and gives an explicit construction. The first assertion was proven in the articles [22] and [105] and the second in [105].

Theorem 4.2 (Borel-Remmert [22], Tits [105]). *Let X be a compact homogeneous complex manifold, G any connected complex Lie group which acts transitively and holomorphically on X, and U the stationary subgroup of some point in X. Then:*

1) *The subgroup $P = N_G(U_0)$ is a parabolic subgroup of G and contains U.*

2) *The bundle $X = G/U$ with base $Y = G/P$, whose projection is defined by the formula $\pi(g \cdot U) = g \cdot P$, is a Tits bundle.*

Corollary 4.3. *The fiber of a Tits bundle is parallelizable.*

To prove the corollary, it suffices to write the fiber in the Klein form

$$P/U = Q/\Gamma, \quad \text{where} \quad Q = P/U_0 \quad \text{and} \quad \Gamma = U/U_0.$$

Proposition 4.4. *A holomorphic bundle of a compact homogeneous complex manifold over a flag manifold which has a parallelizable fiber is a Tits bundle.*

Here, as in Theorem 4.2, it is necessary to use the fact, first observed by Blanchard [16], that a connected complex Lie group of transformations of a complex space X permutes connected components of fibers of a proper holomorphic mapping $X \to Y$.

Example 4.5. If a compact homogeneous complex manifold is simply connected, then the fiber of its Tits bundle is a complex torus (see Corollary 2.7).

Example 4.6. A *homogeneous Hopf manifold* is by definition the quotient space of $\mathbb{C}^n \setminus \{0\}$ relative to the action of a discrete cyclic group generated by the operation of multiplication by d, where $d \in \mathbb{C}^*$ and $|d| \neq 1$. We denote this manifold by X_d. It can be written in the Klein form

$$X_d = \mathrm{SL}(n, \mathbb{C})/U_d,$$

where U_d is the subgroup consisting of matrices of the form

$$\left(\begin{array}{c|c} & 0 \\ * & \vdots \\ & 0 \\ \hline * \ldots * & d^k \end{array}\right) \quad (k \in \mathbb{Z}).$$

Clearly, $P = N((U_d)_0)$ is a maximal parabolic subgroup which preserves the line $\mathbb{C}e_n$, i.e., the stationary subgroup of a point in projective space $\mathbb{P}^{n-1}$. Thus the base of the Tits bundle for X_d is $\mathbb{P}^{n-1}$ and the fiber is the elliptic curve $P/U_d = \mathbb{C}^*/\{d^k\}$. We note that X_d is not simply connected. It is easy to establish the diffeomorphism of smooth manifolds $X_d \approx S^{2n-1} \times S^1$. Conversely, one can show that any complex manifold which is homogeneous and diffeomorphic to $S^{2n-1} \times S^1$ is one of the Hopf manifolds X_d.

A significant complement to Theorem 4.2 is

Proposition 4.7 (Hano [43]). *If in the notation of Theorem 4.2 the group G is semisimple, then U_0 contains the unipotent radical of the parabolic subgroup P.*

Since the structure of parabolic subgroups is well known, this proposition allows us to describe completely the connected component of the identity of a stationary subgroup of a compact homogeneous complex manifold of a semisimple group.

We pause now to consider some applications of the Tits bundle to the calculation of automorphism groups of homogeneous complex manifolds. Let X be a compact complex manifold, G a connected complex Lie group of transformations of X and $n = \dim X$. If we do not require the homogeneity of X, then already for $n = 2$ the dimension of the group G can be arbitrarily large. An example is given by the relatively minimal rational algebraic surfaces. On the other hand, we have the following theorem in whose proof the Tits bundle plays an important role.

Theorem 4.8 (see [6]). *There exists a function $d(n)$ so that for any compact homogeneous complex manifold X of dimension n we have*

$$\dim G \leqslant d(n).$$

We further assume that the group G is transitive on X. There exists a close connection between the topological properties of the manifold X and the algebraic properties of the group G.

Proposition 4.9 (see [4]). *Let X be a compact homogeneous complex manifold and G a connected transitive complex Lie group of transformations of X. We assume that the fundamental group $\pi_1(X)$ is nilpotent of class l. Then G decomposes into the locally direct product of a semisimple group and a nilpotent group, whose nilpotency class does not exceed $l + 1$.*

Examples show that if $\pi_1(X)$ is solvable, then the Levi decomposition of the group G may not be a direct product even locally (see [4], [85]).

To conclude this section we formulate a structure theorem for compact homogeneous Kähler manifolds. We first recall the necessary definitions.

Let X be a compact Kähler manifold and $\omega_1, \ldots, \omega_g$ a basis of the space of holomorphic 1-forms on X. The number g is called the irregularity of the manifold X. Let $\gamma_1, \ldots, \gamma_{2g}$ be a basis of the free part of the group $H_1(X, \mathbb{Z})$. We consider in the space $\mathbb{C}^g$ the vectors

$$\left(\int_{\gamma_j}\omega_1,\ldots,\int_{\gamma_j}\omega_g\right)\quad (j=1,\ldots,2g).$$

They generate a lattice Γ in $\mathbb{C}^g$ of full rank. The complex torus $\mathbb{C}^g/\Gamma$ is called the *Albanese manifold* and is denoted by $A(X)$. The mapping

$$X \ni x \mapsto \left(\int_*^x \omega_1,\ldots,\int_*^x \omega_g\right) \in \mathbb{C}^g,$$

where $*$ is a fixed point of X, is not single-valued but it induces a single-valued holomorphic mapping $\alpha : X \to A(X)$, which is called the *Albanese mapping*. The following universality property is well known (see [16]): if $\beta : X \to T$ is a holomorphic mapping from a manifold X into some complex torus, then there exists a holomorphic mapping $\gamma : A(X) \to T$ such that $\beta = \gamma \circ \alpha$.

Lemma 4.10. *Assume that a compact Kähler manifold X is parallelizable. Then X is a complex torus, and $\alpha : X \to A(X)$ is an isomorphism.*

Proof. Since X is parallelizable, $g = \dim X$ and the forms $\omega_1,\ldots,\omega_g$ are linearly independent on each tangent space. Therefore α is a local isomorphism and $\dim \alpha(X) = g$. The mapping α is a finite unramified covering. Consequently, X is a torus and the universality of the Albanese mapping implies that $\alpha : X \to A(X)$ is an isomorphism of complex tori.

This proves the lemma. A generalization is the following

Theorem 4.11 (Borel-Remmert [22]). *Let X be a compact Kähler manifold. If the complex manifold X is homogeneous, then it is isomorphic to the product of a complex torus and a flag manifold. More precisely, $X = A(X) \times Y$, where $A(X)$ is the Albanese manifold, Y is the base of the Tits bundle for X, and the isomorphism is given by the formula*

$$X \ni x \mapsto (\alpha(x), \pi(x)) \in A(X) \times Y,$$

where α is the Albanese mapping and π is the projection of the Tits bundle.

Chapter 5. Holomorphic Functions on Homogeneous Complex Manifolds

In this section we consider holomorphically separable (see Article I) homogeneous manifolds of complex Lie groups. We denote by $\mathcal{O}(X)$ the algebra of holomorphic functions on a complex space X.

We begin with group manifolds. Let G be a connected complex Lie group and K a maximal compact subgroup of G.

Theorem 5.1 (Matsushima-Morimoto [75]). *G is a Stein manifold if and only if K is a totally real submanifold of G or, in terms of Lie algebras,* $\mathfrak{k} \cap i\mathfrak{k} = \{0\}$.

From this theorem it is easy to obtain

Corollary 5.2. *If a group manifold is holomorphically separable, then it is a Stein manifold.*

Now let U be a closed complex subgroup of G. The question arises: under what conditions on G and U is the quotient space $X = G/U$ a Stein manifold. For certain classes of groups G the answer to this question is known.

A connected complex Lie group G is called *reductive* if G is the complexification of a maximal compact subgroup K in G, i.e., if $\mathfrak{g} = \mathfrak{k} + i\mathfrak{k}$ and $\mathfrak{k} \cap i\mathfrak{k} = \{0\}$. A disconnected complex Lie group is called reductive if it has a finite number of connected components and the connected component G_0 is reductive in the above sense. We note that in the theory of linear algebraic groups a group is called reductive if its unipotent radical is trivial. One can show that this definition agrees with ours in the following sense. A complex Lie group G is reductive if and only if there exists a reductive linear algebraic group over $\mathbb{C}$ which is isomorphic to G as a complex Lie group. Moreover, this algebraic group is unique up to an isomorphism (of algebraic groups).

Theorem 5.3 (Matsushima [73], A.L. Onishchik [82]). *Let G be a connected reductive complex Lie group and $U \subset G$ a closed complex subgroup. In order for the complex manifold $X = G/U$ to be Stein it is necessary*[4] *and sufficient that the subgroup U be reductive. If this condition is satisfied, then the subgroup U is closed in the Zariski topology of the algebraic group G, and X is an affine algebraic variety.*

Proof of Sufficiency. Let L be a maximal compact subgroup in U and $d\mu$ the Haar measure on L. We set

$$(Ef)(g) = \int_L f(gl)\, d\mu(l) \quad (f \in \mathcal{O}(G)).$$

The function Ef is holomorphic and invariant with respect to right translations by elements of L. Since the group U is reductive, this function is invariant also with respect to right translations by elements of U, i.e., $Ef \in \mathcal{O}(X)$.

Let $\{x_n\}$ be a sequence of pairwise distinct points in X having no limit points, and let $x_n = g_n \cdot U$. Since G is a Stein manifold (see Theorem 5.1), there exists a function $f \in \mathcal{O}(G)$ which is equal to n on the subset $g_n \cdot U \subset G$. Then $(Ef)(z_n) = n$, which implies that the manifold X is holomorphically convex and holomorphically separable.

---

[4] In [73], [82] the necessity is proven only for connected subgroups U or for subgroups U with a finite number of connected components. The general case is obtained from this using the result in [11] (see the implication (1) $\Rightarrow$ (2) of Theorem 5.6).

Example 5.4. Let $U = H$ be the Cartan subgroup of $G = \mathrm{SL}(2, \mathbb{C})$ consisting of the diagonal matrices. By Theorem 5.3 the surface $X = \mathrm{SL}(2, \mathbb{C})/H$ is affine algebraic. In this case it is easy to construct a closed imbedding $X \hookrightarrow \mathbb{C}^n$ explicitly. It suffices to take $n = 3$ and to set

$$z_1 = x_{11}x_{12}, \quad z_2 = x_{21}x_{22}, \quad z_3 = x_{21}x_{12} + \tfrac{1}{2},$$

where $x_{ij} = x_{ij}(g)$ are the elements of a matrix $g \in \mathrm{SL}(2, \mathbb{C})$. The functions z_1, z_2, z_3 are invariant with respect to right translations by elements of H, i.e., z_1, z_2, $z_3 \in \mathcal{O}(X)$. They are related by the equation

$$z_3^2 - z_1 z_2 = \tfrac{1}{4},$$

which defines a quadric in $\mathbb{C}^3$. It is not difficult to show that the mapping defined by the functions z_1, z_2, z_3 gives an isomorphism from X onto this quadric.

Up to now we have assumed that the group G is reductive. A sufficient condition for a homogeneous complex manifold G/U of an arbitrary connected complex Lie group G to be a Stein manifold is given in [74]. This condition is especially useful in the theory of nilmanifolds (where G is nilpotent), when it turns out also to be necessary.

As is clear from Theorem 5.3, the analytic problem of whether a homogeneous manifold of a complex Lie group is Stein is closely related to the algebraic-geometric problem of whether a quotient space of linear algebraic group is affine. In the theory of algebraic groups there is a sufficient condition for a quotient space to be affine, which along with Theorem 5.3 often turns out to be useful in various applications. In order to formulate this condition, we denote by $R_u(G)$ the unipotent radical of a linear algebraic group G.

Theorem 5.5 (Bialynicki-Birula [14]). *Let G be a connected linear algebraic group over $\mathbb{C}$ and $U \subset G$ a subgroup closed in the Zariski topology. If $R_u(U) \subset R_u(G)$, then the quotient space G/U is an affine variety. In particular, if G is solvable, then G/U is affine.*

We now consider arbitrary holomorphically separable homogeneous manifolds. Unlike in the cases of Lie groups (see Corollary 5.2) and nilmanifolds (see Theorem 5.11) they are not necessarily Stein manifolds. The simplest example of a holomorphically separable but not holomorphically convex homogeneous manifold is the manifold $\mathbb{C}^n \setminus \{0\}$ on which the group $\mathrm{GL}(n, \mathbb{C})$ acts transitively ($n > 1$). The analog of a holomorphically separable complex space in algebraic geometry is a quasiaffine algebraic variety, i.e., a Zariski open subset of an affine variety. Quasiaffine varieties are holomorphically separable, and our example pertains exactly to this class. This occurs not just by chance: if we restrict to homogeneous manifolds of reductive groups then, as in the case of Stein manifolds, homogeneity implies that a holomorphically separable manifold is algebraic.

Theorem 5.6. *Let G be a connected reductive complex Lie group and $U \subset G$ a closed complex subgroup. The following conditions are equivalent:*

(1) *the homogeneous complex manifold G/U is holomorphically separable;*

(2) *the subgroup U is closed in the Zariski topology and the algebraic variety G/U is quasiaffine;*

(3) *there exist a holomorphic finite-dimensional representation $\varphi : G \to \mathrm{GL}(V)$ and a vector $v \in V$ so that $U = \{g \in G | \varphi(g)v = v\}$;*

(4) *the subgroup U is closed in the Zariski topology, and there exist a holomorphic finite-dimensional representation $\psi : G \to \mathrm{GL}(W)$ and a vector $w \in W$ so that* (a) $P = \{g \in G | \psi(g)w \in \mathbb{C}w\}$ *is a parabolic subgroup;* (b) $U \subset Q$, *where Q is the subgroup of codimension 1 in P defined by the condition $g \in Q \Leftrightarrow \psi(g)w = w$;* (c) $R_u(U) \subset R_u(Q)$.

The implication (1) ⇒ (2) was proven by Barth and Otte [11]; the implications (2) ⇒ (3), (3) ⇒ (2), and (4) ⇒ (2) by Bialynicki-Birula, Hochschild, and Mostow [15]; and the implication (3) ⇒ (4) by A.A. Sukhanov [102]. Condition (4), formulated in terms of weights and roots, gives a rather explicit description of the stationary subgroups of quasiaffine homogeneous varieties [102].

The "antipode" of a holomorphically separable complex space is a complex space on which all holomorphic functions are constant. Homogeneous complex manifolds without holomorphic functions form a natural class, which up to now has been little studied (concerning complex Lie groups without holomorphic functions see [76]). Let G be a connected reductive complex Lie group and U a closed complex subgroup (resp., Zariski closed subgroup). In the general case it is not known for which U all holomorphic (resp., regular) functions on G/U are constant. Under the additional conditions that U is a connected Zariski closed subgroup and that $N_G(U)$ contains a maximal (algebraic) torus, this problem was solved by Pommerening [88]. We note that the analytic and algebraic aspects of the question are closely related. Let U be a closed complex subgroup and $U^\wedge$ its closure in the Zariski topology. We denote by Γ the natural homorphic mapping $G/U \to G/U^\wedge$. As Barth and Otte have shown in [11], any function $f \in \mathcal{O}(G/U)$ on any compact set in G/U can be approximated uniformly by functions of the form $f_n \circ \Gamma$, where f_n is a regular function on $G/U^\wedge$. In particular, $\mathcal{O}(G/U) = \mathbb{C}$ if and only if all regular functions on $G/U^\wedge$ are constant.

We give two examples in which the homogeneous complex manifold X is noncompact but for which $\mathcal{O}(X) = \mathbb{C}$.

Example 5.7. Let X be an abelian group of the form $X = (\mathbb{C}^*)^n/\{\gamma\}$, where $\gamma = (\gamma_1, \ldots, \gamma_n) \in (\mathbb{C}^*)^n$, $|\gamma_1| \neq 1$, and the numbers $\gamma_1, \ldots, \gamma_n$ are multiplicatively independent. The latter condition means that an equality $\gamma_1^{\mu_1} \cdot \ldots \cdot \gamma_n^{\mu_n} = 1$, where $\mu_i \in \mathbb{Z}$, is possible only when $\mu_1 = \cdots = \mu_n = 0$. This condition is equivalent to the density of the cyclic subgroup $\{\gamma\}$ in the Zariski topology of the group $(\mathbb{C}^*)^n$. Therefore it follows from the result of Barth-Otte mentioned above that $\mathcal{O}(X) = \mathbb{C}$. Clearly in the present case this is easy to prove directly using the expansion of a function $f \in \mathcal{O}((\mathbb{C}^*)^n)$ in a Laurent series.

Example 5.8. Let $G = \mathrm{SL}(3, \mathbb{C})$ and $U = T \cdot N$, where

$$T = \left\{ \begin{pmatrix} t & 0 & 0 \\ 0 & 1 & 0 \\ 0 & 0 & t^{-1} \end{pmatrix} \middle| \, t \in \mathbb{C}^* \right\}, \quad N = \left\{ \begin{pmatrix} 1 & x & y \\ 0 & 1 & x \\ 0 & 0 & 1 \end{pmatrix} \middle| \, x, y \in \mathbb{C} \right\}.$$

The adjoint representation defines an action of the group G on the projective space $\mathbb{P}(\mathfrak{g})$. The homogeneous manifold $X = G/U$ can be imbedded equivariantly into $\mathbb{P}(\mathfrak{g})$ so that $X = Y \backslash Z$, where Y and Z are irreducible algebraic submanifolds of $\mathbb{P}(\mathfrak{g})$ with $\dim Y = 5$ and $\dim Z = 3$. It follows already from the existence of such an imbedding that $\mathcal{O}(X) = \mathbb{C}$. Indeed, the singularities of a holomorphic function having codimension > 1 are removable; therefore $\mathcal{O}(X) = \mathcal{O}(Y)$. But Y is compact, so $\mathcal{O}(Y) = \mathbb{C}$.

For the construction of the necessary imbedding $X \hookrightarrow \mathbb{P}(\mathfrak{g})$, we consider the cone in $\mathfrak{g}$ consisting of all nilpotent matrices. We denote by Y its projectivization. The group G preserves Y and has two orbits on Y, which are represented in $\mathfrak{g}$ by the matrices

$$\begin{pmatrix} 0 & 1 & 0 \\ 0 & 0 & 1 \\ 0 & 0 & 0 \end{pmatrix} \quad \text{and} \quad \begin{pmatrix} 0 & 0 & 1 \\ 0 & 0 & 0 \\ 0 & 0 & 0 \end{pmatrix}.$$

Let y_1 and y_2 be the corresponding points in Y. The stationary subgroup of the point y_1 is U; the orbit $G(y_1)$ is isomorphic to $X = G/U$. By identifying X with this orbit, we also obtain the required imbedding. Indeed, $X = Y \backslash Z$, where $Z = G(y_2)$. The stationary subgroup of the point y_2 is a Borel subgroup, which implies that $\dim Z = 3$.

For an arbitrary complex space X it is natural to consider the equivalence relation

$$x_1 \sim x_2 \Leftrightarrow f(x_1) = f(x_2) \quad \forall f \in \mathcal{O}(X).$$

In general, the quotient space $X/\!\sim$ is not even Hausdorff. However, in certain cases we can introduce on $X/\!\sim$ the structure of a complex space so that the quotient mapping $X \to X/\!\sim$ is a holomorphic mapping. A classical result of this type is the reduction theorem of Remmert for holomorphically convex spaces (see Article I, Chapter 4, Section 9).

If $X = G/U$, where G is a connected complex Lie group and U a closed complex subgroup, then the space $X/\!\sim$ is Hausdorff and can be given a natural complex structure. More precisely, there exists a closed complex subgroup $J \subset G$ containing U so that

$$x_1 \sim x_2 \Leftrightarrow \sigma(x_1) = \sigma(x_2),$$

where $\sigma : G/U \to G/J$ and $\sigma(g \cdot U) = g \cdot J$ (see [32]). The homogeneous manifold G/J is holomorphically separable, and the mapping σ induces an algebra isomorphism

$$\sigma^* : \mathcal{O}(G/J) \rightleftarrows \mathcal{O}(G/U).$$

The mapping σ is called the *holomorphic reduction* mapping and the bundle $(G/U, G/J, \sigma)$ the holomorphic reduction bundle.

There is a natural question about holomorphic functions on the fiber J/U. Recall that the fibers of the holomorphic reduction of Remmert are compact and the holomorphic functions on them are constant. If the same were true here, then the study of arbitrary homogeneous manifolds would in some sense reduce to the study of holomorphically separable homogeneous manifolds (the base G/J) and homogeneous manifolds without holomorphic functions (the fiber J/U). Unfortunately in general $\mathcal{O}(J/U) \neq \mathbb{C}$.

Example 5.9 (Barth-Otte [11]). If $G = \mathrm{SL}(2, \mathbb{C})$ and

$$U = \left\{ \begin{pmatrix} 1 & n \\ 0 & 1 \end{pmatrix} \middle| n \in \mathbb{Z} \right\}, \quad \text{then} \quad J = \left\{ \begin{pmatrix} 1 & z \\ 0 & 1 \end{pmatrix} \middle| z \in \mathbb{Z} \right\}.$$

The fiber of the mapping σ is the Stein manifold $J/U = \mathbb{C}^*$.

For certain special classes of homogeneous complex manifolds, in particular, for group manifolds and nilmanifolds, the equality $\mathcal{O}(J/U) = \mathbb{C}$ is true. In these cases the function theoretic classification can be considered to be complete.

Theorem 5.10 (Morimoto [76]). *Let G be a connected complex Lie group. Then,*

$$J = \{g \in G \mid f(g) = f(e) \quad \forall f \in \mathcal{O}(G)\}$$

is a connected central closed complex subgroup with $\mathcal{O}(J) = \mathbb{C}$. The holomorphic reduction mapping $\sigma : G \to G/J$ is a group epimorphism. The group G/J is a Stein manifold.

Theorem 5.11 (Gilligan-Huckleberry [33]). *Let G be a connected nilpotent complex Lie group, $U \subset G$ a closed complex subgroup, and $\sigma : G/U \to G/J$ the holomorphic reduction mapping. Then the base G/J is a Stein manifold, the fiber J/U is connected,* and $\mathcal{O}(J/U) = \mathbb{C}$.

An analogous result holds for homogeneous manifolds of solvable groups under the condition that the stationary subgroup has finitely many connected components [50].

Chapter 6. Meromorphic Functions on Homogeneous Complex Manifolds

Let X be an irreducible compact complex space and $\mathcal{M}(X)$ the field of meromorphic functions on X. The classical Weierstrass-Seigel theorem asserts that the transcendence degree of the field $\mathcal{M}(X)$ does not exceed the dimension of X:

$$\operatorname{tr.d.} \mathcal{M}(X) \leqslant \dim X.$$

For complete algebraic varieties we have an equality, and each function $f \in \mathcal{M}(X)$ is rational. An irreducible compact complex space for which $\operatorname{tr.d.} \mathcal{M}(X) = \dim X$ is called a *Moĭshezon space.* Moĭshezon spaces form a natural class of complex spaces, which includes the complete algebraic varieties but which is not exhausted by them. However, in the category of homogeneous manifolds these classes coincide.

Theorem 6.1 (Grauert-Remmert [39]). *If X is a compact homogeneous complex manifold and* $\operatorname{tr.d.} \mathcal{M}(X) = \dim X$, *then X is a projective algebraic variety. More precisely, X is the product of an abelian variety and a flag manifold.*

The second assertion follows immediately from Theorem 4.11. Theorem 6.1 is proven in [39] with the help of the *meromorphic reduction bundle.* An analogous construction exists also for arbitrary (not necessarily compact) homogeneous complex manifolds [51]. An irreducible complex space is called meromorphically separable if for any two points $x_1, x_2 \in X$ with $x_1 \neq x_2$ there is a function $f \in \mathcal{M}(X)$ which is holomorphic at these two points and such that $f(x_1) \neq f(x_2)$. The meromorphic reduction bundle is related to this concept in the same way as the holomorphic reduction bundle (see Chapter 5) is related to the concept of holomorphic separability.

Theorem 6.2 (Grauert-Remmert [39], Huckleberry-Snow [51]). *Let G be a connected complex Lie group, U a closed complex subgroup, and $X = G/U$. There exists a closed complex subgroup $Q \subset G$ containing U and possessing the following properties:*

(1) *the manifold G/Q is meromorphically separable;*

(2) *the holomorphic mapping $\rho: G/U \to G/Q$ defined by $\rho(g \cdot U) = g \cdot Q$ induces a field isomorphism*

$$\rho^*: \mathcal{M}(G/Q) \rightleftarrows \mathcal{M}(G/U);$$

(3) *for any holomorphic mapping $\varphi: G/U \to Y$, where the complex space Y is meromorphically separable, there exists a holomorphic mapping $\psi: G/Q \to Y$ so that $\varphi = \psi \circ \rho$;*

(4) *the fiber Q/U of the mapping ρ can be written in the Klein form $Q/U = F/\Gamma$, where F is a connected complex Lie group and $\Gamma \subset F$ is a discrete subgroup;*

(5) *if X is compact, then G/Q is a projective variety.*

If X is compact, then in certain cases the mapping ρ coincides with the projection of the Tits bundle. However, this is not always true: for complex tori of dimension g the base of the Tits bundle is a point, while the base of the meromorphic reduction bundle is a complex torus whose dimension can be any number not exceeding g. We now give an example of the coincidence of the two bundles.

Proposition 6.3. *Let G be a connected complex Lie group which coincides with its commutator, and let $X = G/U$ be a compact homogeneous complex manifold of* *this group. Then the subgroup $P = N_G(U_0)$ and the subgroup Q from Theorem 6.2 coincide. In particular, the projection of the Tits bundle $\pi : X \to Y = G/P$ induces an isomorphism between the field $\mathcal{M}(X)$ and the field of rational functions on the flag manifold Y.*

Proof. We show that there exist holomorphic mappings $\psi : G/Q \to G/P$ and $\chi : G/P \to G/Q$ so that $\chi \circ \pi = \rho$ and $\psi \circ \rho = \pi$. Since G/P is projective and therefore a meromorphically separable manifold, the mapping ψ exists by virtue of property (3) (see Theorem 6.2). In view of (5) and the hypothesis $G = G'$ the Albanese manifold $A(G/Q)$ is a point. According to Theorem 4.11, G/Q is a flag manifold and Q is a parabolic subgroup of G. The mapping χ with the required properties exists by the universality of the Tits bundle.

It is clear that ψ and χ are mutually inverse. For $p \in P$ we have $\psi(p \cdot Q) = \psi(\rho(p \cdot U)) = \pi(p \cdot U) = p \cdot P = e \cdot P$. Since ψ is one-to-one, the coset $p \cdot Q$ does not depend on $p \in P$, i.e., $p \in Q$ and $P \subset Q$. It can be proven analogously that $Q \subset P$. This completes the proof.

Along with the field $\mathcal{M}(X)$ it is natural to consider the set of analytic hypersurfaces in a complex manifold X. We denote this set by $\mathcal{H}(X)$. For noncompact homogeneous manifolds the field $\mathcal{M}(X)$ and the set $\mathcal{H}(X)$ have not been studied greatly. For example, the answer to the following question is not known.[5]

Let $G = \mathrm{SL}(2, \mathbb{C})$, let U and J be the subgroups from Example 5.9, and let $X = G/U$. It is clear that the subgroup Q from Theorem 6.2 is contained in J, but it is unknown whether $Q = J$. In other words, is it true that a meromorphic function on $\mathrm{SL}(2, \mathbb{C})$ which is invariant with respect to right translations by matrices of the form $\begin{pmatrix} 1 & n \\ 0 & 1 \end{pmatrix}$, where $n \in \mathbb{Z}$, is invariant with respect to right translations by matrices of the form $\begin{pmatrix} 1 & z \\ 0 & 1 \end{pmatrix}$, where $z \in \mathbb{C}$? The analogous problem has not been solved also for hypersurfaces.

We mention some recent results concerning meromorphic functions on noncompact homogeneous complex manifolds and analytic hypersurfaces in these manifolds.

Example 6.4 (see [1]). Let $X = \mathrm{SL}(n, \mathbb{C})/\mathrm{SL}(n, \mathbb{Z})$. Then $\mathcal{H}(X) = \varnothing$ and $\mathcal{M}(X) = \mathbb{C}$.

A generalization is the following

Theorem 6.5 (Huckleberry-Margulis [47]). *Let G be a complex semisimple Lie group and $\Gamma \subset G$ a subgroup which is dense in the Zariski topology. Then there does not exist in G an analytic hypersurface which is invariant with respect to right translations by the elements of Γ. In particular, all Γ-invariant meromorphic*

[5] A positive answer to this question was recently obtained by F. Berteloot, A.T. Huckleberry and K. Oejeklaus (see: F. Berteloot, K. Oeljeklaus, Invariant Plurisubharmonic Functions and Hypersurfaces on Semisimple Complex Lie Groups, to appear in "Mathematische Annalen")

functions on G are constant. If Γ is a closed complex subgroup, then $\mathscr{H}(G/\Gamma) = \varnothing$ *and* $\mathscr{M}(G/\Gamma) = \mathbb{C}$.

We note that Theorem 6.5 is not true for reductive groups G.

Theorem 6.6 (see [3]). *Let G be a connected complex nilpotent Lie group and* $\Gamma \subset G$ *an arbitrary subgroup, and suppose that on G there are no nonconstant holomorphic functions which are invariant with respect to right translations by the elements of Γ. Then any Γ-invariant analytic hypersurface in G (thus any Γ-invariant meromorphic function on G) is invariant also with respect to the commutator group G′. If Γ is a closed complex subgroup and V is the smallest closed complex subgroup in G which contains Γ and G′, then the natural mapping* $G/\Gamma \to G/V$ *induces a bijection* $\mathscr{H}(G/V) \xrightarrow{\sim} \mathscr{H}(G/\Gamma)$ *and a field isomorphism* $\mathscr{M}(G/V) \xrightarrow{\sim} \mathscr{M}(G/\Gamma)$.

This theorem is, as is the preceding, a negative result. The classical theory of multiperiodic functions deals with the fields of meromorphic functions on $\mathbb{C}^n$ which are invariant with respect to an additive subgroup of the group $\mathbb{C}^n$. Theorem 6.6 shows that when $\mathbb{C}^n$ is replaced by a nilpotent group then no new fields of similar type arise. We have been advised by Huckleberry that Theorem 6.6 has recently (1985) been generalized by Loeb to solvable groups.

Chapter 7. Almost Homogeneous Complex Spaces

Let X be an irreducible complex space and G a connected complex Lie group acting holomorphically on X.

The complex space X is called an *almost homogeneous space* of the group G if the orbit $G(x_0)$ of some point $x_0 \in X$ is open in X.

It is not hard to show that the complement $E_G = X \setminus G(x_0)$ is an analytic subset of X. To do this, it is necessary to apply Remmert's theorem about fibers to the mapping $G \times X \to X \times X$ which sends (g, x) to (gx, x). It follows from this theorem that the set

$$S = \{x \in X \mid \dim G(x) < \dim X\}$$

is analytic. The group G is transitive in a neighborhood of each point of $X \setminus S$, and since $X \setminus S$ is connected, this action is globally transitive, which implies that $X \setminus S = G(x_0)$ and $E_G = S$.

The concept of an almost homogeneous complex space was introduced by Remmert and Van de Ven [94]. Almost homogeneous spaces are a generalization of homogeneous manifolds; for the latter $E_G = \varnothing$. Almost homogeneous spaces occur in various situations, for example, in the study of homogeneous bundles over homogeneous manifolds or by the restriction of transitive actions to a subgroup (for interesting examples of this type see [61]). A special case of an almost homogeneous complex space is a complex space on which a group G acts with a finite number of orbits. To this class belong, in particular, the so-called complex symmetric spaces introduced by Borel [20], which generalize Hermitian symmetric spaces. Finally, in algebraic geometry almost homogeneous spaces arise naturally in the study of actions of algebraic groups: the closure of an orbit in the Zariski topology is an almost homogeneous space (in complex analysis this is generally not true).

The assertion of Theorem 4.11 carries over in weakened form to compact almost homogeneous Kähler manifolds.

Theorem 7.1. *Let X be a compact almost homogeneous Kähler manifold and $\alpha: X \to A(X)$ its Albanese mapping. Then*:

(1) *The mapping α is the projection of a locally trivial bundle with base $A(X)$ and with connected fiber* (*we denote it by Y*);

(2) *Y is a simply connected projective algebraic variety*;

(3) *There exist finite unramified coverings $X^* \to X$ and $A^* \to A(X)$ such that X^* is an almost homogeneous Kähler manifold which is C^∞-diffeomorphic to $Y \times A^*$. The projection $X^* \to A^*$ is precisely the Albanese mapping for X^**, i.e., $A^* = A(X^*)$.

Assertion (1) was proven by Remmert and Van de Ven (see [91]); assertion (2) by Oeljeklaus [80]; and assertion (3) by Barth and Oeljeklaus [10]. The necessity of passing to a finite covering in (3) is evident from the following example.

Example 7.2 (Barth-Oeljeklaus [10]). Let $Z = \mathbb{C} \times \mathbb{P}^1 \times \mathbb{P}^1$, Γ the (free abelian) group of transformations of Z with the two generators

$$(z, p, q) \mapsto (z + 1, p, q) \quad \text{and} \quad (z, p, q) \mapsto (z + \omega, q, p),$$

where $\omega \in \mathbb{C}\setminus\mathbb{R}$, $\Lambda = \mathbb{Z} + \mathbb{Z}\omega$ a lattice in $\mathbb{C}$, $A = \mathbb{C}/\Lambda$, $X = Z/\Gamma$, and $\alpha: X \to A$ the mapping induced by the projection of Z onto the first factor. Then X is a Kähler manifold and $\alpha: X \to A$ is its Albanese mapping. The group $G = \mathbb{C} \times \mathrm{PSL}(2, \mathbb{C})$ acts on Z by the formula

$$(z, p, q) \xrightarrow{(w, g)} (z + w, gp, gq),$$

where $\omega \in \mathbb{C}$ and $g \in \mathrm{PSL}(2, \mathbb{C})$; this action commutes with the action of Γ. Since G has an open orbit on Z, namely, $\mathbb{C} \times (\mathbb{P}^1 \times \mathbb{P}^1\setminus\Delta)$, where Δ is the diagonal, the induced action $G \times X \to X$ also has an open orbit, and X is an almost homogeneous manifold of the group G. The fiber Y of the Albanese mapping $\alpha: X \to A$ is isomorphic to $\mathbb{P}^1 \times \mathbb{P}^1$; however, X is not even homeomorphic to $A \times \mathbb{P}^1 \times \mathbb{P}^1$, since

$$H^2(X, \mathbb{Z}) = \mathbb{Z} \oplus \mathbb{Z}; \quad H^2(A \times \mathbb{P}^1 \times \mathbb{P}^1, \mathbb{Z}) = \mathbb{Z} \oplus \mathbb{Z} \oplus \mathbb{Z}.$$

If in the hypotheses to Theorem 7.1 we also require that the first Betti number $b_1(X) = 0$, then according to assertion (2) the manifold X is a projective variety.

It is known that this variety is unirational [7]. Its rationality has been proven for $\dim X \leqslant 7$ (see [50]).

One of the first results in the theory of almost homogeneous complex spaces was the classification of compact nonsingular surfaces X which are almost homogeneous with respect to the group $(\mathrm{Aut}\, X)_0$ (see Potters [91]). It turns out that such surfaces are either Kähler or *generalized Hopf surfaces*, i.e., they have as universal covering $\mathbb{C}^2 \setminus \{0\}$. In the Kähler case the surface X either is rational, or is a topologically trivial $\mathbb{P}^1$-bundle over an elliptic curve, or else is isomorphic to a complex torus. We have in these cases $\dim A(X) = 0$, 1, or 2 respectively.

A generalized Hopf surface is almost homogeneous if and only if its fundamental group is abelian.

Example 7.3. We show that any nonsingular complex surface X which is homeomorphic to $S^1 \times S^3$ is almost homogeneous. According to a theorem of Kodaira (1966), X is a generalized Hopf surface. Moreover, X is obtained from $\mathbb{C}^2 \setminus \{0\}$ by taking the quotient relative to the cyclic group of transformations with generator

$$\gamma : (z_1, z_2) \mapsto (az_1 + \lambda z_2^m, bz_2),$$

where $0 < |a| \leqslant |b| < 1$ and $(b^m - a)\lambda = 0$. We consider the two possible cases.

a) $\lambda = 0$. As G we take the group of automorphisms of $\mathbb{C}^2 \setminus \{0\}$ of the form $(z_1, z_2) \mapsto (sz_1, tz_2)$ $(s, t \in \mathbb{C}^*)$. Then G commutes with γ and therefore acts on X. One of the orbits of this action is open. Its complement E_G consists of two nonintersecting elliptic curves (the images of the coordinate axes in X).

b) $\lambda \neq 0$. As G we take the group of all automorphisms of $\mathbb{C}^2 \setminus \{0\}$ of the form

$$(z_1, z_2) \mapsto (t^m z_1 + \tau z_2^m, tz_2) \quad (t \in \mathbb{C}^*, \tau \in \mathbb{C}).$$

Then, as in case a), G acts on X with an open orbit. Here E_G is an elliptic curve (the image of the z_1 axis).

Along with the classification of smooth compact almost homogeneous complex surfaces, there is a classification of almost homogeneous affine algebraic surfaces (see [37], [89], [13]). The almost homogeneous three-dimensional affine algebraic varieties with E_G empty or finite have also been enumerated (see [90]).

In the high dimensional case it is natural to pose the following problem. Suppose that we know certain properties of the set E_G, for example, suppose that we have information about its analytic structure or its orbit decomposition. What can we deduce about the structure of an almost homogeneous space X or, more precisely, about the structure of the pair (X, G)? The first result in this direction is the next

Theorem 7.4 (Oeljeklaus [79]). *Let X be a compact complex manifold which is almost homogeneous with respect to a connected complex Lie group G, and assume that E_G contains at least one isolated point. Then X is isomorphic to $\mathbb{P}^n$; in particular, X is a homogeneous manifold (of a larger group). Here E_G is either a point or the disconnected union of a point and a hyperplane, and the corresponding groups G can be easily enumerated.*

Up to now we have encountered examples of compact almost homogeneous spaces for which E_G is either connected or has two connected components. It turns out that under certain hypotheses E_G cannot have more than two connected components. Let $\Omega = G(x_0)$ be the open orbit in X and U the stationary subgroup of the point x_0. If U has a finite number of connected components, then by a theorem of Borel (1953) the homogeneous manifold $\Omega = G/U$ has not more than two ends in the sense of Freudenthal. Let N be the number of connected components in E_G and ν the number of ends of Ω. Then $\nu \geqslant N$. Therefore if $N \geqslant 2$ we have $\nu = N = 2$. In other words, in order to describe the compact almost homogeneous spaces with disconnected E_G we need to know which homogeneous manifolds have two ends and to understand the structure of their equivariant compactifications. The answer to the first question is especially simple to formulate under the hypothesis that the group G and the subgroup U are linear algebraic.

Theorem 7.5 (see [5]). *Let G be a connected linear algebraic group over $\mathbb{C}$ and $U \subset G$ a subgroup which is closed in the Zariski topology. The homogeneous manifold G/U has two ends if and only if there exist a parabolic subgroup $P \subset G$ and a nontrivial rational character $\varphi : P \to \mathbb{C}^*$ so that $U = \operatorname{Ker} \varphi$.*

Geometrically Theorem 7.5 says that the homogeneous manifold G/U is the total space of a principal $\mathbb{C}^*$-bundle over the flag manifold G/P. It has a natural G-equivariant completion $M(G, P, \varphi)$, which is obtained by adjoining to the space of the bundle the zero and infinitely distant sections. More formally, $M(G, P, \varphi)$ is the quotient space of product $G \times \mathbb{P}^1$ relative to the following action of the group P: an element $p \in P$ acts via the formula

$$(g, \zeta) \overset{p}{\mapsto} (gp^{-1}, \varphi(p)\zeta) \quad (\zeta \in \mathbb{C} \cup \{\infty\}).$$

A partial answer to the question about the structure of the equivariant compactifications with disconnected E_G is given by the next result.

Theorem 7.6 (see [5]). *Assume that a connected linear algebraic group G acts regularly and effectively on a complete irreducible algebraic variety X.*[6] *Assume that there exists a Zariski open orbit $\Omega \subset X$ and that the complement $E_G = X \setminus \Omega$ is disconnected in the complex topology. Then the set E_G consists of precisely two components E_1 and E_2, each of which is homogeneous with respect to G. The group G is reductive. For some parabolic subgroup $P \subset G$ and character $\varphi : P \to \mathbb{C}^*$ there exists a G-equivariant regular mapping*

$$M(G, P, \varphi) \to X,$$

under which the zero section is mapped onto E_1, the infinitely distant section is mapped onto E_2, and the open orbit is mapped isomorphically onto Ω.

[6] The algebraic variety X is defined over C. It can have singular points.

If the space X is normal, then Theorem 7.6 can be made more precise. In this case the sets $X\backslash E_i$ for $i = 1, 2$ admit a simple description in terms of algebraic groups and their representations, and the manifold X is equivariantly imbedded into projective space.

After an extension of the group acting with an open orbit, the latter can increase. For example, this is the case in Theorem 7.4. In the notation of Theorem 7.6 we can ask if there exists an automorphism $g: X$ such that $gE_1 \cap \Omega \neq \varnothing$ (clearly $g \notin G$). There is no such automorphism automatically if E_1 consists of singular points or if there exists a regular birational mapping $X \to X'$ which lowers the dimension of E_1. As Lescure has shown [67], if these obstructions are absent then an automorphism with the required properties exists.

Although Theorem 7.5 does have an analog in the complex analytic case (see [60], [34], [31]), the geometry becomes more complicated and the equivariant completions have been investigated only partially. For example, there is the following result generalizing Theorem 7.4 to spaces with singular points (the algebraic version can be obtained from Theorem 7.6).

Theorem 7.7 (Huckleberry-Oeljeklaus [48]). *Let X be an irreducible complex space and G a connected complex Lie group acting on X with an open orbit Ω so that $E_G = X\backslash\Omega$ contains an isolated point. If* $\dim X > 1$, *then X is locally irreducible, and the normalization of X is a one-to-one mapping. We assume further that X is normal. Then X is a projective or affine cone over a flag manifold Y, which is imbedded equivariantly into a projective space. The isolated point of E_G is the vertex of a cone, v. If X is noncompact, then $E_G = \{v\}$; and if X is compact, then either $E_G = \{v\}$ or $E_G = \{v\} \cup Y$. It is not difficult to enumerate the corresponding groups. If X has no singular points, then either $X = \mathbb{C}^n$ or $X = \mathbb{P}^n$.*

As is evident from Theorems 7.4 and 7.7, almost homogeneous spaces for which $\dim E_G = 0$ are organized especially simply. Next in complexity is the case of $\dim E_G = 1$. Here under the hypothesis that the space X is compact a complete classification has also been obtained (in the nonsingular case by Huckleberry and Oeljeklaus [49] and in the singular case by Lescure).

Another type of restriction considered in the theory of almost homogeneous complex spaces is a condition on the orbit decomposition. The simplest orbit decomposition is obtained when E_G is homogeneous with respect to G, i.e., when X is a two-orbit space. In this case we have the following result.

Theorem 7.8 (see [53], [2]). *Let X be a complete nonsingular algebraic variety and G a connected linear algebraic group acting regularly on X. We assume that* $\operatorname{codim} E_G = 1$ *and that G acts on E_G transitively. Then there is a G-equivariant regular mapping $\tau: X \to Y$, where $Y = G/P$ is a flag manifold of the group G, and the fiber of τ is isomorphic to $\mathbb{P}^n$, $\mathbb{P}^n \times \mathbb{P}^n$, $\mathbb{Q}_n$, $\mathbb{G}_{2,2n}$ or to a Hermitian symmetric space of the group E_6.*

Let us assume in addition that the group G is semisimple and that the stationary subgroup of a point in the open orbit Ω is reductive, i.e., that Ω is an affine manifold. In this case τ maps to a point, and X is one of the standard two-orbit manifolds enumerated in Theorem 7.8. They have been studied in detail. The open orbit Ω has the form $\Omega = K^{\mathbb{C}}/L^{\mathbb{C}}$, where $A = K/L$ is a compact Riemannian symmetric space of rank 1 (here $G = K^{\mathbb{C}}$). Thus A is imbedded in X as a totally real submanifold. It turns out that some noncompact real form of the group $(\mathrm{Aut}\, X)_0$ has two orbits on X, namely, A and $X \setminus A$. The manifold $X \setminus A$ is *strictly pseudoconcave*, i.e., it admits an exhaustion by relatively compact open sets with smooth strictly pseudoconcave boundaries.

Example 7.9. Let X be the quadric in $\mathbb{P}^{n+1}$ defined by the equation $z_0^2 = z_1^2 + \cdots + z_{n+1}^2$. The group $G = \mathrm{SO}(n+1, \mathbb{C})$ acts on X by the formula

$$(z_0 : z') \overset{g}{\mapsto} (z_0 : gz'),$$

where $g \in G$ and $z' = (z_1 : \ldots : z_{n+1})^t$. Here G has the open orbit $\Omega = \{z \in X \mid z_0 \neq 0\}$ whose complement $E_G = \{z \in X \mid z_0 = 0\} = \mathbb{Q}_{n+1}$ is homogeneous with respect to G. Let A be the set of real points of the manifold X. Then $A \subset \Omega$, and this imbedding is the standard imbedding of the sphere $A = \mathrm{SO}(n+1)/\mathrm{SO}(n) = S^n$ into the affine quadric $\Omega = \mathrm{SO}(n+1, \mathbb{C})/\mathrm{SO}(n, \mathbb{C})$. In the affine coordinates $\zeta_k = z_k/z_0$, where $k = 1, \ldots, n+1$, we have:

$$\Omega = \left\{ \zeta \in \mathbb{C}^{n+1} \,\middle|\, \sum_1^{n++1} \zeta_k^2 = 1 \right\},$$

$$A = \left\{ \zeta \in \mathbb{R}^{n+1} \,\middle|\, \sum_1^{n+1} \zeta_k^2 = 1 \right\} = \left\{ \zeta \in \Omega \,\middle|\, \sum_1^{n+1} |\zeta_k|^2 = 1 \right\}.$$

The open sets

$$\left\{ \zeta \in \Omega \,\middle|\, \sum_1^{n+1} |\zeta_k|^2 > 1 + \varepsilon \right\} \cup E_G \quad (\varepsilon > 0)$$

are relatively compact in $X \setminus A$, have smooth strictly pseudoconcave boundaries, and define an exhaustion of $X \setminus A$. Thus $X \setminus A$ is a strictly pseudoconcave manifold. The real group $\mathrm{SO}_0(1, n+1)$ acts in the natural way on X and has two orbits: A and $X \setminus A$.

Any complex manifold obtained from a compact complex space by discarding a finite number of points has the property of strict pseudoconcavity. In particular, projective homogeneous cones with vertices removed are of this sort (see Theorem 7.7). Aside from them and the manifolds $X \setminus A$ described above there exists exactly one strictly pseudoconcave homogeneous complex manifold, namely, $\mathbb{P}^n \setminus \mathbb{B}^n$, where $\mathbb{B}^n$ is the ball in $\mathbb{C}^n$ (see [51], [52]).

To conclude Chapter 7, we consider one algebraic direction in the theory of almost homogeneous complex spaces. Let G be an algebraic group defined over $\mathbb{C}$, U a closed Zariski subgroup of G, and $G \times X \to X$ a regular action, one of whose orbits is open and isomorphic to G/U. Thus we have an equivariant open imbedding $G/U \hookrightarrow X$, and we can pose the problem of describing all such imbeddings for fixed G and U. Assume that G is a connected reductive linear group. It turns out that the classification of imbeddings $G/U \hookrightarrow X$ is more complex the larger the minimal codimension of an orbit of a Borel subgroup $B \subset G$ in G/U (*complexity* of the manifold G/U in the terminology of Luna-Vust [68]). The most complete results have been obtained under the hypothesis that the complexity of G/U is equal to zero, i.e., that B has an open orbit in G/U. In this case the number of orbits on X is always finite, and the orbit decomposition of X in many important cases admits a combinatorial description [109], [68], [86]. The starting point here is the theory of toroidal varieties [59], [25], where $G = B = (\mathbb{C}^*)^n$ and $U = \{e\}$.

Homogeneous manifolds of complexity 0 arise naturally in the theory of representations of compact Lie groups. Namely, if K is a connected compact Lie group, L a closed subgroup, $G = K^{\mathbb{C}}$ and $U = L^{\mathbb{C}}$, then the complexity of G/U is equal to zero if and only if each irreducible K-module occurs in the space of continuous functions on K/L not more than once (see [108]).

We note that certain homogeneous manifolds with complexity 1 are also investigated in [68]. For example, the equivariant open imbeddings of the manifold $G/\{e\}$ are described, where $G = \mathrm{SL}(2, \mathbb{C})$.

Chapter 8. Siegel Domains

In [24] E. Cartan determined all symmetric bounded domains in $\mathbb{C}^n$ and also all homogeneous bounded domains in $\mathbb{C}^2$ and $\mathbb{C}^3$. Putting together these results, he discovered that any homogeneous bounded domain in $\mathbb{C}^2$ or $\mathbb{C}^3$ is symmetric. "This curious result," wrote E. Cartan, "sets forth a new problem, namely, to determine whether there are any nonsymmetric homogeneous bounded domains. It is clear that there is no basis for a negative answer. The discovery of a nonsymmetric homogeneous bounded domain would clearly solve this problem, and possibly it would indicate a path along which all homogeneous bounded domains should be looked for."

Borel [19] and Koszul [66] proved that any bounded domain which is homogeneous with respect to a semisimple group of automorphisms is symmetric. Their result was strengthened by Hano [42], who proved the same for a unimodular group of matrices.

The answer to E. Cartan's question was given by I.I. Pyatetskiĭ-Shapiro [92], who constructed examples of nonsymmetric homogeneous bounded domains in dimensions 4 and 5. One of these examples is reproduced below (see Examples 8.4 and 8.7). Later on it turned out that there are continuous families of pairwise nonisomorphic homogeneous bounded domains (beginning with dimension 7). Thus, symmetric domains are in a sense exceptional, since the number of them (up to isomorphism) in each dimension is finite.

As E. Cartan predicted, the discovery of a nonsymmetric bounded homogeneous domain led to the classification of all homogeneous bounded domains. A decisive role in this classification was played by the concept of a Siegel domain, which was introduced by I.I. Pyatetskiĭ-Shapiro [93]. Before giving the definition of a Siegel domain, we consider some examples.

Example 8.1 (*the generalized Siegel upper halfplane*). We call a classical symmetric domain of type III (see Chapter 1) a generalized unit disc:

$$D_p = \{Z \in M_{p,p} | Z = Z', Z\bar{Z} < I\}.$$

If $Z \in D_p$, then $\det(I - Z) \neq 0$, and we can set

$$W = i(I + Z)(I - Z)^{-1} \quad (Z \in D_p).$$

Then $W = W^t$ and moreover,

$$\begin{aligned}\operatorname{Im} W &= \tfrac{1}{2}\{(I - Z)^{-1}(I + Z) + (I + \bar{Z})(I - Z)^{-1}\}\\ &= \tfrac{1}{2}(I - Z)^{-1}\{(I + Z)(I - \bar{Z}) + (I - Z)(I + \bar{Z})\}(I - \bar{Z})^{-1}\\ &= (I - Z)^{-1}(I - Z\bar{Z})(I - \bar{Z})^{-1} > 0.\end{aligned}$$

The unbounded domain in $M_{p,p}$ consisting of the symmetric matrices whose imaginary parts are positive definite is called the generalized Siegel upper halfplane. We denote it by S_p. We have shown that the mapping $Z \mapsto i(I + Z)\cdot(I - Z)^{-1}$ sends D_p to S_p. Indeed this mapping establishes an isomorphism between D_p and S_p. It is not hard to verify that the inverse mapping is given by $Z \mapsto (Z - iI)\cdot(Z + iI)^{-1}$.

Now let Ω be an open convex cone in $\mathbb{R}^n$ (i.e., $x, y \in \Omega$, $\lambda, \mu \geqslant 0$, $\lambda + \mu \neq 0 \Rightarrow \lambda x + \mu y \in \Omega$) and assume that Ω contains no lines. A domain $\mathbb{C}^n = \mathbb{R}^n + i\mathbb{R}^n$ consisting of all points $z = x + iy$ with $y \in \Omega$ is a special case of a tube domain. It is called a *Siegel domain of the first kind*. In Example 8.1 Ω is the cone of positive definite symmetric matrices of order p in the space of all symmetric matrices $\mathbb{R}^{p(p+1)/2}$. The classical symmetric domains of types I (with $p = q$), II, III, and IV can be realized as Siegel domains of the first kind. However, the ball $\mathbb{B}^n = \left\{z \in \mathbb{C}^n | \sum_1^n |z_k|^2 < 1\right\}$ (i.e., a domain of type I with $p = n$, $q = 1$) does not admit such a realization.

Example 8.2. We consider the holomorphic mapping $\mathbb{B}^{m+1} \to \mathbb{C}^{m+1}$ defined by

$$u_k = \frac{iz_k}{1 - z_{m+1}} \quad (k = 1, \ldots, m), \quad w = i\frac{1 + z_{m+1}}{1 - z_{m+1}},$$

where $u_1, \ldots, u_m, w$ are coordinates in $\mathbb{C}^{m+1}$. Then

$$\operatorname{Im} w - \sum_{k=1}^m |u_k|^2 = \frac{1 - \sum_{k=1}^m |z_k|^2}{|1 - z_{m+1}|^2},$$

which implies that the image of $\mathbb{B}^m$ is contained in the domain defined by the inequality

$$\operatorname{Im} w - \sum_{k=1}^{m} |u_k|^2 > 0.$$

In reality our mapping is an isomorphism from the ball onto this domain. The inverse mapping is defined by

$$z_k = \frac{2u_k}{w+i} \quad (k = 1,\dots,m), \quad z_{m+1} = \frac{w-i}{w+i}.$$

As above let Ω be an open convex cone in $\mathbb{R}^n$ containing no lines. The mapping $F: \mathbb{C}^m \times \mathbb{C}^m \to \mathbb{C}^n$ is called Ω-Hermitian if 1) $F(u, v)$ is linear with respect to u; 2) $\overline{F(u, v)} = F(v, u)$; 3) $F(u, u) \in \bar{\Omega}$, where $\bar{\Omega}$ is the closure of Ω, and 4) $F(u, u) = 0$ only when $u = 0$. The domain

$$D = D(\Omega, F) = \{(z, u) \in \mathbb{C}^n \times \mathbb{C}^m | \operatorname{Im} z - F(u, u) \in \Omega\}.$$

is called a *Siegel domain of the second kind.* In Example 8.2 $n = 1$, $\Omega = \mathbb{R}^+$, and $F(u, v) = \sum_1^m u_k \bar{v}_k$. Another example is any Siegel domain of the first kind. In this case $m = 0$ and $F = 0$.

It is not difficult to show that a Siegel domain is isomorphic to a bounded domain of holomorphy and is homeomorphic to a cell.

For any domain $D \subset \mathbb{C}^N$ we denote by $\operatorname{Aff} D$ the subgroup of the group $\operatorname{Aut} D$ consisting of the affine transformations of the space $\mathbb{C}^N$. For a Siegel domain $D = D(\Omega, F) \subset \mathbb{C}^N = \mathbb{C}^n \times \mathbb{C}^m$ the transformations of the group $\operatorname{Aff} D$ have the form

$$(z, u) \mapsto (Az + a + 2iF(Bu, b) + iF(b, b), Bu + b),$$

where $a \in \mathbb{R}^n$, $b \in \mathbb{C}^m$, $A \in \mathrm{GL}(n, \mathbb{R})$, $A(\Omega) = \Omega$, $B \in \mathrm{GL}(m, \mathbb{C})$, and $\mathrm{A}F(u, v) = F(Bu, Bv)$.

Theorem 8.3 (Kaup, Matsushima, Ochiai [58]). *If the Siegel domain $D = D(\Omega, F) \subset \mathbb{C}^N = \mathbb{C}^n \times \mathbb{C}^m$ is homogeneous, then it is affinely homogeneous, i.e., the group $\operatorname{Aff} D$ acts on D transitively. If D is isomorphic to another Siegel domain $D' = D(\Omega', F') \subset \mathbb{C}^N = \mathbb{C}^{n'} \times \mathbb{C}^{m'}$, then the domains D and D' are affinely equivalent. More precisely, $n' = n$, $m' = m$, and there exist linear transformations $A \in \mathrm{GL}(n, \mathbb{R})$ and $B \in \mathrm{GL}(m, \mathbb{C})$ so that $\Omega' = A(\Omega)$, $F'(u, v) = A(F(B^{-1}u, B^{-1}v))$, and $D' = \{(z, u) \in \mathbb{C}^n \times \mathbb{C}^m | (A^{-1}z, B^{-1}u) \in D\}$.*

To prove this theorem, it is necessary to make a detailed study of the automorphism group of the Siegel domain D. In particular, it turns out that the coefficients of velocity fields of the group $\operatorname{Aut} D$ are polynomials in the coordinates of $\mathbb{C}^N$ of degree not higher than 2. In the group $\operatorname{Aut} D$ there is the one-parameter subgroup

$$(z, u) \mapsto (tz, t^{1/2}u) \quad (t > 0).$$

We denote by $\mathfrak{g}$ the Lie algebra of the group $\operatorname{Aut} D$, and by δ the velocity field of this one-parameter subgroup: $\delta = \sum z_i \frac{\partial}{\partial z_i} + \frac{1}{2} \sum u_j \frac{\partial}{\partial u_j}$. The element δ defines in $\mathfrak{g}$ the grading

$$\mathfrak{g} = \mathfrak{g}_{-1} \oplus \mathfrak{g}_{-1/2} \oplus \mathfrak{g}_0 \oplus \mathfrak{g}_{1/2} \oplus \mathfrak{g}_1,$$

where $\mathfrak{g}_\lambda = \{x \in \mathfrak{g} | [\delta, x] = \lambda x\}$. The subalgebra $\mathfrak{g}_{-1} \oplus \mathfrak{g}_{-1/2} \oplus \mathfrak{g}_0$ corresponds to the subgroup $\operatorname{Aff} D$, and the subalgebra $\mathfrak{g}_{-1} \oplus \mathfrak{g}_0 \oplus \mathfrak{g}_1$ to the subgroup which preserves the flat section $D \cap \{u = 0\}$, which is a Siegel domain of the first kind $D(\Omega, 0) \subset \mathbb{C}^n$.

A large number of articles are devoted to Siegel domains and, in particular, to their automorphisms (see [36], [58], [104], [55], [77], [95], [27]).

We are especially interested in homogeneous Siegel domains. A cone $\Omega \subset \mathbb{R}^n$ is called a *homogeneous cone*, if the group

$$\operatorname{Aff} \Omega = \{A \in \mathrm{GL}(n, \mathbb{R}) | A(\Omega) = \Omega\}$$

is transitive on Ω. In view of Theorem 8.3 the following conditions together are necessary and sufficient for the homogeneity of the Siegel domain $D(\Omega, F)$:

1) the cone Ω is homogeneous;

2) for each element $A \in T$, where $T \subset \operatorname{Aff} \Omega$ is a subgroup which is transitive on Ω, there exists a matrix $B \in \mathrm{GL}(m, \mathbb{C})$ so that $AF(u, v) = F(Bu, Bv)$.

Example 8.4. We identify $\mathbb{C}^3$ (resp., $\mathbb{R}^3$) with the space of complex (resp., real) symmetric matrices of order two. Let Ω be the cone consisting of all positive definite symmetric matrices, and let the Ω-Hermitian mapping $F : \mathbb{C}^1 \times \mathbb{C}^1 \to \mathbb{C}^3$ be defined by the formula

$$F(u, v) = \begin{pmatrix} u\bar{v} & 0 \\ 0 & 0 \end{pmatrix}.$$

The Siegel domain $D(\Omega, F) \subset \mathbb{C}^4 = \mathbb{C}^3 \times \mathbb{C}^1$ is defined by the inequalities

$$\begin{vmatrix} y_{11} - |u|^2 & y_{12} \\ y_{12} & y_{22} \end{vmatrix} > 0; \quad y_{11} - |u|^2 > 0 \quad (y_{ij} = \operatorname{Im} z_{ij}).$$

The group $\operatorname{Aff} \Omega$ consists of all transformations $Y \mapsto gYg^t$, where $g \in \mathrm{GL}(2, \mathbb{R})$ and $Y \in \Omega$. As T we may take the subgroup consisting of transformations with an upper-triangular matrix g. The domain $D(\Omega, F)$ is homogeneous.

The role of Siegel domains in the theory of homogeneous complex manifolds is explained by the following fundamental theorem.

Theorem 8.5 (È.B. Vinberg, S.G. Gindikin, I.I. Pyatetskiĭ-Shapiro [107]). *Every homogeneous bounded domain in $\mathbb{C}^N$ is isomorphic to a homogeneous Siegel domain of the second kind.*

We note that a Siegel domain, which is the realization of a given homogeneous bounded domain, is by the construction in [107] affinely homogeneous. Under the hypothesis of affine homogeneity it is proven in [107] that the realization is unique up to affine equivalence. By Theorem 8.3 the assertion of uniqueness of the realization is also true without the indicated hypothesis.

The transformation rule for the Bergman kernel function under an isomorphism of domains enables us to define a kernel function for unbounded domains which are isomorphic to bounded ones, in particular, for Siegel domains. As in the case of bounded domains, we obtain an invariant Kähler metric (the Bergman metric) and an invariant volume form. In [35] S.G. Gindikin obtained an explicit expression for the kernel function $K(z, u; \zeta, \eta)$ of a homogeneous Siegel domain $D = D(\Omega, F)$. In particular, on the diagonal (i.e., when $\zeta = z$ and $\eta = u$) we have

$$K = \text{const} \cdot \lambda(\operatorname{Im} z - F(u, u)),$$

where $\lambda : \Omega \to \mathbb{R}^+$ is some special function in the cone Ω (a multi-dimensional generalization of the power function).

The definition of a symmetric domain given in Chapter 1 for bounded domains carries over word for word to domains isomorphic to bounded domains. General conditions for a Siegel domain to be symmetric are obtained in [96]. The following simple result is sometimes useful to check that a Siegel domain is nonsymmetric.

Proposition 8.6. *If a Siegel domain $D(\Omega, F)$ possesses a symmetry at the point $(z_0, 0)$, then this symmetry has the from*

$$(z, u) \mapsto (\varphi(z), \psi(z) \cdot u),$$

where $z \mapsto \varphi(z)$ is the symmetry of the Siegel domain of the first kind $D(\Omega, 0) = D(\Omega, F) \cap \{u = 0\}$ at the point z_0 and $\psi(z)$ is a holomorphic matrix-valued function in $D(\Omega, 0)$.

Example 8.7. Let $D = D(\Omega, F) \subset \mathbb{C}^4$ be the domain from Example 8.4. The symmetry at the point $(iI, 0)$, if it exists, must be given by the formula

$$(Z, u) \overset{s}{\mapsto} (-Z^{-1}, \psi(Z)u).$$

Here $Z = (z_{ij})$ is a symmetric matrix of order 2, $u \in \mathbb{C}$, and $\operatorname{Im} Z - \begin{pmatrix} |u|^2 & 0 \\ 0 & 0 \end{pmatrix} > 0.$ Using the simply transitive solvable automorphism group, it is easy to find an explicit form for the Bergman kernel function

$$K = \text{const} \cdot y_{22}[(y_{11} - |u|^2)y_{22} - y_{12}^2]^{-4} \quad (y_{ij} = \operatorname{Im} z_{ij}).$$

One verifies immediately that s does not preserve the Bergman metric. Therefore the domain D is nonsymmetric.

For homogeneous Siegel domains of the first kind the symmetry condition can be formulated especially simply. For this it is necessary and sufficient that the homogeneous cone $\Omega \subset \mathbb{R}^n$ from which the Siegel domain is constructed be *self-dual*. This means that for some choice of scalar product the cone Ω coincides with the dual cone

$$\Omega^* = \{x \in \mathbb{R}^n | (x, y) > 0 \, \forall y \in \bar{\Omega} \setminus \{0\}\}.$$

È.B. Vinberg (1963) has constructed examples of homogeneous convex cones which do not contain entire lines and are not self-dual. The simplest of these is the cone consisting of all positive definite symmetric matrices $Y = (y_{ij})$ of order 3 for which $y_{32} = y_{23} = 0$. To it there corresponds a nonsymmetric homogeneous Siegel domain of the first kind in $\mathbb{C}^5$ (see [36]).

Chapter 9. Homogeneous Kähler Manifolds

Let X be a Kähler manifold with Kähler metric h. The manifold X is called a *homogeneous Kähler manifold* of a group G if G is a transitive Lie group of transformations of X which preserve the complex structure and the metric h. To classify homogeneous Kähler manifolds, we introduce some geometric concepts.

We assume that on the complex manifold X there is a fixed volume form ω. In local coordinates it is given as $\omega = i^n K dz_1 \wedge d\bar{z}_1 \wedge \ldots \wedge dz_n \wedge d\bar{z}_n$, where K is a positive function. The Hermitian form $\eta = \sum \frac{\partial^2 \log K}{\partial z_i \partial \bar{z}_j} dz_i \otimes d\bar{z}_j$, which does not depend on the choice of the coordinate system, is called the *canonical Hermitian form* of the manifold X. If X is a Kähler manifold with Kähler metric h and ω is the corresponding volume form, then η is the Ricci tensor of the metric h.

Now let X be a homogeneous complex manifold with a transitive Lie group G. If there exists on X a G-invariant volume form ω, then it is unique up to constant multiple. In this case the canonical Hermitian form η is G-invariant and uniquely defined. In particular, if the homogeneous manifold X is Kähler, then it may admit several G-invariant Kähler metrics, but they all have the same Ricci tensor. For homogeneous bounded domains the form η coincides with the Bergman metric and, in particular, is positive definite.

In the compact case the classification of homogeneous Kähler manifolds is obtained from Theorem 4.11. It is necessary only to observe that in the decompositon $X = A(X) \times Y$ the tangent spaces to the factors are orthogonal with respect to the invariant Kähler metric on X (see [72]).

Theorem 9.1. *Every compact homogeneous Kähler manifold X is the product of the flat complex torus $A(X)$ and a simply connected compact homogeneous Kähler manifold Y, where the Kähler metric on X is the product of the Kähler metrics on $A(X)$ and Y. The manifold Y can be written in the Klein form $Y = G/U$, where G is a connected compact semisimple Lie group and U is the centralizer of a torus in G.*

On the other hand, the structure of homogeneous Kähler manifolds of semisimple Lie groups is completely described by the following two theorems.

Theorem 9.2. (Borel [19]). *Let G be a connected semisimple Lie group and $X = G/U$ a homogeneous Kähler manifold of the group G. Then the stationary subgroup U is connected and compact and is the centralizer of a torus in G. Let $G = G_1 \cdot \ldots \cdot G_s$ be the decomposition of the group G into a product of simple normal factors, and let K be a maximal compact subgroup of G containing U. Then $K = K_1 \cdot \ldots \cdot K_s$ and $U = U_1 \cdot \ldots \cdot U_s$, where $K_i = K \cap G_i$, $U = U \cap G_i$, and for each i there is the alternative: either $G_i = K_i$ or G_i/K_i is a Hermitian symmetric space. Here $G/U = (G_1/U_1) \times \cdots \times (G_s/U_s)$ and G/U is the total space of a holomorphic bundle whose base is a bounded symmetric domain isomorphic to G/K and whose fiber is the compact homogeneous Kähler manifold K/U.*

Theorem 9.3 (Koszul [66]). *Under the hypotheses of the preceding theorem the canonical Hermitian form η of the Kähler manifold G/U is nondegenerate. The number of negative squares in it is equal to* $\dim K - \dim U$. *In particular, if the group G is compact, then the form η is negative definite.*

The last statement admits a generalization.

Theorem 9.4 (Shima [99]). *If a homogeneous Kähler manifold $X = G/U$ has a negative definite canonical Hermitian form η, then G is a compact semisimple Lie group.*

We now formulate a conjecture concerning the structure of an arbitrary homogeneous Kähler manifold, which is analogous to the theorem on the decomposition of a Hermitian symmetric space into a product of Hermitian symmetric spaces of compact type, of flat type, and of noncompact type.

Conjecture (È.B. Vinberg, S.G. Gindikin (1967)). *Any homogeneous Kähler manifold is the total space of a holomorphic bundle whose base is a homogeneous bounded domain and whose fiber, viewed with the induced Kähler metric, decomposes into the direct product of a flat homogeneous Kähler manifold and a homogeneous Käher manifold which is simply connected and compact.*

Theorems 9.1 and 9.2 show that the conjecture is true if the homogeneous Kähler manifold $X = G/U$ is compact or if the group G is semisimple. In [106] the conjecture is proven for the case when G is solvable and the operators $\operatorname{Ad} g$ for $g \in G$ have only real eigenvalues. The latter restriction has recently been removed (see Dorfmeister [28]). Thus the conjecture is true for homogeneous Kähler manifolds of solvable Lie groups.

On the other hand, the validity of the conjecture has been verified under some additional hypotheses on the canonical Hermitian form η. Theorem 9.4 shows that the conjecture is true if the form η is negative definite. If $\eta = 0$ then X is a flat homogeneous Kähler manifold (D.V. Alekseevskiĭ, B.N. Kimel'fel'd [8]). If η is positive definite, then the complex manifold X is isomorphic to a homogeneous bounded domain (Nakajima, 1984). Thus the conjecture is true also in these cases. Finally, if the form η is nondegenerate and the group G is unimodular, then by a theorem of Hano (see [42]) G is semisimple and the validity of the conjecture follows from Theorem 9.2.

We deduce a few consequences from this conjecture. The bundle referred to in it is unique, which means it is preserved under all automorphisms of the complex manifold X. Indeed, its fibers can be determined as the maximal subsets on which all bounded holomorphic functions are constant. Further, the base of the bundle is a Stein manifold homeomorphic to a cell, and the structure group can be viewed as the group of all biholomorphic transformations of the fiber, which is a complex Lie group. It follows from a theorem of Grauert (1958) (see Article II) that the bundle is analytically trivial. Therefore, if the conjecture is true, then every homogeneous Kähler manifold is isomorphic as a complex manifold to a direct product of homogeneous Kähler manifolds of the three basic types, i.e., simply connected and compact, flat, and a homogeneous bounded domain (the Kähler structure here is in general not the product structure).

Added in proof. Recently J. Dorfmeister and K. Nakajima announced that they have proven the conjecture of È.B. Vinberg and S.G. Gindikin in full generality. ("La Conjecture fondamentale pour les variétés homogènes kähleriennes," CiR. Acad. Sc. Paris, t. 303, Série 1, n° 8, 1986, p. 335–338).

Bibliography*

Some information about complex homogeneous manifolds can be found in books on differential geometry (see, for example, [62], Vol. 2, Supplement 24). Concerning Hermitian symmetric spaces see [44], Chap. 8; [62], Vol. 2, Chap. 11; and [29]. Homogeneous bounded domains are studied in the books [93], [55], [77]. In the first of these, as also in [100], special attention is given to the theory of automorphic functions. Aside from the indicated monographs the most essential sources are the classical works [110], [21] (for Chap. 2); [23] (for Chap. 3); [105], [22] (for Chap. 4); [39] for Chap. 6); [36] (for Chaps. 8 and 9); [58] (for Chap. 8); and also the recent surveys [50], [32] (for Chaps. 5–7); and the article [27] which contains, in particular, a complete bibliography on homogeneous bounded domains (for Chap. 8).

1. Ahiezer, D.N. (Akhiezer): Invariant meromorphic functions on complex semisimple Lie groups, Invent. Math. 65, 325–329 (1982). Zbl. 479.32010
2. Ahiezer, D.N. (Akhiezer): Equivariant completions of homogeneous algebraic varieties by homogeneous divisors. Ann. Global Anal. and Geom. 1, No. 1, 49–78 (1983). Zbl. 537.14033
3. Ahiezer, D.N. (Akhiezer): Invariant analytic hypersurfaces in complex nilpotent Lie groups. Ann. Global Anal. and Geom. 2, 129–140 (1984). Zbl. 576.32039
4. Akhiezer, D.N.: Compact complex homogeneous spaces with solvable fundamental group. Izv. Akad. Nauk SSSR, Ser. Mat., 38, 59–80 (1974); English transl.: Math. USSR, Izv.8, 61–83 (1975).
5. Akhiezer, D.N.: Dense orbits with two ends. Izv. Akad. Nauk SSSR, Ser. Mat. 41, 308–324 (1977); English transl.: Math. USSR, Izv. 11, 293–307 (1977). Zbl. 373.14016
6. Akhiezer, D.N.: A bound for the dimension of the automorphism group of a compact complex

*For the convenience of the reader, references to reviews in Zentralblatt für Mathematik (Zbl.), compiled using the MATH database, have, as far as possible, been included in this bibliography.

homogeneous space. Soobshch. Akad. Nauk Grus. SSR (Russian) 110, 469–472 (1983) Zbl. 579.32049
7. Akao, K.: On prehomogeneous compact Kähler manifolds. Proc. Japan Acad., Ser. A, 49, 483–485 (1973). Zbl. 272.32009
8. Alekseevskiĭ, D.V., Kimel'fel'd, B.N.: The structure of homogeneous Riemann spaces with zero Ricci curvature. Funkts. Anal. Prilozh. 9, No. 2 5–11 (1975); English transl.: Funct. Anal. Appl. 9, 97–102 (1975). Zbl. 316.53041
9. Aribaud, Fr.: Une nouvelle démonstration d'un théorème de R. Bott et B. Kostant. Bull. Soc. Math. Fr. 95, 205–242 (1967). Zbl. 155.69
10. Barth, W., Oeljeklaus, E.: Über die Albanese-Abbildunge einer fasthomogenen Kählermannigfaltigkeit. Math. Ann. 211, 47–62 (1974). Zbl. 276.32022
11. Barth, W., Otte, M.: Invariante holomorphe Funktionen auf reduktiven Liegruppen. Math. Ann. 201, 97–112 (1973). Zbl. 253.32018
12. Berezin, F.A., Pyatetskiĭ-Shapiro, I.I.: Homogeneous extensions of a complex space. (Russian) Dokl. Akad. Nauk SSSR 99, 889–892 (1954). Zbl. 58.308
13. Bertin, J.: Pinceaux de droites et automorphismes des surfaces affines. J. Reine Angew. Math. 341, 32–53 (1983). Zbl. 501.14028
14. Bialynicki-Birula, A.: On homogeneous affine spaces of linear algebraic groups. Am. J. Math. 85, 577–582 (1963). Zbl. 116.382
15. Bialynicki-Birula, A., Hochschild, G. Mostow, G.D.: Extensions of representations of algebraic linear groups. Am. J. Math. 85, 131–144 (1963). Zbl. 116.23
16. Blanchard, A.: Sur les variétés analytiques complexes. Ann. Sci. Ec. Norm. Supér., III. Sér. 73, 157–202 (1956). Zbl. 73.375
17. Boothby, W.M.: Homogeneous complex contact manifolds. Proc. Symp. Pure Math. 3, 144–154 (1961). Zbl. 103.387
18. Boothby, W.M.: A note on homogeneous complex contact manifolds. Proc. Am. Math. Soc. 13, 276–280 (1962). Zbl. 103.387
19. Borel, A.: Kählerian coset spaces of semisimple Lie groups. Proc. Natl. Acad. Sci. USA 40, 1147–1151 (1954). Zbl. 58.160
20. Borel, A.: Symmetric compact complex spaces. Arch. Math. 33, 49–56 (1979). Zbl. 423.32015
21. Borel A., Hirzebruch, F.: Characteristic classes and homogeneous spaces I. Am. J. Math. 80, 458–538 (1958); II. 81, 315–382 (1959); III. 82, 491–504 (1960).
22. Borel A., Remmert, R.: Über kompakte homogene Kählersche Mannigfaltigkeiten. Math. Ann. 145, 429–439 (1962). Zbl. 111.180
23. Bott, R.: Homogeneous vector bundles. Ann. Math. 66, 203–248 (1957). Zbl. 94.357
24. Cartan, E.: Sur les domaines bornés homogènes de l'espace de n variables complexes. Abh. Math. Semin. Univ. Hamb. 11, 116–162 (1935). Zbl. 11.123
25. Danilov, V.I.: The geometry of toric varieties. Usp. Mat. Nauk 33, No. 2 85–134 (1978); English transl.: Russ. Math. Surv. 33, No. 2 97–154 (1978).
26. Demazure, M.A.: A very simple proof of Bott's theorem. Invent. Math. 33, 271–272 (1976). Zbl. 383.14017
27. Dorfmeister, J.: Homogeneous Seigel domains. Nagoya Nath. J. 86, 39–83 (1982).
28. Dorfmeister, J.: Homogeneous Kähler manifolds admitting a transitive solvable group of automorphisms. Ann. Sci. Ec. Norm. Supér, IV. Sér. 18, 143–180 (1985).
29. Drucker, D.: Exceptional Lie algebras and the structure of hermitian symmetric spaces. Mem. Am. Math. Soc. 208, 207 p. (1978). Zbl. 395.17009
30. Drucker, D.: Simplified descriptions of the exceptional bounded symmetric domains. Geom. Dedicata 10, 1–29 (1981). Zbl. 486.32020
31. Gilligan, B.: Ends of complex homogeneous manifolds having nonconstant holomorphic functions. Arch. Math. 37, 544–555 (1981). Zbl. 459.32013
32. Gilligan, B.: Holomorphic reductions of homogeneous spaces, (Springer) Lect. Notes Math. 1014, 27–36 (1983). Zbl. 529.32013

33. Gilligan, B.: Huckleberry, A.T.: On non-compact complex nil-manifolds. Math. Ann. 238, 39–49 (1978). Zbl. 405.32009
34. Gilligan, B., Huckleberry, A.T.: Complex homogeneous manifolds with two ends. Mich. J. Math. 28, 183–198 (1981). Zbl. 452.32022
35. Gindikin, S.G.: Analysis in homogeneous domains. Usp. Mat. Nauk 19, 3–92 (1964); English transl.: Russ. Math. Surv. 19, 1–89 (1964) Zbl. 144.81
36. Gindikin S.G., Pyatetskiĭ-Shapiro, I.I., Vinberg, E.B.: Homogeneous Kähler manifolds, in Geometry of homogeneous bounded domains, Edizioni Cremonese, Rome 1–87 (1968). Zbl. 183.354
37. Gizatullin, M. Kh.: Affine surfaces which are quasihomogeneous with respect to an algebraic group. Izv. Akad. Nauk SSSR, Ser. Mat. 35, 738–753 (1971); English transl.: Math. USSR, Izv. 5, 754–769 (1971). Zbl. 218.14019
38. Goto, M.: On algebraic homogeneous spaces. Am. J. Math. 76, 811–818 (1954).
39. Grauert, H., Remmert, R.: Über kompakte homogene komplexe Mannigfaltigkeiten. Arch. Math. 13, 498–507 (1962). Zbl. 118.374
40. Griffiths, P.A.: Some geometric and analytic properties of homogeneous complex manifolds. Acta Math. 110, 115–155, 157–208 (1963). Zbl. 171.446
41. Griffiths P.A., Schmid, W.: Locally homogeneous complex manifolds. Acta Math. 123, 253–302 (1970). Zbl. 209.257
42. Hano, J.-I.: On Kählerian homogeneous spaces of unimodular Lie groups. Am. J. Math. 79, 885–900 (1957). Zbl. 96.162
43. Hano, J.-I.: On compact complex coset spaces of reductive Lie groups. Proc. Am. Math. Soc. 15, 159–163 (1969). Zbl. 129.157
44. Helgason, S.: Differential geometry and symmetric spaces, Academic Press, N.Y. and London (1962). Zbl. 111.181
45. Hua, Lo-Keng: Harmonic analysis of functions of several complex variables in classical domains. Am. Math. Soc., Providence (1963). Zbl. 112.74
46. Huckleberry, A.T., Livorni, E.L.: A classification of homogeneous surfaces. Can. J. Math. 33, 1097–1110, (1981). Zbl. 504.32025
47. Huckleberry, A.T., Margulis, G.A.: Invariant analytic hypersurfaces. Invent. Math. 71, 235–240 (1983). Zbl. 507.32024
48. Huckleberry A.T., Oeljeklaus, E.: A characterization of complex homogeneous cones. Math. Z. 170, 181–194 (1980). Zbl. 412.32030
49. Huckleberry, A.T., Oeljeklaus, E.: Classification theorems for almost homogeneous spaces. Institute E. Cartan. Nancy, No. 9 (1984). Zbl. 549.32024
50. Huckleberry, A.T., Oeljeklaus, E.: Homogeneous spaces from a complex analytic viewpoint, in Manifolds and Lie groups, Progress in Mathematics, Vol 14, [Birkhäuser-Verlag, Basel, Boston, Stuttgart] 159–186 (1981). Zbl. 527.32020
51. Huckleberry A.T., Snow, D.M.: Pseudoconcave homogeneous manifolds. Ann. Sc. Norm. Super Pisa, Cl. Sci., IV. Ser. 7, 29–54 (1980). Zbl. 506.32015
52. Huckleberry A.T., Snow, D.M.: A classification of strictly pseudoconcave homogeneous manifolds. Ann. Sc. Norm. Super. Pisa, Cl. Sci., IV. Ser. 8, 231–255 (1981). Zbl. 464.32019
53. Huckleberry, A.T., Snow, D.M.: Almost homogeneous Kähler manifolds with hypersurface orbits. Osaka J. Math. 19, 763–786 (1982). Zbl. 507.32023
54. Ise, M.: Bounded symmetric domains of exceptional type, J. Fac. Sci., Univ. Tokyo, Sect. IA 23, 75–105 (1976). Zbl. 357.32016
55. Kaneyuki, S.: Homogeneous bounded domains and Siegel domains. (Springer) Lect. Notes Math. 241 (1971). Zbl. 241.32011
56. Kantor, I.L.: The cross ratio of points and other invariants on homogeneous spaces with parabolic stationary subgroups I. Tr. Semin. Vektorn. Tenzorn. Anal. [Prilozh. Geom. Mekh. Fiz.] (Russian) 17, 250–313 (1974). Zbl. 307.53029
57. Kaup, W.: Reelle Transformationsgruppen und invariante Metriken auf komplexen Räumen. Invent. Math. 3, 43–70 (1967). Zbl. 157.134

58. Kaup W., Matsushima, Y., Ochiai, T.: On the automorphisms and equivalences of generalized Siegel domains. Am. J. Math. 92, 475–498 (1970). Zbl. 198.425
59. Kempf, G., Knudsen, F., Mumford, D., Saint-Donat, B.: Toroidal embeddings I. (Springer) Lect. Notes Math. 339 (1973). Zbl. 271.14017
60. Khosrovyan, O.M.: On complex homogeneous spaces with two ends, in Geometric methods in problems of analysis and algebra. (Russian) Yaroslavl' Univ. 35–42 (1978). Zbl. 413.32013
61. Kimel'fel'd, B.N.: Reductive groups which are locally transitive on the flag manifolds of the orthogonal groups. (Russian) Tr. Tbilis. Mat. Inst. Razmadze 62, 49–75 (1979). Zbl. 459.32014
62. Kobayashi, S., Nomizu, K.: Foundations of differential geometry, Vol. I. Interscience, N.Y. (1963), Vol 2. (1969). Zbl. 119.375 Zbl. 175.485
63. Koecher, M.: An elementary approach to bounded symmetric domains. Rice Univ. Houston (1969). Zbl. 217.109
64. Korányi, A., Wolf, J.A.: Realization of Hermitian symmetric spaces as generalized half-planes. Ann. Math., II. Ser. 81, 265–288 (1965). Zbl. 137.274
65. Kostant, B.: Lie algebra cohomology and the generalized Borel-Weil theorem. Ann. Math., II. Ser. 74, 329–387 (1961). Zbl. 134.35
66. Koszul, J.L.: Sur la forme hermitienne canonique des espaces homogènes complexes. Can. J. Math. 7, 562–576 (1955). Zbl. 66.161
67. Lescure, F.: Elargissement du groupe d'automorphismes pour des varietés quasi-homogènes. Math. Ann. 261, 455–462 (1982). Zbl. 531.32018
68. Luna, D., Vust. Th.: Plongements d'espaces homogènes. Comment. Math. Helv. 58, 186–245 (1983). Zbl. 545.14010
69. Malyshev, F.M.: Complex homogeneous spaces of semisimple Lie groups of the first category. Izv. Akad. Nauk SSSR, Ser. Mat. 39, 992–1002 (1975); English transl.: Math. USSR, Izv. 9, 939–949 (1977). Zbl. 322.53024
70. Malyshev, F.M.: Complex homogeneous spaces of semisimple Lie groups of type D_n. Izv. Akad. Nauk SSSR, Ser. Mat. 41, 829–852 (1977); English transl.: Math. USSR, Izv. 11, 783–805 (1978). Zbl. 362.53043
71. Malyshev, F.M.: Complete complex structures on homogeneous spaces of semisimple Lie groups. (Russian) Izv. Akad. Nauk SSSR, Ser. Mat. 43, 1294–1318 (1979). Zbl. 428.32021
72. Matsushima, Y.: Sur les espaces homogènes Kähleriens d'un groupe de Lie réductif. Nagoya Math. J. 11, 53–60 (1957). Zbl. 99.375
73. Matsushima, Y.: Espaces homogènes de Stein des groupes de Lie complexes I. Nagoya Math. J. 16, 205–218 (1960). Zbl. 94.282
74. Matsushima, Y.: Espaces homogènes de Stein des groups de Lie complexes II. Nagoya Math. J. 18, 153–164 (1961). Zbl. 143.302
75. Matsushima, Y., Morimoto, A.: Sur certains espaces fibrés holomorphes sur une varieté de Stein. Bull. Soc. Math. Fr. 88, 137–155 (1960). Zbl. 94.281
76. Morimoto, A.: Non-compact complex Lie groups without non-constant holomorphic functions, in Proceedings of the conference on complex analysis at the Univ. of Minneapolis. Springer-Verlag, Berlin–Heidelberg–New York. 256–272 (1965). Zbl. 144.79
77. Murakami, S.: On automorphisms of Siegel domains. (Springer) Lect. Notes Math. 286 (1972). Zbl. 245.32001
78. Narasimhan, M.S., Okamoto, K.: An analogue of the Borel-Weil-Bott theorem for hermitian symmetric pairs of non-compact type. Ann. Math., II. Ser. 91, 486–511 (1970). Zbl. 257.22013
79. Oeljeklaus, E.: Ein Hebbarkeitssatz für Automorphismengruppen kompakter komplexer Mannigfaltigkeiten. Math. Ann. 190, 154–166 (1970). Zbl. 201.102
80. Obljeklaus, E.: Fasthomogene Kählermannigfaltigkeiten mit verschwindender ersten Bettizahl. Manuscr. Math. 7, 175–183 (1972). Zbl. 248.32021
81. Oeljeklaus, K., Richthofer, W.: Homogeneous complex surfaces. Math. Ann. 268, 273–292 (1984). Zbl. 551.32023

82. Onishchik, A.L.: Complex hulls of compact homogeneous spaces. Dokl. Akad. Nauk SSSR 130, 726–729 (1960); English transl.: Sov. Math. 1, 88–93 (1960). Zbl. 90,94
83. Onishchik, A.L.: Inclusion relations between transitive compact groups of transformations. Tr. Mosk. Mat. O.-va. 11, 199–242 (1962). Zbl. 192.126
84. Onishchik, A.L.: Decompositions of reductive Lie groups. Mat. Sb., Nov. Ser. 80 (122), 553–599 (1969). English transl. II, Sev., Am. Math. Soc. 50, 5–58 (1966). Zbl. 222.22011 Engl. transl.: Math. USSR, Sb, 9 (1969), 515–554 (1970)
85. Otte, M., Potters, J.: Beispiele homogener Mannigfaltigkeiten. Manuscr. Math. 10, 117–127 (1973). Zbl. 264.32013
86. Pauer, F.: Normale Einbettungen von G/U. Math. Ann. 257, 371–396 (1981). Zbl. 461.14013
87. Pauer, F.: Glatte Einbettungen von G/U. Math. Ann. 262, 421–429 (1983). Zbl. 512.14029
88. Pommerening, K.: Observable radizielle Untergruppen von halbeinfachen algebraischen Gruppen. Math. Z. 165, 243–250 (1979). Zbl. 379.20040
89. Popov, V.L.: Classification of affine algebraic surfaces that are quasihomogeneous with respect to an algebraic group. Izv. Akad. Nauk SSSR, Ser. Mat. 37, 1038–1055 (1973); English transl.: Math. USSR, Izv. 7, 1039–1055 (1975). Zbl. 251.14018
90. Popov, V.L.: Classification of three-dimensional affine algebraic varieties that are quasihomogenous with respect to an algebraic group, Izv. Akad. Nauk SSSR, Ser. Mat. 39, 566–609, (1975); English transl.: Math. USSR, Izv. 9, 535–576 (1976).
91. Potters, J.: On almost homogeneous compact complex analytic surfaces. Invent. Math. 8, 244–266 (1969). Zbl. 205.251
92. Pyatetskiĭ-Shapiro, I.I. (Piatetski-Shapiro, I.I.) On a problem of E. Cartan. Dokl. Akad. Nauk SSSR (Russian) 124, 272–273 (1959). Zbl. 89,62
93. Pyatetskiĭ-Shapiro, I.I. (Piatetski-Shapiro, I.I.) Automorphic functions and the geometry of classical domains. GIFML, Moscow (1961); English transl.: Gordon and Breach, N.Y. (1969).
94. Remmert, R., van de Ven, A.: Zur Funktionentheorie homogener komplexer Mannigfaltigkeiten. Topology 2, 137–157 (1963). Zbl. 122.86
95. Rothaus, O.S.: Automorphisms of Siegel domains. Am. J. Math. 101, 1167–1179 (1979). Zbl. 443.32022
96. Satake, I.: On classification of quasi-symmetric domains. Nagoya Math. J. 62, 1–12 (1976). Zbl. 331.32026
97. Schmid, W.: On a conjecture of Langlands. Ann. Math., II. Ser. 93, 1–42 (1971). Zbl. 291.43013
98. Schmid, W.: L^2-cohomology and discrete series. Ann. Math., II. Ser. 103, 375–394 (1976). Zbl. 333.22009
99. Shima, H.: On homogeneous complex manifolds with negative definite canonical hermitian form. Proc. Japan Acad. 46, 209–211, (1970). Zbl. 207.204
100. Siegel, C.L.: Topics in complex function theory, Vol. III. Abelian functions and modular functions of several complex variables. Wiley-Interscience, N.Y., London, Sydney, Toronto (1973). Zbl. 257.32002
101. Steinsiek, M.: Transformation groups on homogeneous-rational manifolds. Math. Ann. 260, 423–435 (1982). Zbl. 503.32017
102. Sukhanov, A.A.: On observable subgroups of linear algebraic groups, Candidate Dissertation, MGU, Moscow (1978).
103. Takeuchi, M.: On orbits in a compact hermitian symmetric space. Am. J. Math. 90, 657–680 (1968). Zbl. 181.243
104. Tanaka, N.: On infinitesimal automorphisms of Sieigel domains. J. Math. Soc. Japan 22, 180–212 (1970). Zbl. 188.81
105. Tits, J.: Espaces homogènes complexes compacts. Comment. Math. Helv. 37, 111–120 (1962). Zbl. 108.363
106. Vinberg, È.B., Gindikin, S.G.: Kählerian manifolds admitting a transitive solvable automor-

phism group. Mat. Sb., Nov. Ser. 74, 357–377 (1967); English transl.: Math USSR, Sb. 3, 333–351 (1967). Zbl. 153.399

107. Vinberg, È.B., Gindikin, S.G., Pyatetskiĭ-Shapiro, I.I.: On the classification and canonical realization of complex homogeneous bounded domains. (Russian) Tr. Mosk. Mat. O.-va 12, 359–388 (1963). Zbl. 137.56
108. Vinberg, È.B., Kimel'fel'd, B.N.: Homogeneous domains on flag manifolds and spherical subgroups of semisimple Lie groups. Funkts. Anal. Prilozh. 12, No. 3, 12–19 (1978); English transl.: Funct. Anal. Appl. 12, 168–174 (1979). Zbl. 439.53055
109. Vinberg, È.B., Popov, V.L.: On a class of quasihomogeneous affine varieties. Izv. Akad. Nauk SSSR, Ser. Mat. 36, 749–764 (1972); English transl.: Math. USSR, Izv. 6, 743–758 (1973). Zbl. 248.14014.
110. Wang, H.-C.: Closed manifolds with homogeneous complex structures. Am. J. Math. 76, 1–32 (1954). Zbl. 55.166
111. Wang, H.-C.: Complex parallelizable manifolds. Proc. Am. Math. Soc. 5, 771–776 (1954). Zbl. 56.154
112. Wolf, J.A.: On the classification of Hermitian symmetric spaces. J. Math. Mech. 13, 489–495 (1964). Zbl. 245.32011
113. Wolf, J.A.: The action of a real semisimple group on a complex flag manifold I. Orbit structure and holomorphic arc components. Bull. Am. Math. Soc. 75 1121–1237 (1969) Zbl. 183.509
114. Wolf, J.A., Korányi, A.: Generalized Cayley transformations of bounded symmetric domains. Am. J. Math. 87, 899–939 (1965). Zbl. 137.274

Author Index

Subject Index